Smart Sensors, Measurement and Instrumentation

Volume 6

Series Editor

Subhas Chandra Mukhopadhyay
School of Engineering and Advanced Technology (SEAT)
Massey University (Manawatu)
Palmerston North
New Zealand
E-mail: S.C.Mukhopadhyay@massey.ac.nz

For further volumes:
http://www.springer.com/series/10617

Càndid Reig · Susana Cardoso de Freitas,
Subhas Chandra Mukhopadhyay

Giant Magnetoresistance (GMR) Sensors

From Basis to State-of-the-Art Applications

Springer

Càndid Reig
Department of Electronic Engineering
University of Valencia
Burjassot
Spain

Dr. Subhas Chandra Mukhopadhyay
School of Engineering and
Advanced Technology
Massey University (Manawatu)
Palmerston North
New Zealand

Susana Cardoso de Freitas
INESC-MN
Rua Alves Redol
Lisboa
Portugal

ISSN 2194-8402 ISSN 2194-8410 (electronic)
ISBN 978-3-642-37171-4 ISBN 978-3-642-37172-1 (eBook)
DOI 10.1007/978-3-642-37172-1
Springer Heidelberg New York Dordrecht London

Library of Congress Control Number: 2013933304

Printed on acid-free paper

Springer is part of Springer Science+Business Media (www.springer.com)

Giant Magnetoresistance (GMR) Sensors

Editorial

The *spintronics* concept was stated during the eighties in order to describe charge transport mechanisms in which the spin of the electrons play an important rule. The Giant Magnetoresistance effect, reported at the end of this decade, was the definitive step in pushing this concept towards technological applications. Initially, read heads in hard disk drives were benefited of this emerging technology. The bits magnetically saved onto the surface of the disk are read by the binary changing of the resistance of GMR devices. With the time, and taking advantage of the gained knowledge from this research, new multilayered structures were engineered in order to make use of the linear capabilities of these devices. Their inherent properties regarding high sensitivity, large scale of integration and compatibility with standard CMOS circuitry have proved definite advantages and have placed GMR devices as the preferable magnetic sensors in most of the current applications.

This manuscript is dedicated to draw the scope of the state of the art of current magnetic sensing applications based on GMR sensors. From the definition of the concepts to the analysis of several selected cases of success, the different topics related to this novel technology are described.

In the first chapter, the theoretical fundamentals are introduced. The concepts of GMR, spin valve and tunneling magnetoresistance are presented. The descriptions of the physical mechanisms are analyzed, including the sensors based linear regime.

The second chapter is dedicated to the involved microfabrication processes. Being multilayered micro-structures, the design and realization of GMR devices share some techniques with standard CMOS processes. Lithography, pattern transfer by lift-off or etching, ... steps are commonly applied. Other processes such as magnetic layers deposition are specific to GMR technology and their compatibility with standard CMOS processes is a matter of concern.

The range of detectivity of the GMR sensors is basically limited by the signal to noise ratio (SNR) instead of the signal level. A detailed description of the noise mechanisms in GMR sensors is presented in the third chapter. The analysis is particularized to spin valves and magnetic tunnel junctions. The experimental setup for performing noise measurements is provided. Biasing strategies for minimizing noise effects are also proposed.

A GMR sensor can be basically understood as a magnetic resistive sensor. In this sense, the traditional approaches developed for typical resistive sensors can be directly considered. For example, Wheatstone bridges are commonly used in GMR based sensors applications. Novel approaches such as current based biasing or resistive array read-out interfaces need to be specifically described. Chapter four analyses the state of the art of resistive sensors interfaces, focusing on GMR particularities and ranging from realization with discrete components to microelectronic designs.

Microelectronics in general and the measurement of electric current at the integrated circuit (IC) level is one of the fields in which GMR sensors are applied. After the presentation of the concept (to measure the electric current by means of the associated magnetic field) some cases of success are revised. Particular issues such as bandwidth limitations or self-heating effect are analyzed. Some applications are proposed. The compatibility with non-dedicated CMOS processes is also studied.

The automotive world has also incorporated GMR sensors to its portfolio. In this chapter some developments regarding the use of GMR sensors in automobiles at Infineon are presented. Such sensors can be used in steering angle measurement, rotor position measurement, wheel speed measurement and crank shaft speed. Data acquisition and processing in these systems is also analyzed.

Being magnetic sensors and due to their inherent advantages related to miniaturization, GMR sensors are also applied in compass magnetometers as those included in portable navigation devices such as global positioning systems (GPS) and other mobile gadgets, mainly smartphones. The compass indicates the static orientation of the user which, with the addition of gyroscopic information along with GPS data can give precise location and orientation to the users. In this chapter, the fundamental concepts and the current trends of GMR based compass magnetometers are described.

To give an idea of its maturity, GMR sensors have already left the earth and have been included as part of the instrumentation in experimental satellites. In this chapter, the use of commercial off-the-shelf GMR based 3D magnetic sensors on board a picosatellite is described. After a comparison of currently available alternative, the design of the considered system is presented. The system passed all the qualification processes.

Highly promising results are being obtained when using GMR sensors in non-destructive evaluation (NDE) processes. Two complimentary chapters are dedicated to this topic. In the first one, the inspection of failures in printed circuit boards (PCBs) is approached. The explanation of the underlying physics (eddy-current testing, ECT), the description of the GMR based scanning probe as well as the associated electronic interface and the analysis of the results, also including finite element modeling (FEM) are treated. In the other chapter, an example of GMR sensors used in biomedicine is described. A needle probe, including GMR sensors is presented. An experimental setup and procedure with agar injected with magnetic fluid to simulate actual clinical process was developed.

For improving the performance of NDE processes based on GMR sensors, the use of arrays can be considered. In the last chapter, this approach is analyzed; comparing figures of merit with previously stated systems such as SQUIDs, Hall effect or giant magnetoimpedance (GMI) based ones. Then the use of magnetoresistive sensors in imaging and scanning microscopy is particularly demonstrated for die level fault isolation.

From all described work we can conclude that we can definitively change our appreciation on GMR sensors. In few years they have moved from being 'potential' to become the first option in much of the applications requiring the sensing of low magnetic fields.

Càndid Reig
Susana Cardoso de Freitas
Subhas Chandra Mukhopadhyay

Dr. Càndid Reig
Department of Electronic Engineering
Superior Technical School of Engineering
University of Valencia

Av. de la Universitat
E46100- Burjassot
Spain

Tlf. (Office):	(+34) 9635 44038
Tlf. (Lab):	(+34) 9635 43426
Fax:	(+34) 9635 44353
e-mail:	Candid.Reig@uv.es

Dr. Càndid Reig received the B.Sc. and Ph.D. degrees in Physics from the University of Valencia, Valencia, Spain, in 1994 and 2000, respectively.

In 1997, he joined the Department of Applied Physics, University of Valencia, where he developed his doctoral thesis, dealing on properties of semimagnetic semiconductors. After one year (2001) as an Assistant Professor within the University Miguel Hernández (Elche, Spain), in 2002, he was an Assistant Professor with the University of Valencia. Then, he was granted a postdoctoral stay at the INESC-MN (Lisbon, Portugal), working on GMR sensors, focusing on electric current measurement.

He is currently an Associate Professor with the Department of Electronic Engineering, University of Valencia, where he teaches and does research. His current research interests include the design of novel GMR-based sensors for low field applications, their characterization, modeling and the development of integrated microelectronic interfaces.

Dr. Reig has participated as researcher and principal investigator in national and international research projects related to magnetoresistive devices and their industrial applications. He has also advised PhD students within these topics. As results, he has published more than 40 journal papers, contributed more than 50 works to conferences and coauthored 4 book chapters.

Dr. Reig actively participates in periodic international conference program committees (ICST, SENSORDEVICES, . . .), in research journals' editorial boards (Sensors and Transducers, Microelectronics and Solid State Electronics, . . .) as well as acting as a referee for several research journals.

Dr. Susana Cardoso de Freitas
INESC Microsistemas e Nanotecnologias

Rua Alves Redol, 9
1000-029 Lisboa
Portugal

Tlf. (direct line):	(+351) 21 3100380
Tlf. (secretary):	(+351) 21 3100237
Fax:	(+351) 21 3145843
e-mail:	scardoso@inesc-mn.pt

Dr. Susana Cardoso de Freitas (born in Sintra, Portugal in 1973) received her Ph.D in Physics from Instituto Superior Técnico (IST, Lisbon) in 2002, from her studies of GMR multilayers and magnetic tunnel junctions for read head and MRAM applications. The research was done at INESC (Lisbon). In 2000 she was a "Co-op Pre-Professional Engineer" at IBM, T.J.Watson Research Center (USA). Since 2002 she has been at INESC-Microsystems and Nanotechnologies (INESC-MN), initially as post-doctoral fellow and presently as Senior Researcher, developing ion beam deposition of materials (tunnel junctions and spin valves) on up to 200mm wafers, for application in sensors, biochips and MRAMs. She is the responsible for technology transfer and services related to magneto-resistive sensors and thin film materials and has been involved in several collaborations for magneto-resistive materials optimization and sensor fabrication (down to 50nm dimensions), being co-author of several papers on thin film materials, magnetoresistive sensors for biomedical applications, biochips with integrated MR sensors, hybrid MEMS/MR devices or nano-devices for spintransfer.

Since 2003 she is also an Invited Auxiliary Professor at the Physics Department of IST (Lisbon), and is responsible for student coordination (over 12 master students, 3 PhD student and 5 post-doctoral fellows).

She has coordinated 4 research projects funded by Portuguese Ministry of Science and Technology (FCT) related to magnetoresistive device optimization, noise characterization, and has been the portuguese responsible for 3 EU Marie-Curie RTN (SPINSWITCH, ULTRASMOOTH, SpinICur) and 1 ICT project (Magnetrodes). She is co-author of 140 publications, 25 conference proceedings and 5 book chapters.

Dr. Subhas Chandra Mukhopadhyay
Professor of Sensing Technology, School of Engineering and Advanced Technology,

DISTINGUISHED LECTURER, IEEE SENSORS COUNCIL
FIEEE (USA), FIEE (UK) and Topical Editor, IEEE Sensors Journal,
Associate Editor, IEEE Transactions on Instrumentation and Measurements,
Technical Editor, IEEE Transactions on mechatronics,
Postgraduate Advisor, School of Engineering and Advanced Technology,
Editor-in-Chief, International Journal on Smart Sensing and Intelligent Systems,
(www.s2is.org)

AG-HORT Building AH 3.77
Massey University (Manawatu)
Palmerston North
New Zealand
Tel : +64-6-3505799 ext. 2480 Fax : +64-6-350 2259
email : S.C.Mukhopadhyay@massey.ac.nz
Personal web: http://seat.massey.ac.nz/personal/s.c.mukhopadhyay

Dr. Subhas Chandra Mukhopadhyay graduated from the Department of Electrical Engineering, Jadavpur University, Calcutta, India in 1987 with a Gold medal and received the Master of Electrical Engineering degree from Indian Institute of Science, Bangalore, India in 1989. He obtained the PhD (Eng.) degree from Jadavpur University, India in 1994 and Doctor of Engineering degree from Kanazawa University, Japan in 2000.

Currently, he is working as a Professor of Sensing Technology with the School of Engineering and Advanced Technology, Massey University, Palmerston North, New Zealand. His fields of interest include Smart Sensors and Sensing Technology, Wireless Sensors Network, Electromagnetics, control, electrical machines and numerical field calculation etc.

He has authored/co-authored over 270 papers in different international journals and conferences, edited nine conference proceedings. He has also edited ten special issues of international journals and twelve books with Springer-Verlag as guest editor.

He is a **Fellow** of **IEEE**, a **Fellow** of **IET** (UK), an associate editor of IEEE Sensors journal and IEEE Transactions on Instrumentation and Measurements. He is also a Technical Editor of IEEE Transactions on Mechatronics. He is a Distinguished Lecturer of IEEE Sensors council. He is in the editorial board of many international journals. He has organized many international conferences either a General Chair or Technical programme chair.

Contents

Spintronic Phenomena: Giant Magnetoresistance, Tunnel Magnetoresistance and Spin Transfer Torque

C. Baraduc, M. Chshiev, and B. Dieny

SPINTEC UMR8191, CEA-CNRS-UJF-INPG, 17 rue des martyrs, Grenoble, France
{claire.baraduc,mair.chshiev,bernard.dieny}@cea.fr

Abstract. This introduction to spintronic phenomena deals with the three major physical effects of this research field: giant magnetoresistance, tunnel magnetoresistance and spin transfer torque. This presentation aims at describing the concepts in the simplest way by recalling their historical development. The correlated technical improvements mostly concerning material issues are also described showing their evolution with time.

In recent decades, progress in fabrication and characterization of systems with reduced dimensionality has stimulated fundamental research on a wide range of quantum phenomena and has enabled development of nanomaterials with new functionalities related to new information technologies. The most remarkable event, in this context, is the discovery of giant magnetoresistance (GMR) in magnetic multilayered structures in 1988 by the groups of A. Fert [1] and P. Grünberg [2]. They observed a significant change in the resistance of multilayers when the magnetizations of adjacent ferromagnetic layers separated by a nonmagnetic spacer were brought into alignment by an applied magnetic field. In other words, they showed that the electron flow through the structure is controled by the relative orientation of magnetizations in adjacent layers similarly to a polarizer/analyzer optical experiment (Fig. 1). The discovery of this *spin filtering* effect opened new ways of exploring magnetic properties of materials by means of spin-dependent transport and generated a new field of research called spin electronics or spintronics [3, 4, 5, 6], which combines two traditional fields of physics: magnetism and electronics. In other words, it is not only the electron charge but also the electron spin that is used to operate a device. Spin is the intrinsic angular momentum of a particle which, in the case of the electron, is characterized by a quantum number equal to 1/2 with two possible states called "spin-up" and "spin-down" (or "majority" and "minority"). In ferromagnetic materials, the Coulomb interaction and Pauli exclusion principle cause a long-range ordering of the unpaired up (or down) spins leading to the finite magnetic moment μ per unit volume (magnetization M) resulting from the difference of majority and minority density of states (DOS). Furthermore, such inequality of the DOS for two spin states at the Fermi surface leads to significantly different conductivities for the spin up and the spin down electrons as was demonstrated by A. Fert and I. Campbell in the late 1960s [7, 8]. Along with the existence

C. Reig et al.: *Giant Magnetoresistance (GMR) Sensors*, SSMI 6, pp. 1–30.
DOI: 10.1007/978-3-642-37172-1_1

of the long range interlayer coupling between two ferromagnets separated by a nonmagnetic spacer [9], these observations were the key steps in the discovery of GMR suggesting that the transport in ferromagnetic materials is spin-dependent and can be considered within the two current model [10]. Giant magnetoresistance became the supreme manifestation of spin-dependent transport and was recognized by the award of the Nobel Prize 2007 to A. Fert and P. Grünberg.

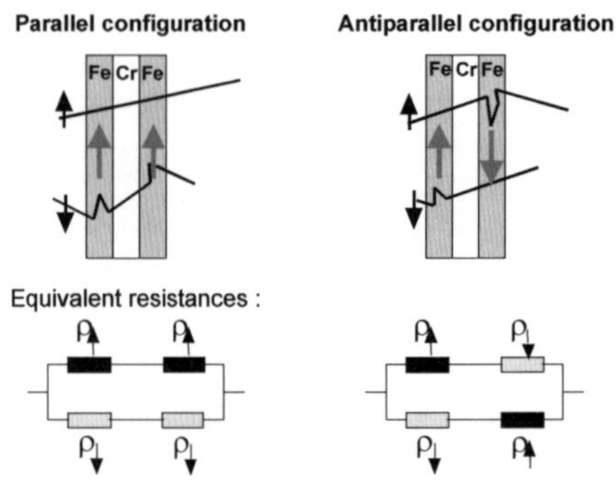

Fig. 1. In the parallel magnetic configuration, one spin species easily flows through the device, leading to a low electrical resistance. By contrast, in the antiparallel magnetic configuration, resistance is high: this case is similar to extinction in a optical polarizer-analyzer set-up.

The advent of GMR has renewed the interest in spin dependent tunneling [11] across semiconductors and insulators and led to demonstrations of high tunnel magnetoresistance ratios (TMR) at room temperature [12, 13] in magnetic tunnel junctions (MTJs). MTJs consist of two semi-infinite ferromagnetic electrodes separated by an insulating barrier and have been objects of great interest from both fundamental and applied perspectives. High sensitivity to magnetic fields makes these structures good candidates for hard drive magnetoresistive read heads, logic devices and magnetic random access memories [14, 15, 16, 17]. Various aspects have been addressed, both theoretically and experimentally, regarding spin dependent properties in MTJs such as the role of disorder and impurities at the interfaces between ferromagnetic and oxide layers, the impact of the junction composition on TMR ratio, temperature dependence of the latter, etc [18, 19, 20, 21, 22, 23, 24, 6, 25, 26, 27, 28, 29, 30]. A major breakthrough in the area of spin-dependent tunneling was the prediction of extremely high tunnel magnetoresistance ratios for certain epitaxially grown magnetic tunnel junctions [31, 32, 33, 34]. This prediction was based on a spin filtering effect that may arise from the symmetry of the wave functions [35]. Recent experiments [36, 37, 38, 39] largely confirmed predictions made by W. H. Butler et al. [32] and J. Mathon et al. [33] for MgO based tunnel junctions. This discovery also has a large impact on the technologies of MRAM and of read sensors for hard drives.

Another major phenomenon which plays a crucial role in spintronics was proposed by J. Slonczewski [40] and L. Berger [41] who predicted that a spin-polarized current can induce magnetic switching and dynamic excitations in ferromagnetic thin films. This *spin transfer torque* effect can be viewed as the reciprocal effect of spin filtering which is responsible for GMR and TMR. It turns out that the relative magnetic configuration can be controlled by the spin-polarized currents themselves. These predictions inspired a great deal of work aimed at understanding the interactions between spin-polarized currents and ferromagnetic nanostructures [42, 43, 44, 45]. Experiments later confirmed that this effect can lead to current-controlled hysteretic switching in magnetic nanostructures in moderate (or even zero) applied magnetic field [46, 47, 48, 49, 5, 50, 51, 52, 53, 54]. This behavior is not only of scientific interest but also finds potential applications in spintronic devices such as current controlled switching of magnetic random access memories (MRAM) elements and has implications for the signal-to-noise ratio of magnetic hard-disk read heads [55].

This chapter is divided into three parts. The first one is devoted to transport in GMR: it briefly presents the historical development of GMR and the related materials issue as well as the question of spin dependent transport in ferromagnets, then it addresses transport properties of GMR in the two geometrical configurations (CIP and CPP) and the spin accumulation effect. The second section is devoted to TMR: it presents the evolution of the understanding of the spin dependent transport in magnetic tunnel junction and the specificity of MgO-based tunnel junctions linked to symmetry filtering. Finally, the third section addresses transport properties in non-collinear magnetizations configuration and spin transfer torque in both systems, GMR and TMR.

1 Giant Magnetoresistance

1.1 Short History of GMR

After the discovery of GMR by A. Fert [1] and P. Grünberg [2], this magnetoresistive effect has been observed in many multilayered structures of the form $(F/NM)_n$ where F is a transition metal magnetic layer (Fe, Co, Ni or their alloys) and NM is a non-ferromagnetic transition metal or a noble metal ($V, Cr, Nb, Mo, Ru, Re, Os, Ir, Cu, Ag, Au$). In such magnetic multilayers, the thickness of each individual layer is typically in the nanometer range and the thickness of the normal metal layer is chosen in order to get spontaneous antiparallel alignment of magnetizations in successive ferromagnetic layers. So the zero-field resistance is high and significantly decreases when a large applied magnetic field aligns magnetizations of all layers. The measured magnetoresistance is much larger than the anisotropic magnetoresistance and was therefore called "giant". It ranges from a tenth of a percent in V or Mo based multilayers to more than 100% in Fe/Cr [56] or Co/Cu multilayers [57, 58]. The GMR effect was also observed in multilayers with double coercivity such as $(Ni_{80}Fe_{20}/Cu/Co/Cu)_n$ [59]: at high field magnetizations in successive magnetic layers are parallel, whereas they are in antiparallel alignment when the applied field is comprised between the two coercive fields. These systems also exhibit a large magnetoresistive ratio (MR): 16% variation in resistance was observed between 0 Oe and 50 Oe at 300 K in

$NiFeCo/Cu/Co/Cu$ multilayers [60]. Nevertheless both structures, either the antiferromagnetically coupled multilayers or the double coercivity ones, present some technical drawbacks for magnetic sensors application. The first ones are usually only sensitive to very high field (of the order of kOe), necessary to overcome the antiferromagnetic coupling, and the latter are sensitive to the magnetic history of the structure: when the soft layer switches, the resistance either increases or decreases depending on the direction of the hard layer magnetization. All these drawbacks were overcome with the invention of a third type of structure called spin-valve [61]. It essentially consists of two magnetic layers separated by a non-magnetic spacer layer. The magnetization of one of the layer is pinned by exchange interaction with an adjacent antiferromagnetic layer, whereas the magnetization of the other layer is free to rotate in an applied field. Since the two layers are very weakly coupled, the change from parallel to antiparallel magnetic configurations can occur in small fields giving these systems a large sensitivity. Since their discovery in 1990, the magnetic and transport properties of spin-valves have been considerably improved especially in terms of magnetic stability of the pinned layer and MR amplitude. The overall structural quality of the stack has been highly enhanced by using appropriate buffer layer such as $NiFeCr$ alloys. The pinning strength of the antiferromagnetic layer has been improved thanks to the use of Mn based alloys, in particular ordered

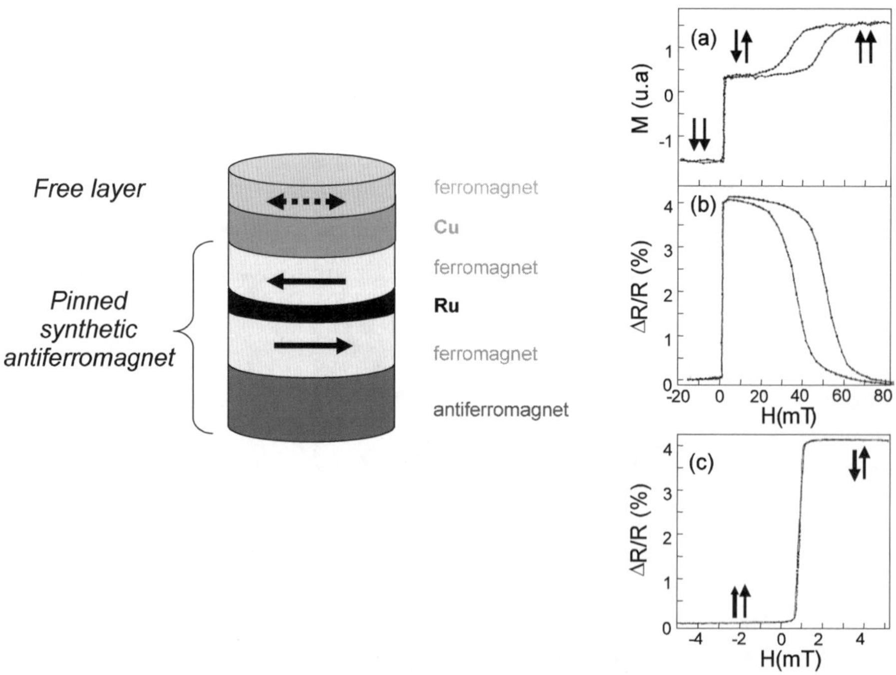

Fig. 2. Left: structure of a spin-valve where the pinned layer is a synthetic antiferromagnet. Right: first result on the GMR effect in spin-valves [61]; (a) room-temperature hysteresis loop ; (b) magnetoresistance ;(c) MR response when the free layer switches between the parallel and antiparallel configuration.

alloys such as $Pt_{50}Mn_{50}$, which provide a large bias field with a high blocking temperature.

Moreover the magnetic stability of the pinned layer has also been enhanced by the introduction of so-called synthetic antiferromagnetic pinned layer (Fig. 2) composed of two $1.5\ nm - 3\ nm$ thick ferromagnetic layers (typically $CoFe$ alloys) antiferromagnetically coupled through a thin Ru layer [62]. The strong antiparallel coupling between the two ferromagnetic layers is due to RKKY interaction when the Ru layer thickness is about $0.5 - 1.0\ nm$. The antiparallel alignment of the magnetizations of the two ferromagnetic layers result in a reduced net moment compared to a single pinned layer. Therefore the torque exerted by the applied field on the synthetic antiferromagnet is reduced compared to a single pinned layer thus improving its magnetic stability. Another advantage of using synthetic antiferromagnetic pinned layers is that, in a microscopic device, the antiparallel alignment of the magnetization in the synthetic antiferromagnet reduces the magnetostatic stray field created by the pinned layer on the sensing layer [63]. The GMR amplitude has also been significantly enhanced (up to 20%) by improving the overall structural quality of the stack and by adding Co-rich interfacial layers. Starting from $NiFe/Cu/NiFe/FeMn$ spin-valves, it was observed that introducing very thin Co or Co-rich $CoFe$ alloys at the interfaces between $NiFe$ and Cu can lead to a doubling of the MR amplitude at room temperature [64]. This result is explained by the fact that Co is much less miscible in copper than Ni. Intermixing at the $NiFe/Cu$ interfaces reduces the exchange stiffness leading to magnetic excitations around the interfaces that are responsible for spin-flip scattering of the conduction electrons. This spin memory loss has a detrimental impact on the MR amplitude. In contrast, when Co is introduced at the interface between $NiFe$ and Cu, it constitutes a good barrier against diffusion between $NiFe$ and Cu. Furthermore Co has a much higher Curie temperature than $NiFe$ ($T_{c,Co} = 1400\ K$, $T_{c,NiFe} = 800\ K$). Therefore, the Co insertion leads to interfacial magnetic stiffening thereby reducing the amount of magnetic fluctuations along the interfaces.

1.2 Spin Dependent Transport in Ferromagnetic Transition Metals

Let us now discuss the physical origin of the GMR effect. In fact, all spintronic phenomena (GMR, TMR and spin transfer torque) are related to the fact that electronic transport is spin dependent. In 1971, it was experimentally demonstrated using superconductor / ferromagnetic tunnel junctions that conduction electrons in ferromagnetic transition metals are spin polarized [65]. This experiment was then theoretically explained by considering that itinerant s-electrons are hybridized with the spin-polarized localized d-electrons [66]. More precisely, band structure calculations show exchange splitting between spin-up and spin-down electron bands with a very complex pattern: depending on the k-vector, a given band may have an itinerant character for some k values with a nearly parabolic shape and a localized character (nearly flat band) for other k values. However dealing with the real band structure is quite cumbersome and simple models are widely used by the spintronic community since they capture most of the physics of GMR and TMR. These models state that the spin polarized current observed in ferromagnetic transition metals is due to spin polarized free electrons. By convention and regardless of the minus sign, spin and spin momentum are considered equivalent: spin-up (resp. spin-down) electrons have their momentum in the same (resp. opposite) direction as the local magnetization and are called majority (resp. minority) electrons. These two species are considered to carry current in parallel. This two currents model proposed by Mott [10] is justified because the spin-orbit coupling is weak in these relatively light elements and because magnon scattering is negligible at temperatures much below the Curie temperature. Two descriptions are mainly used (see Fig. 3). The first one is the Stoner model where the free electrons bands present exchange splitting. In that case, the spin polarized current is due to the fact that the density of state and the electron velocity at Fermi level are different for spin-up and spin-down electrons. This description is usually used to explain tunnel magnetoresistance. The second model consider that the spin dependent current is mostly due to the fact that scattering rates are different for spin-up and spin-down electrons. In most cases, scattering toward *d*-states is most frequent for minority electrons since the number of available states is much larger. Within this description, generally used for GMR, the density of state and the velocity of conduction electrons are the same for spin-up and spin-down electrons and the current polarization is only ascribed to a strong difference in the diffusion constant [67].

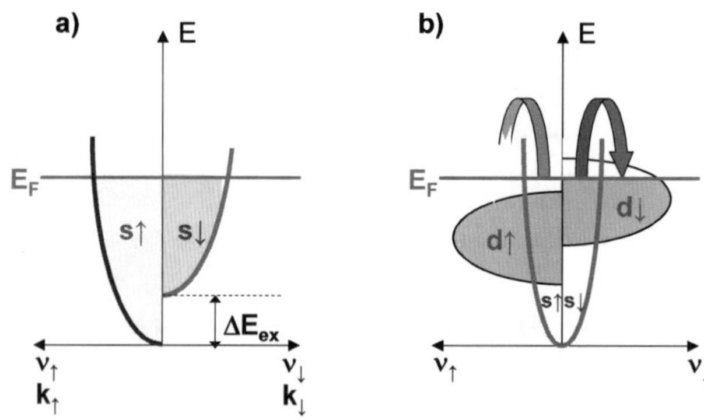

Fig. 3. Two simple models used in spintronics: a) Stoner model: exchange splitting of the free electron bands; b) large difference in the s-d scattering rate between spin-up and spin-down electrons

1.3 Current in Plane GMR

Historically GMR were electrically connected at the top surface, thus allowing the current to flow along the layers of the structure (see Fig. 4). In this current in plane (CIP) geometry, current flows in each layer in parallel, which could *a priori* not lead to magnetoresistance. In fact, the spacer between adjacent ferromagnetic layers is so thin that electrons successively travel in both layers

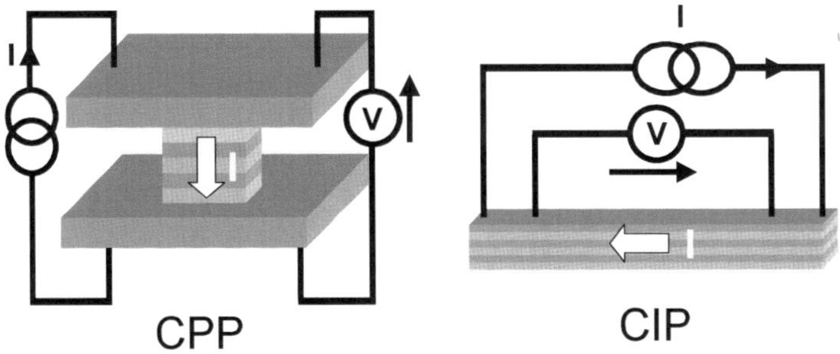

Fig. 4. Schematic representation of the difference between CPP and CIP transport geometries. In the CPP geometry, the multilayered structure is sandwiched between bottom and top electrodes.

thus alternatively probing the two ferromagnets. Magnetoresistance is therefore a second order effect that appears because of finite size effect. It can be modeled in a semi-classical approach using the Blotzmann equation in thin film, taking into account the probability p of specular reflection at the interfaces (Fuchs-Sondheimer theory). This theory is extended to the two currents model including spin dependent reflection and transmission coefficients at the interfaces. The total conductance of the structure is then calculated as the conductance of all layers in parallel corrected by a term that contains the interfaces effect. It is this last term that is responsible for the magnetoresistance. At this stage, it is worth noting that the conductance variation $\Delta G = G_P - G_{AP}$ between the parallel and the antiparallel state is the macroscopic quantity most directly related to the CIP-GMR of spin-valve structures. Other quantities, such as $\Delta R/R$ that is proportional to the signal to noise ratio, are subjected to extrinsic effects: for example, shunting of the current by outer layers affect the overall conductance without changing ΔG. A significant magnetoresistance signal can only be obtained if electrons travel easily from one ferromagnetic layer to the other which means that the non-magnetic layer must be thin, smaller than the elastic mean free path. However, at low spacer layer thickness, the parallel coupling between ferromagnetic layers increases, so that a compromise must be found between a large GMR amplitude and a coupling that is too large. This compromise is usually obtained for spacer layer thickness between 1.8 and 2.2nm of Cu. Moreover, the magnetoresistance can be optimized by a judicious choice of the ferromagnetic layers thickness. Let us first suppose that there is no specular reflection at the interfaces (rough interfaces, $p = 0$). In that case, for a very thin ferromagnetic layer, scattering will be mostly due to the interfaces whatever the electron spin, thus reducing the scattering contrast between the two electron species. When the ferromagnetic thickness is of the order of the majority electron mean free path, only minority electrons are mostly scattered within the ferromagnetic layer. This condition ensures an optimum scattering contrast, whereas a thicker ferromagnetic layer would be responsible for current shunting. The situation is different for smooth interfaces leading to specular reflection ($p = 1$). In that case, the outer interface plays the role of a mirror and the structure is equivalent to $(F/NM)_n$ multilayer with an infinite number of repeats. Then the optimum magnetoresistance is obtained when majority electrons can fully cross one ferromagnetic layer without being scattered whereas minority electrons are scattered. Therefore the optimum thickness of the ferromagnetic layer is now of the order of the mean free path of minority electrons.

1.4 Current Perpendicular to Plane GMR

Simple Resistor Network

Let us now consider the CPP geometry where the current flows perpendicular to the plane of the layers as illustrated in Fig. 4. In the simplest approximation, the various layers can be considered as carrying the current in series. In the absence of spin-flip mechanism, the transport properties can be described in a very simple two-channel resistor model. It is interesting to note that this simplest approximation can already explain the CPP-GMR, which was not the case for the CIP-GMR. Whereas consideration of finite size effects and in particular, of elastic electron mean free paths, are required for CIP-GMR, these physical lengths are no longer characteristic lengths in CPP transport. They play an indirect role through the spin-dependent resistivities but do not determine how the CPP-GMR varies as a function of the thickness of the various layers.

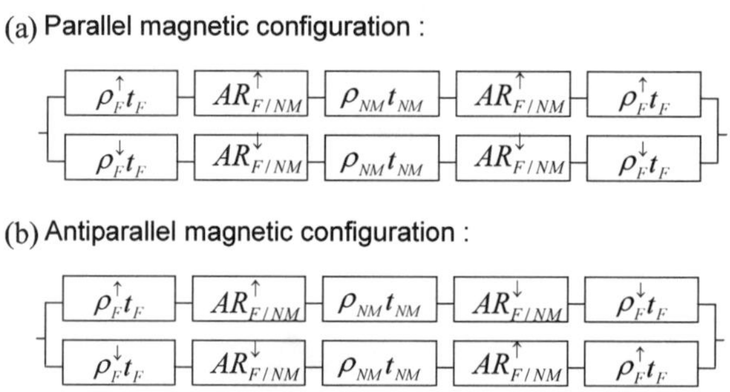

Fig. 5. Two channel resistance network which can be used to model the CPP-GMR in $F(t_F)/NM(t_{NM})/F(t_F)$ sandwiches. (a) Parallel magnetic configuration; (b) antiparallel magnetic configuration.

The serial resistance network sketched in Fig. 5 has been successfully used to interpret a number of experimental results obtained primarily at low temperature [68, 69, 70, 71, 72, 73, 74, 75, 76]. This simple model considers the spin-dependent resistivities in the ferromagnetic layers ($\rho^{\uparrow(\downarrow)}$) and the spin-dependent interfacial resistances, per unit area, at the F/NM interfaces ($AR^{\uparrow(\downarrow)}_{F/NM}$):

$$\rho_F^{\uparrow,(\downarrow)} = 2\rho_F^* \left[1 - (+)\beta\right] \quad (1)$$

$$\rho_N^{\uparrow,(\downarrow)} = 2\rho_N^* \tag{2}$$
$$AR_{F/NM}^{\uparrow,(\downarrow)} = 2AR_{F/NM}^* \left[1 - (+)\gamma\right] \tag{3}$$

Parameters β and γ, with values between -1 and 1, characterize the scattering asymmetry in the bulk and at the interface between the successive layers, respectively. The measurable resistivity of the ferromagnetic metal ρ_F and of the non-magnetic metal ρ_N and the measurable F/NM interfacial resistances are related to these parameters by the following relations:

$$\rho_F = \rho_F^*(1-\beta^2) \tag{4}$$
$$\rho_N = \rho_N^* \tag{5}$$
$$AR_{F/NM} = AR_{F/NM}^*(1-\gamma^2) \tag{6}$$

The resistor model explained a large number of results, especially at low temperature, in Co based multilayers with relatively thin layers. However, strong deviations from this model were observed in $NiFe$ based systems. Similarly, the resistor model could not explain the different magnetoresistive properties obtained in multilayers in which the ordering of the layers was changed, for example interleaved $Co\ 1nm/Cu\ 2nm/Co\ 6nm/Cu\ 2nm)_4$ multilayers compared to separated $(Co\ 1nm/Co\ 20nm)_4/(Co\ 6nm/Cu\ 20nm)_4$ multilayers [77]. The main difference between these two situations appears in the antiparallel magnetic configuration: every other Co layer has a down magnetization in the interleaved multilayers whereas in the separated multilayers, one half of the stack has magnetization up and the other half has magnetization down. The origin of the deviations from the simple resistor network were ascribed to spin relaxation and spin accumulation effects, which are presented below since they play a very important role in CPP transport [67].

Spin Accumulation

Interfacial spin accumulation can be easily understood by considering an F/NM interface such as the one depicted in Fig. 6-a. In a ferromagnet, the two spin channels do not carry the same amount of current due to the strong difference in scattering rates: the spin-up current density $J_\uparrow = \frac{J}{2}(1+\beta_j)$ is larger than the spin-down current density $J_\downarrow = \frac{J}{2}(1-\beta_j)$. By contrast, in the non-magnetic layer, the two spin currents are equal. Let us now consider a fictitious box around the F/NM interface: a large spin-up current enters this box whereas a medium spin-up current leaves this box, thus leading to accumulation of spin-up electrons within the box. These extra spin-up electrons will fill empty states above the Fermi level, so that their chemical potential $\mu_\uparrow$ is larger than the Fermi level. Similarly by considering incoming and outgoing spin-down electrons, we observe a net loss of spin-down electrons, thus leading to a chemical potential of spin-down electrons below the Fermi level (see Fig. 6-b). This excess (resp. defect) of spin-up (resp. spin-down) electrons would continuously increase with time if there were no spin-flip. In fact, a stationary state is reached that corresponds to a

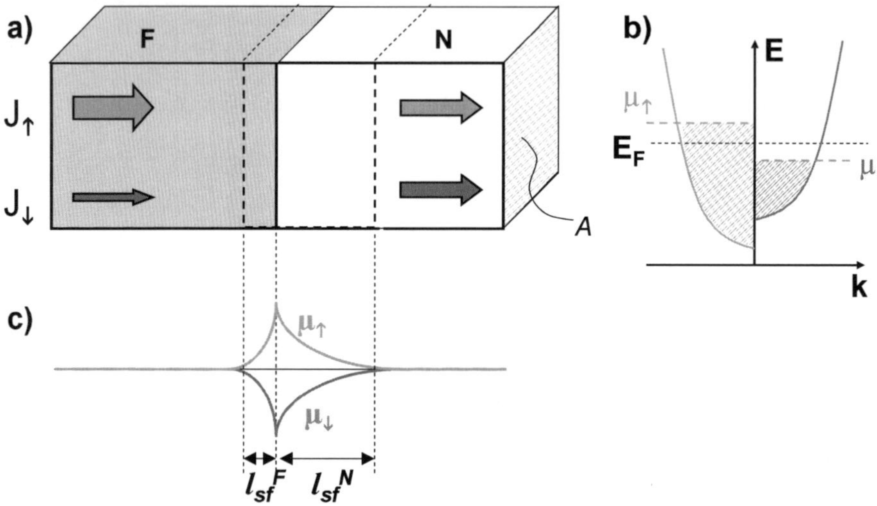

Fig. 6. Spin accumulation: (a) schematics of the current at the F/NM interface; (b) chemical potential of the two spin species due to spin accumulation; (c) evolution in space of the chemical potential (the global slope due to the voltage bias is not represented)

balance between the accumulation, which rate depends on the current value, and the spin-flip process which rate is linked to the spin-flip time τ_{sf}. The amount of excess spin-up electrons is therefore equal to the number of spin-up electrons accumulated during the first time interval τ_{sf}.

$$n_{\uparrow,acc} = \frac{JA}{2e}\beta_j \tau_{sf} \tag{7}$$

where A is the area of the multilayer structures. These accumulated spins at the interface will diffuse from the interface until they finally flip. The equilibrium will be recovered at a certain distance from the interface equal to the spin diffusion length $l_{sf} = \sqrt{D\tau_{sf}}$ where D is the mean diffusion constant $1/D = 1/D_{\uparrow} + 1/D_{\downarrow}$. The spin diffusion length l_{sf} is usually longer in non-magnetic metals than in ferromagnets. The difference in the chemical potentials is then directly related to the amount of spin accumulation in the box of volume $A(l_{sf}^F + l_{sf}^N)$:

$$\Delta\mu = \mu_{\uparrow} - \mu_{\downarrow} = \frac{1}{\nu_F}\frac{\beta_j J \tau_{sf}}{2e(l_{sf}^F + l_{sf}^N)} \tag{8}$$

where ν_F is the density of states at the Fermi level.

Taking into account spin accumulation gives a satisfactory description of transport in CPP-GMR much beyond the inaccurate model of simple resistor network. (Let us note that an accurate description must also include the interfacial spin memory loss). Calculations are performed using the two equations proposed by the Valet-Fert theory where σ is the spin, and z the direction of the current flow:

$$2e\rho^* \frac{\partial j^\sigma}{\partial z} = -\frac{\mu^\sigma - \mu^{-\sigma}}{l_{sf}^2} \tag{9}$$

$$j^\sigma = \frac{1}{e\rho^\sigma} \frac{\partial \mu^\sigma}{\partial z} \tag{10}$$

The first equation is the continuity equation and the second one the Ohm law. The characteristic length of the CPP transport is therefore the spin diffusion length. When the interfacial spin dependent scattering does not play a dominant role, the maximum magnetoresistance is obtained for ferromagnetic layer thickness of the order of the spin diffusion length. The spacer thickness must also be smaller than the spin diffusion length of the non magnetic metal, which is always the case for nanometric layers.

In concluding this section, the above discussion has shown the intrinsic differences between CIP and CPP transport. The CIP GMR cannot be modeled by a simple parallel resistor network. The lowest order of approximation requires inclusion of finite size effects for which the characteristic lengths are the spin-dependent elastic mean free paths. In contrast, CPP transport can be described at the lowest order of approximation by a two channel serial resistance network. Nevertheless, finite size effects play also an important role in CPP transport because in most practical situations, the thickness of the layers (particularly with NiFe) is of the order of the spin-diffusion length and interfacial spin-flip is significant.

2 Tunnel Magnetoresistance

2.1 Amorphous Magnetic Tunnel Junctions

The phenomenon of tunnel magnetoresistance takes its origin from pioneering experiments on spin dependent tunneling by Meservey and Tedrow [65, 78, 11] and was first reported at low temperatures by Julliere [79] in magnetic tunnel junctions comprising *Co* and *Fe* layers separated by *Ge* spacer. The resistance of this structure was increasing about ~14 % when relative configuration of adjacent ferromagnetic layers changed from parallel to antiparallel. The discovery of GMR inspired a significant breakthrough in TMR observations two decades later after Julliere's work when large room temperature TMR values were reported in amorphous Al_2O_3-based magnetic tunnel junctions simultaneously by J. Moodera et al [12] and T. Miyazaki et al [13]. First descriptions of TMR were usually based on Julliere's model which is based on two current model [10] with assumption that the tunneling conductance is proportional to the density of states of two ferromagnetic electrodes at the Fermi energy denoted as $\nu_{L(R)}^{\uparrow(\downarrow)}$ for spin up(down) electrons in the left(right) electrode, respectively (see Fig. 7). In this case, the conductance for the parallel and antiparallel magnetization configuration can be written respectively as:

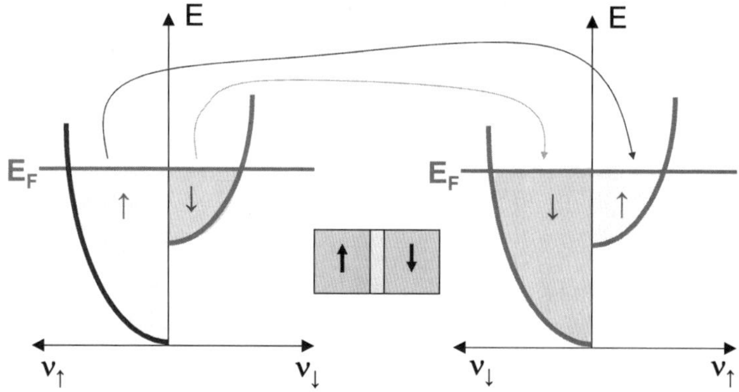

Fig. 7. Transport in the antiparallel state of a magnetic tunnel junction considering the Fermi golden rule. The conductance is proportional to the product of the densities of initial and final states.

$$G_P \propto \nu_L^{\uparrow}\nu_R^{\uparrow} + \nu_L^{\downarrow}\nu_R^{\downarrow} \tag{11}$$

$$G_{AP} \propto \nu_L^{\uparrow}\nu_R^{\downarrow} + \nu_L^{\downarrow}\nu_R^{\uparrow}. \tag{12}$$

If we now introduce a spin polarization in terms of spin polarized DOS for the left(right) electrode as

$$P_{L(R)} = \frac{\nu_{L(R)}^{\uparrow} - \nu_{L(R)}^{\downarrow}}{\nu_{L(R)}^{\uparrow} + \nu_{L(R)}^{\downarrow}}, \tag{13}$$

the TMR can be expressed directly in terms of spin polarization according to the following expression:

$$TMR = \frac{2P_L P_R}{1 - P_L P_R}. \tag{14}$$

However, this expression is not satisfactory since the spin polarization of the tunneling conductance measured with Meservey-Tedrow technique was found positive for transition metals and their alloys implying that $\nu_{L(R)}^{\uparrow} > \nu_{L(R)}^{\downarrow}$. This is not true at least for transition metals and their alloys on the face centered cubic (fcc) side of the Slater-Pauling curve representing dependence of transition metal magnetic moment as a function of electron-to-atom ratio. Indeed, for fcc Ni,Co and Fe as well as for permalloy the majority d-band is filled yielding thereby higher DOS at the Fermi level for minority electrons, i.e. $\nu_{L(R)}^{\uparrow} < \nu_{L(R)}^{\downarrow}$ leading to negative spin polarization defined according to Eq. (13). The contradiction was resolved by M. B. Stearns [66] who suggested that the tunneling current is due to parabolic bands of the conduction electrons of these ferromagnets and depends on wave vector k which becomes spin dependent as a consequence of exchange splitting between spin-up and spin-down bands (cf 1.2). The definition of the polarization becomes:

$$P_{L(R)} = \frac{k_{L(R)}^{\uparrow} - k_{L(R)}^{\downarrow}}{k_{L(R)}^{\uparrow} + k_{L(R)}^{\downarrow}} \tag{15}$$

which is equivalent to the former definition in the case of parabolic electron bands.

This simple model could not explain, however, experimental observations where the sign of the TMR ratio depends not only on the electrode properties but also on the insulator suggesting that the spin polarization is not an intrinsic property of the ferromagnet alone but depends on the entire junction's composition comprising the insulator and the ferromagnet/insulator interfaces [23,24,21,80]. It was therefore necessary to generalize the definition of spin polarization and it was redefined in the framework of free electron model as [81,82,83]:

$$P_{L(R)} = \frac{k^{\uparrow}_{L(R)} - k^{\downarrow}_{L(R)}}{k^{\uparrow}_{L(R)} + k^{\downarrow}_{L(R)}} \cdot \frac{q_0^2 - k^{\uparrow}_{L(R)} k^{\downarrow}_{L(R)}}{q_0^2 + k^{\uparrow}_{L(R)} k^{\downarrow}_{L(R)}} \tag{16}$$

with $k^{\sigma}_{L(R)} = \sqrt{(2m/\hbar^2)(E_F - V^{\sigma}_{L(R)})}$ and $q_0 = \sqrt{(2m/\hbar^2)(U - E_F)}$ where U (resp. $V^{\sigma}_{L(R)}$)) is the barrier height (resp. spin dependent electron potential in electrodes). According to Eq. (16) the spin polarization of the tunneling current is not anymore uniquely defined by the wave vector of the ferromagnetic electrodes and contains a factor which depends on the barrier height. Let us note that for very large barrier, i.e. when $q_0 >> k^{\sigma}_{L(R)}$, one recovers the polarization defined by Eq. (15) which in its turn reduces back to Eq. (13) if one considers the DOS is proportional to the wave vector at the Fermi energy.

One can generalize the definition of the spin polarization to the case of magnetic tunnel junctions under a finite applied bias V. In this case the wave vectors become $k^{\sigma}_{L(R)} = \sqrt{(2m/\hbar^2)\left[E - V^{\sigma}_{L(R)} - (+)\mathrm{eV}/2\right] - k_{||}^2}$ and $q_{L(R)} = \sqrt{(2m/\hbar^2)\left[U - E + (-)\mathrm{eV}/2\right] + k_{||}^2}$ where $k_{||}$ indicates transverse to the transport direction wave vector and the evanescent wave vector $q(z)$ depends on position z inside the barrier of thickness a. The tunnel current density for respectively parallel and antiparallel magnetization configurations is then defined as $j_P = j^{\uparrow\uparrow} + j^{\downarrow\downarrow}$ and $j_{AP} = j^{\uparrow\downarrow} + j^{\downarrow\uparrow}$ where:

$$j^{\sigma\sigma'} = \frac{e}{2\pi h} \int dE \left[f(E) - f(E + \mathrm{eV})\right] \int T^{\sigma\sigma'}(E, \mathrm{V}, k_{||}) k_{||} dk_{||} \tag{17}$$

with transmission probabilities given by (we omit the arguments for convenience):

$$T^{\sigma\sigma'} = \frac{8 q_L q_R k^{\sigma}_L k^{\sigma}_R}{\left[q_L^2 + {k^{\sigma}_L}^2\right]\left[q_R^2 + {k^{\sigma}_R}^2\right] \cosh\left(2\int_0^a q(z)dz\right) + 4 q_L q_R k^{\sigma}_L k^{\sigma}_R - \left[q_L^2 - {k^{\sigma}_L}^2\right]\left[q_R^2 - {k^{\sigma}_R}^2\right]}.$$

This expression becomes considerably simplified in case of thick/high barrier, i.e. when the positive exponent within cosh function becomes large compared to its negative counterpart, yielding:

$$T^{\sigma\sigma'} = T^{\sigma}_L T^{\sigma'}_R \exp\left(-2\int_0^a q(z)dz\right) \tag{18}$$

where we conveniently introduced so called spin dependent interfacial transmission probabilities at the left(right) interfaces defined as:

$$T^{\sigma}_{L(R)} = \frac{4q_{L(R)}k^{\sigma}_{L(R)}}{q^{2}_{L(R)} + k^{\sigma}_{L(R)}{}^{2}}$$

and we get finally spin polarization at the left(right) interface for a tunnel junction under a finite bias:

$$P_{L(R)} = \frac{T^{\uparrow}_{L(R)} - T^{\downarrow}_{L(R)}}{T^{\uparrow}_{L(R)} + T^{\downarrow}_{L(R)}}.$$

2.2 TMR in Crystalline Tunnel Junctions

Magnetic tunnel junctions attracted an even growing interest when huge tunnel magnetoresistive effects which can reach thousands of percents were obtained in crystalline magnetic tunnel junctions comprising body centered cubic (bcc) ferromagnetic electrodes (layers) separated by rocksalt MgO insulating barriers. The nature of spin dependent transport in these systems goes beyond the free electron model and is directly related to electronic structure properties of crystalline materials involved which can be calculated using first principles calculations. For reviews on this topic one can see, for example, Refs. [35, 84, 85]. The spin filtering effect arises from the symmetry of the Bloch states in the bulk ferromagnetic electrodes and the complex band structure of the insulator [31,86]. The electrons carrying the current are described by Bloch wave functions classified with respect to their symmetry. The symmetry of the Bloch states depends on the crystal atomic structure and can be described by the allowed combination of the s, p, and d orbital characters contributing to the corresponding wave function. In this picture, electrons tunnel through the barrier as evanescent Bloch states which match the symmetry of the incident Bloch states in the electrode. In coherent tunneling picture, an independent transport channel may be associated to each symmetry of the Bloch functions characterized with a specific attenuation in the insulator barrier. Thus, a multi-channel description of transport may be proposed where each channel is described by its spin polarization, symmetry and attenuation within the barrier. The total conductivity is then the sum of the conductivities of all available channels. This description requires that the wave function symmetry is conserved as the electron traverses the interface, which in its turn is possible only if lateral two-dimensional (2-D) periodicity is preserved. In other words, it is necessary that the insulating barrier fits epitaxially to the electrodes and that 2-D order parallel to the interfaces between the barrier and the electrodes is maintained. Finally, if among Bloch states of the ferromagnetic electrodes the highly symmetric one at the Fermi energy is available only for one of two spin states, a huge conductance will appear when magnetizations of adjacent FM layers are in parallel state. At the same, due to unavailability of high symmetry Bloch state in opposite spin channel, the antiparallel conductance will be very low yielding very high TMR

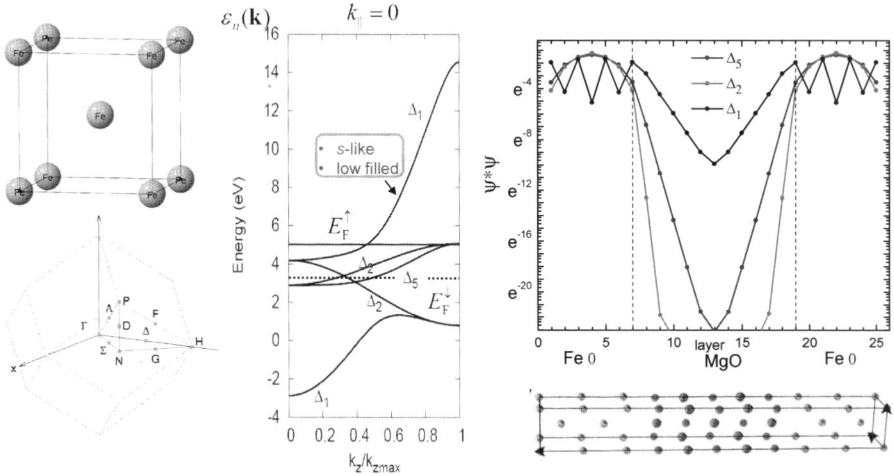

Fig. 8. (Left panel) The bcc cell and its 1st Brillouin zone with high symmetry k-points. (Middle panel) Band structure of bcc Fe along the (001) direction. The majority(minority) Fermi energy is denoted $E_F^{\uparrow}(\downarrow)$. (Right panel) The Fe(001)/MgO(001) supercell and dependence of the Bloch state character with different symmetries within the supercell along (001) direction.

ratio. Indeed, Fig. 8 shows the bcc unit cell, the 1st Brillouin zone with high symmetry k-points (left panel) and the band structure for Fe-bcc in the (001) direction (middle panel) where one can see that at the Fermi energy there are Bloch states with symmetries Δ_1, Δ_5 and Δ_2' for the majority and Δ_2, Δ_5 and Δ_2' for the minority electrons. One can see that only the Δ_1 majority-spin state (formed by s, p_z and d_{z^2}) possesses s-character at the Fermi energy. By contrast, the minority-spin states tend to have mainly p and d-character only. And this property holds for bcc $Co(001)$ and $FeCo(001)$ as well. It turns out that the Δ_1 wave function which is present in both the valence and the conduction bands of certain insulators (MgO, GaAs, ZnSe etc.) continues through the gap as an evanescent state with a relatively small decay rate [31,86,32,33,87,88,89]. This is demonstrated in the right panel of Fig. 8 where one can see that there is indeed a huge difference in the way Bloch states that live primarily on the Fe decay into the MgO. While the Δ_1 state decays relatively slowly, the other Bloch functions experience a strong decay rate within the insulator and will thus provide a small contribution to the conductance. As a result, the tunneling conductance is huge for the parallel alignment of the magnetizations since it is governed by the majority Δ_1 states, while in the antiparallel configuration this channel is almost closed since there are no available Δ_1 minority states [32, 33]. This theoretical description is largely confirmed in recent experiments [36, 37, 38, 39] for $CoFe(001)|MgO(001)$ crystalline tunnel junctions. Nevertheless, despite of a good qualitative agreement, the TMRs ratio measured in single crystal MTJs are lower than the theoretical expectations. The early results were obtained with

MBE grown tunnel junctions but higher values of TMR have been experimentally reported for systems elaborated by sputtering. In MBE grown samples, the TMR of 67% at room temperature was reported in 2002 [90]. The reduced TMR was attributed to the structural quality since it was altered by the plastic relaxation induced by strain. In sputtered samples, the plastic relaxation disappears. The MgO barrier is grown on initially amorphous electrodes subsequently re-crystallized by annealing, the ferromagnetic electrodes adopting the structure of the insulator during annealing. By using different stoichiometries for the CoFeB amorphous electrodes re-crystallized by subsequent annealing steps, the filtering efficiency reflected by the TMR ratio has been continuously enhanced. In 2004, the groups of S.S. Parkin et al in IBM Almaden and S. Yuasa et al in Japan reported simultaneously a 220% TMR for both sputtered and epitaxially grown $CoFe/MgO$ MTJs [36, 37]. Further enhancement of the TMR ratio in epitaxially grown systems is possible by increasing the spin polarization of injected electrons using "alternative" bcc ferromagnetic systems. In 2006, a 410% TMR ratio has been reported in $bcc - Co/MgO/Fe$ MTJ by Yuasa [91] and the current room temperature TMR record was reported by Ikeda et al [39]. The structural quality of the MTJ stacks as a factor impacting TMR ratios in MgO-based tunnel junctions have been widely discussed. In particular, the presence of oxygen vacancies or oxygen interstitial atoms at the ferromagnetic/insulator interface may have significant impact on the transport properties and their role has been discussed as potential interfacial resonant states [92].

3 Angular Dependence of Transport and Spin Transfer Torque

3.1 Transport for Non-collinear Magnetizations Configuration

Up to now, we only considered the case of parallel or antiparallel magnetization configuration. The case of non-collinear magnetization is more complex since the ferromagnetic layers have different quantization axis leading to spin channels mixing. When the tunnel barrier is thick/large enough, the angular conductance simply depends linearly on the cosine of the magnetizations angle:

$$G = \frac{G(0) + G(\pi)}{2} + \frac{G(0) - G(\pi)}{2} \cos\theta \tag{19}$$

Deviations from this formula only occurs for thin tunnel barrier: in that case a $\sin^2\theta$ term appears in the denominator. For current-in-plane GMR, the angular dependence of transport can approximatively be described as a $\cos\theta$ variation of the resistance:

$$r = \frac{R(\theta) - R(0)}{R(\pi) - R(0)} = \frac{1 - \cos(\theta)}{2} \tag{20}$$

In the case of CIP-GMR, experiments are consistent with this simple formula (see Fig. 9) that can be theoretically demonstrated provided there is no potential step at the interfaces. Otherwise the angular dependence is expected to be

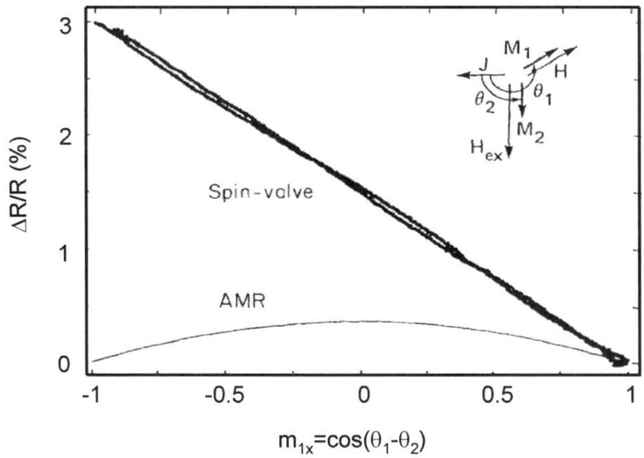

Fig. 9. Experimental demonstration that CIP GMR varies as the cosine of the angle between magnetizations in free and fixed layer [96]. This result implies that GMR sensors have a linear response around $\theta = \pi/2$.

more complex [94]. However the correction is proportional to $(\frac{k_\uparrow - k_\downarrow}{k_\uparrow - k_\downarrow})^2$ which is so small that this deviation was not experimentally observed [95]. By contrast, a deviation from the simple formula Eq. (20) was experimentally observed in CPP-GMR [97]. It was theoretically ascribed to the particular angular variation of spin-accumulation effects taking place in magnetic metallic multilayers, and described by a resistance network by Slonczewski [103]. For symmetric CPP-GMR, experimentalists usually apply the simple formula proposed by Slonczewski [103]:

$$r = \frac{1 - \cos^2(\theta/2)}{1 + \chi \cos^2(\theta/2)} \tag{21}$$

where χ is a fitting parameter.

3.2 Spin Transfer Torque

In such non-collinear magnetizations configuration, the question of spin accumulation must be reconsidered. Let us remember that in the collinear case, the current flow is responsible for an excess (or default) of one spin species leading to the appearance of a small magnetization in the paramagnetic metal [99] and a negligible modification of the magnetization amplitude in the ferromagnetic layer. The situation is quite different in the non-collinear configuration. In 1996 J. Slonczewski [40] and L. Berger [41] independently predicted that the current flowing perpendicular to the plane in a metallic multilayer can generate a spin transfer torque strong enough to reorient the magnetization in one of the layers due to a transfer of angular momentum between the propagating electrons and the background magnetization of that layer. This *spin transfer torque* could either induce steady state precession or reversal of the magnetization. In addition to its intrinsic scientific interest, this phenomenon may lead to several possible applications since it offers the possibility to monitor magnetization directly with an electrical current and not with an applied field. This effect can be easily understood by considering a GMR where the two ferromagnetic layers F_1 and F_2 have their magnetizations misaligned by an angle θ (see Fig. 10-a). Let us consider the fictitious box that include both ferromagnetic/spacer interfaces. The electrons flowing from F_1 to F_2 enter the fictitious box with a spin polarization along the direction of the magnetization $\boldsymbol{M_1}$ and leave the box with a spin polarization along $\boldsymbol{M_2}$. Since the incoming ($\boldsymbol{\mu_1}$) and outgoing ($\boldsymbol{\mu_2}$) magnetic moment is not the same, some magnetic moment is transferred to the system.

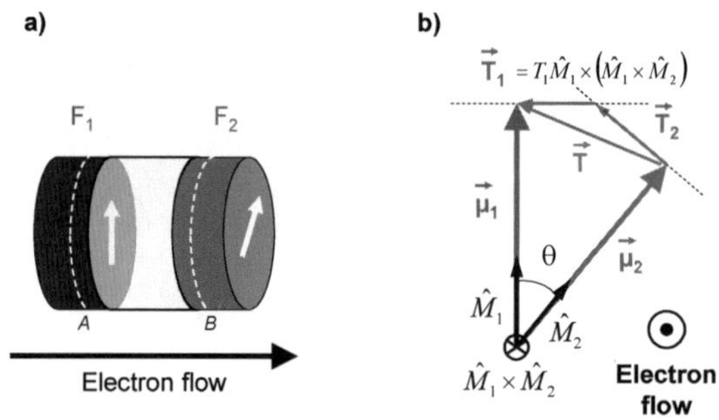

Fig. 10. Spin transfer torque: (a) schematics of a F/NM interface in non-collinear magnetic configuration; (b) in-coming ($\boldsymbol{\mu_1}$) and out-going ($\boldsymbol{\mu_2}$) magnetic moment due to electron flow and total momentum ($\boldsymbol{T} = \boldsymbol{T_1} + \boldsymbol{T_2}$) left within the fictitious box

However magnetization cannot significantly increase, so any temporal variation of $\boldsymbol{M}$ has to be perpendicular to $\boldsymbol{M}$:

$$\frac{d|\boldsymbol{M}|^2}{dt} = 2\boldsymbol{M} \cdot \frac{d\boldsymbol{M}}{dt} = 0 \tag{22}$$

Consequently only the component perpendicular to $\boldsymbol{M}$ of the electrons magnetic moment can be transferred to the bulk magnetization. The total transferred moment per time unit $\boldsymbol{T}$ is then decomposed in two parts $\boldsymbol{T_1}$ and $\boldsymbol{T_2}$ (Fig. 10-b). One part acts on $\boldsymbol{M_1}$ and the other on $\boldsymbol{M_2}$. The magnetic moment transferred to $\boldsymbol{M_1}$ is then a vector perpendicular to $\boldsymbol{M_1}$ which lies in the plane defined by $(\boldsymbol{M_1}, \boldsymbol{M_2})$. Since the unit vector normal to this plane is $\hat{M}_1 \times \hat{M}_2$, $\boldsymbol{T_1}$ and $\boldsymbol{T_2}$ are necessarily written as:

$$\boldsymbol{T_{1,2}} = T_{1,2}\hat{M}_{1,2} \times (\hat{M}_1 \times \hat{M}_2) \tag{23}$$

By analogy to magnetization dynamics, where the time derivative of the magnetization is equal to the torque exerted by the magnetic field, this transferred moment per time unit was originally called "pseudo-torque" by Slonczewski [100]. It is now universally called spin transfer torque. Nevertheless we must keep in mind that this term is not exactly a torque but stems directly from transverse spin accumulation. In a typical spin-valve setup, one of the magnetic layers is pinned by exchange bias [101, 102, 103] to an antiferromagnet while another is allowed

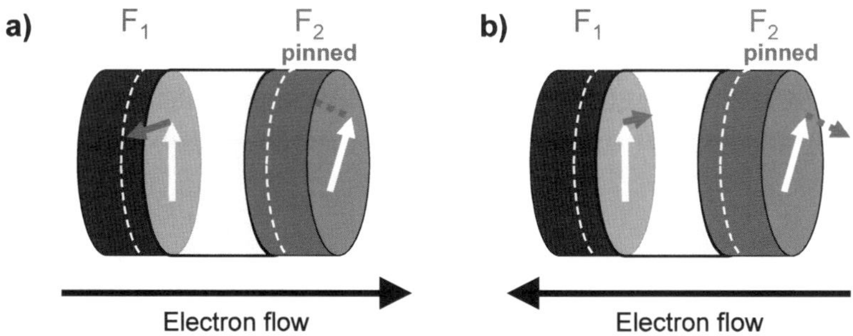

Fig. 11. Spin transfer torque under current reversal: (a) the anti-parallel configuration is favored when electrons flow from the free to the fixed layer; (b) the parallel configuration is favored when the current flow is reversed

to switch. In the case being considered, pinning the magnetization $\boldsymbol{M_2}$ favors anti-parallel (AP) alignment of magnetization of the "free" layer $\boldsymbol{M_1}$. When the current direction is reversed, the spin transfer torque will now align $\boldsymbol{M_1}$ and $\boldsymbol{M_2}$ in parallel (P) configuration (see Fig. 11). In a magnetic tunnel junction, spin transfer torque depends on $\sin\theta$ as expected from Eq.(23). However in metallic multilayer, the angular dependence of the torque is more complex since the angular dependence of the current is also asymmetric (cf 3.1). Therefore, in the most general case, the spin torque on the free-layer magnetization is written:

$$\boldsymbol{T}_{//} = \gamma_0 \frac{a_j}{M_s} \boldsymbol{M} \times (\boldsymbol{M} \times \hat{p}) \tag{24}$$

where γ_0 is the gyromagnetic ratio, M_s is the saturation magnetization, $\hat{p}$ the magnetic polarization of the fixed layer and the subscript $//$ recalls that the torque is parallel to the plane that contains the two magnetization vectors. In the case of a metallic multilayer, the prefactor a_j is angle dependent [98]:

$$\gamma_0 a_j = g(\theta) \frac{\mu_B I}{e M_s V} \tag{25}$$

where I is the DC applied current, V is the ferromagnetic layer volume and $g(\theta) = q_+/(B_0 + B_1 \cos\theta) + q_-/(B_0 - B_1 \cos\theta)$. The angular dependence of the torque is represented in Fig. 12. However, in some specific cases where the GMR is asymmetric, one may observe a much complex angular dependence [104]. In this so-called "wavy-structures" containing two different ferromagnetic layers such as $NiFe/Cu/Co$, one layer is smaller than the spin diffusion length and one is larger. The specific angular dependence observed in these systems may be explained with a simple geometrical argument [105] as shown in Fig. 12.

Most of the existing theories of spin transfer torque in metallic spin-valves use semiclassical approaches [106, 50, 107, 108, 109, 110] based on a generalization of the Valet-Fert theory for CPP-GMR [67] to the case of noncollinear magnetizations since the transport in such structures is usually considered to be diffusive

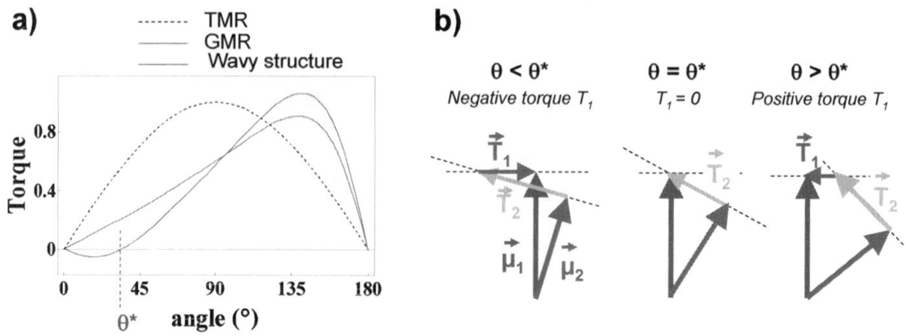

Fig. 12. Spin transfer torque: (a) angular dependence of the torque in the case of magnetic tunnel junction, giant magnetoresistance and wavy-structures; (b) simple geometrical argument explaining the origin of the specific angular dependence of the torque in wavy-structures (adapted from [105])

and quantum coherent effects are not important. To account for the influence of quantum coherence effects on spin transfer torque in the ballistic transport regime, several groups used approaches based on non-equilibrium Green functions or scattering formalism [42, 111, 112, 113]. The majority of these theories including the original ones by J. Slonczewski [40] and L. Berger [41] considered only the in-plane component of spin transfer torque discussed above assuming that the electron spin remains in the plane formed by $\boldsymbol{M_1}$ and $\boldsymbol{M_2}$. However, there exists another "out-of-plane" term [44,45,111] of spin transfer torque which acts perpendicularly to the plane of the magnetizations $\boldsymbol{M_1}$ and $\boldsymbol{M_2}$ which is variously called "perpendicular" or "field-like" and written as $T_\perp \hat{M}_1 \times \hat{M}_2$. Contrary to the in-plane torque, this term also exists in the absence of current flow. It can be viewed as a magnetic coupling between the two ferromagnets similarly to the RKKY coupling observed in the collinear magnetizations configuration in $F_1/N/F_2$ structures. In the RKKY case, the magnetic coupling between the two ferromagnets is mediated by the conduction electrons that "feel" both interfaces [114, 115]. A similar origin may explain the magnetic coupling in the non-collinear configuration. At first sight, it may seem strange to consider magnetic coupling as a torque. In fact, it is formally the same. From the coupling energy $E_{ex} = J\boldsymbol{M_1} \cdot \boldsymbol{M_2}$, one may extract the magnetic field applied on layer i: $\boldsymbol{h_i} = \partial E_{ex}/\partial \boldsymbol{M_i} \propto \boldsymbol{M_j}$ $(i \neq j)$. And this field exerts a torque on the magnetization $\boldsymbol{M_i}$, in each ferromagnetic layer $i = 1, 2$:

$$\boldsymbol{T}_{i,\perp} = -\gamma_0 \boldsymbol{M_i} \times \boldsymbol{h_i} \propto -\boldsymbol{M_i} \times \boldsymbol{M_j} \qquad (\, i \neq j) \tag{26}$$

Here the subscript $\perp$ emphasizes the fact that this torque is transverse to the plane that contains $\boldsymbol{M_1}$ and $\boldsymbol{M_2}$, contrary to the longitudinal spin transfer torque. Let us also notice that the torques on both magnetizations are equal and opposite: $\boldsymbol{T}_{1,\perp} = -\boldsymbol{T}_{2,\perp}$ since no net out-of-plane moment is deposited by the current flow. The appearance of a perpendicular moment on each interface is due

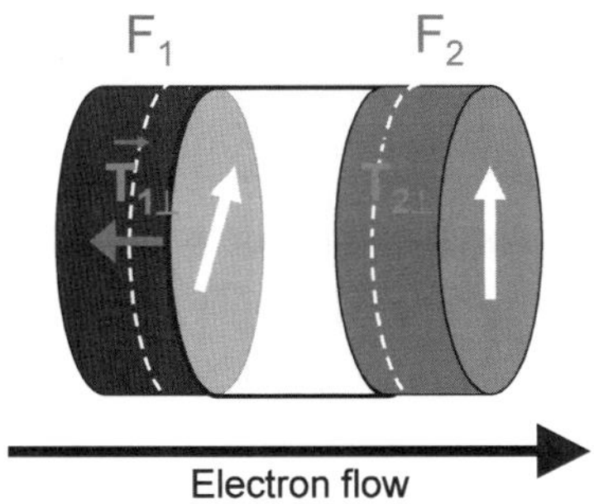

Fig. 13. The out-of-plane torque is equal and opposite on the two interfaces. It stays the same under current reversal in the case of symmetric structures [116].

to the spin precession of the electron that enters the ferromagnet as a vanishing wave before being reflected [117]. Since the reflected electron gained some magnetic moment in the perpendicular direction, electrons indefinitely reflected by the two interfaces contribute to a transverse torque on both ferromagnets. However this term is usually small in the case of metallic spin valves [44, 118], of the order of a few percents of the longitudinal torque. By contrast, the out-of-plane torque plays a particularly important role in the case of magnetic tunnel junctions under finite applied voltages [119,120,110,121,122,123,124], and can reach 20% of the longitudinal torque. It is often called *interlayer exchange coupling* or *conservative exchange coupling* [81]. The difference in the out-of-plane torque amplitude between GMR and TMR is related to the fact that in magnetic tunnel junctions, only the directions of incidence close to the perpendicular are selected, since the effective barrier thickness is much larger for grazing incidence. In that case, there is less classical dephasing between precessing electrons, thus allowing a significant out-of-plane moment to build up at the interfaces. By contrast to longitudinal torque, the out-of-plane torque does not change sign by current reversal. This property can be easily understood by considering Fig. 13. For a symmetric structure, current reversal can be imagined by rotating the picture: the current now flows in the opposite direction and the out-of-plane torques still point outwards [116]. This was theoretically demonstrated (see Fig. 14), showing that the transverse torque is an even function of the applied voltage [120]:

$$T_{\perp}(V) = T_0 + T_1 V^2 \tag{27}$$

In asymmetric junctions however, a linear term in V appears in the expression of the transverse torque. This theoretical expression was recently confirmed by experimental data [125].

The effect of spin transfer torque on magnetization dynamics can be described by adding both STT terms to the Landau-Lifshitz-Gilbert(LLG) equation [45]:

$$\frac{d\mathbf{M}}{dt} = -\gamma_0 \mathbf{M} \times \mathbf{H_{eff}} + \frac{\alpha}{M_s} \mathbf{M} \times \frac{d\mathbf{M}}{dt} - \frac{|g_e|\mu_B}{\hbar} \mathbf{T} \tag{28}$$

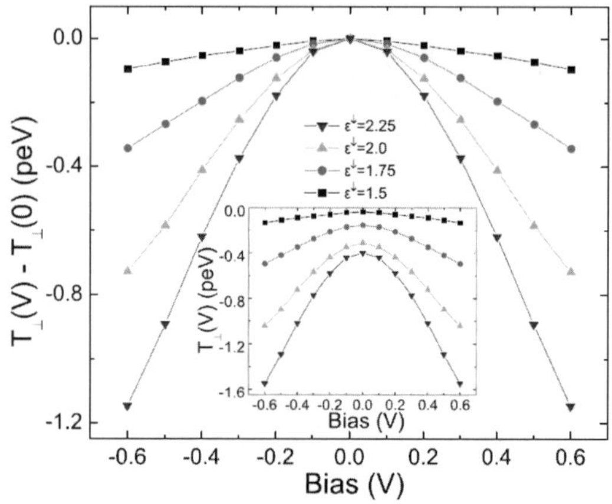

Fig. 14. Bias dependence of the current-induced perpendicular component of the net spin torque per unit area, for $\theta = \pi/2$, and for various values of the on-site energy for spin down [120]

where

$$\mathbf{T} = T_{||}\hat{\mathbf{M}} \times (\hat{\mathbf{M}} \times \hat{\mathbf{M}}') + T_{\perp}\hat{\mathbf{M}} \times \hat{\mathbf{M}}' \tag{29}$$

and γ_0 is the gyromagnetic ratio, g_e and α represent the electron g-factor and the Gilbert damping, respectively, and $\mathbf{H_{eff}}$ is the effective field which includes the anisotropy field, the demagnetizing field and the external applied field. From Eqs. (29) and (28), the out-of-plane torque acts as an effective field while the in-plane torque acts as an effective (anti-)damping. As a function of its sign, $\mathbf{T}_{||}$ may excite or damp the magnetization $\mathbf{M}$, whereas $\mathbf{T}_{\perp}$ only affects the energy surface of the ferromagnetic layer. Different magnetic behavior may be observed: magnetization switching from a stable state to another, stabilization of magnetic states at low energy minima, or coherent and incoherent precessions [126, 127, 128, 129, 130]. In particular, at low field when the energy cost to reverse magnetization is not too high, spin transfer torque may overcome the damping and produce magnetization switching. This effect is thoroughly studied for MRAM writing as an alternative writing scheme which would be downsize scalable. At larger field, steady-state precessions of the magnetization may pave the way towards tunable RF-oscillators in the GHz frequency range. A considerable scientific activity in this field is devoted to improving the output power and the quality factor of such oscillators. For the magnetic sensors application however, the existence of spin transfer torque is less favourable: if the bias current overcomes the critical value, spin transfer torque induces magnetic excitations that are responsible for supplementary noise. In this area, a significant effort is devoted to develop materials with a larger Gilbert damping.

4 Conclusion

These last few examples show the interaction between fundamental research and applicative development that is so characteristic of spintronics. Less than

25 years have elapsed since the first discovery of GMR, and many applications of spintronics already exist, from the first GMR read-heads for hard-disk drive now replaced by TMR, to the more recent MRAMs already launched into the market. Spintronics is also a very active research field that has shown a rapid evolution, with many decisive steps : the discovery of GMR, the development of TMR, the experimental evidence of the spin transfer torque in nanometric structures and the variety of effects it causes. This rapid evolution seems not to dim: new effects are currently investigated such as Rashba effect, spin-dependent thermoelectric effects or anisotropy controled by electric field; new materials are being studied like magnetic semiconductors, organic GMR or graphene based GMR; and finally new applications are foreseen such as racetrack memories or RF-oscillators. Spintronics still offers many promising routes to explore.

References

[1] Baibich, M.N., Broto, J.M., Fert, A., Nguyen Vandau, F., Petroff, F., Etienne, P., Creuzet, G., Friederich, A., Chazelas, J.: Giant Magnetoresistance of (001)Fe/(001) Cr Magnetic Superlattices. Physical Review Letters 61(21), 2472–2475 (1988)

[2] Binasch, G., Grunberg, P., Saurenbach, F., Zinn, W.: Enhanced Magnetoresistance In Layered Magnetic-Structures With Antiferromagnetic Interlayer Exchange. Physical Review B 39(7), 4828–4830 (1989)

[3] Ohno, H.: Making nonmagnetic semiconductors ferromagnetic. Science 281(5379), 951–956 (1998)

[4] Wolf, S.A., Awschalom, D.D., Buhrman, R.A., Daughton, J.M., von Molnar, S., Roukes, M.L., Chtchelkanova, A.Y., Treger, D.M.: Spintronics: A spin-based electronics vision for the future. Science 294(5546), 1488–1495 (2001)

[5] Fert, A., Barthelemy, A., Ben Youssef, J., Contour, J.P., Cros, V., De Teresa, J.M., Hamzic, A., George, J.M., Faini, G., Grollier, J., Jaffres, H., Le Gall, H., Montaigne, F., Pailloux, F., Petroff, F.: Review of recent results on spin polarized tunneling and magnetic switching by spin injection. Materials Science and Engineering B-Solid State Materials For Advanced Technology 84(1-2), 1–9 (2001)

[6] Barthelemy, A., Fert, A., Contour, J.P., Bowen, M., Cros, V., De Teresa, J.M., Hamzic, A., Faini, J.C., George, J.M., Grollier, J., Montaigne, F., Pailloux, F., Petroff, F., Vouille, C.: Magnetoresistance and spin electronics. Journal of Magnetism and Magnetic Materials 242(Part 1), 68–76 (2002)

[7] Campbell, I.A., Fert, A., Pomeroy, A.R.: Evidence For 2 Current Conduction in Iron. Philosophical Magazine 15(137), 977 (1967)

[8] Fert, A., Campbell, I.A.: 2-Current Conduction In Nickel. Physical Review Letters 21(16), 1190 (1968)

[9] Grunberg, P., Schreiber, R., Pang, Y., Brodsky, M.B., Sowers, H.: Layered Magnetic-Structures - Evidence For Antiferromagnetic Coupling of Fe Layers Across Cr Interlayers. Physical Review Letters 57(19), 2442–2445 (1986)

[10] Mott, N.F., Wills, H.H.: The Resistance and Thermoelectric Properties of the Transition Metals. Proceedings of the Royal Society A 156, 368–382 (1936)

[11] Meservey, R., Tedrow, P.M.: Spin-Polarized Electron-Tunneling. Physics Reports-Review Section of Physics Letters 238(4), 173–243 (1994)

[12] Moodera, J.S., Kinder, L.R., Wong, T.M., Meservey, R.: Large Magnetoresistance At Room-Temperature In Ferromagnetic Thin-Film Tunnel-Junctions. Physical Review Letters 74(16), 3273–3276 (1995)
[13] Miyazaki, T., Tezuka, N.: Giant Magnetic Tunneling Effect In Fe/Al2O3/Fe Junction. Journal of Magnetism and Magnetic Materials 139(3), L231–L234 (1995)
[14] Parkin, S.S.P., Roche, K.P., Samant, M.G., Rice, P.M., Beyers, R.B., Scheuerlein, R.E., O'Sullivan, E.J., Brown, S.L., Bucchigano, J., Abraham, D.W., Lu, Y., Rooks, M., Trouilloud, P.L., Wanner, R.A., Gallagher, W.J.: Exchange-biased magnetic tunnel junctions and application to nonvolatile magnetic random access memory (invited). Journal of Applied Physics 85(8, Part 2b), 5828–5833 (1999)
[15] Tehrani, S., Chen, E., Durlam, M., DeHerrera, M., Slaughter, J.M.: High density submicron magnetoresistive random access memory (invited). Journal of Applied Physics 85(8, Part 2b), 5822–5827 (1999)
[16] Ikeda, S., Hayakawa, J., Lee, Y.M., Matsukura, F., Ohno, Y., Hanyu, T., Ohno, H.: Magnetic tunnel junctions for spintronic memories and beyond. IEEE Transactions on Electron Devices 54(5), 991–1002 (2007)
[17] Matsunaga, S., Hayakawa, J., Ikeda, S., Miura, K., Hasegawa, H., Endoh, T., Ohno, H., Hanyu, T.: Fabrication of a nonvolatile full adder based on logic-in-memory architecture using magnetic tunnel junctions. Applied Physics Express 1(9), 091301 (2008)
[18] Tsymbal, E.Y., Pettifor, D.G.: Spin-polarized electron tunneling across a disordered insulator. Physical Review B 58(1), 432–437 (1998)
[19] Tsymbal, E.Y., Pettifor, D.G.: The influence of impurities within the barrier on tunneling magnetoresistance. Journal of Applied Physics 85(8, Part 2b), 5801–5803 (1999)
[20] Sharma, M., Wang, S.X., Nickel, J.H.: Inversion of spin polarization and tunneling magnetoresistance in spin-dependent tunneling junctions. Physical Review Letters 82(3), 616–619 (1999)
[21] LeClair, P., Swagten, H.J.M., Kohlhepp, J.T., van de Veerdonk, R.J.M., de Jonge, W.J.M.: Apparent spin polarization decay in Cu-dusted Co/Al2O3/Co tunnel junctions. Physical Review Letters 84(13), 2933–2936 (2000)
[22] LeClair, P., Swagten, H.J.M., Kohlhepp, J.T., de Jonge, W.J.M.: Tunnel conductance as a probe of spin polarization decay in Cu dusted Co/Al2O3/Co tunnel junctions. Applied Physics Letters 76(25), 3783–3785 (2000)
[23] De Teresa, J.M., Barthelemy, A., Fert, A., Contour, J.P., Lyonnet, R., Montaigne, F., Seneor, P., Vaures, A.: Inverse tunnel magnetoresistance in Co/SrTiO3/La0.7Sr0.3MnO3: New ideas on spin-polarized tunneling. Physical Review Letters 82(21), 4288–4291 (1999)
[24] De Teresa, J.M., Barthelemy, A., Fert, A., Contour, J.P., Montaigne, F., Seneor, P.: Role of metal-oxide interface in determining the spin polarization of magnetic tunnel junctions. Science 286(5439), 507–509 (1999)
[25] Bagrets, D., Bagrets, A., Vedyayev, A., Dieny, B.: Influence of s-d interfacial scattering on the magnetoresistance of magnetic tunnel junctions. Physical Review B 65(6), 064430 (2002)
[26] Vedyayev, A., Bagrets, D., Bagrets, A., Dieny, B.: Resonant spin-dependent tunneling in spin-valve junctions in the presence of paramagnetic impurities. Physical Review B 63(6), 064429 (2001)
[27] Dimopoulos, T., Da Costa, V., Tiusan, C., Ounadjela, K., van den Berg, H.A.M.: Interfacial phenomena related to the fabrication of thin Al oxide tunnel barriers and their thermal evolution. Applied Physics Letters 79(19), 3110–3112 (2001)

[28] Dimopoulos, T., Gieres, G., Colis, S., Wecker, J., Luo, Y., Samwer, K.: Magnetic tunnel junctions with yttrium oxide barrier. Applied Physics Letters 83(16), 3338–3340 (2003)
[29] Dimopoulos, T., Henry, Y., Da Costa, V., Tiusan, C., Ounadjela, K.: Influence of barrier overoxidation and annealing on the inelastic spin-dependent tunneling in AlOx-based junctions. Journal of Applied Physics 95(11, Part 2), 6936–6938 (2004)
[30] Tsymbal, E.Y., Mryasov, O.N., LeClair, P.R.: Spin-dependent tunnelling in magnetic tunnel junctions. Journal of Physics-Condensed Matter 15(4), R109–R142 (2003)
[31] MacLaren, J.M., Zhang, X.G., Butler, W.H., Wand, X.D.: Layer KKR approach to Bloch-wave transmission and reflection: Application to spin-dependent tunneling. Physical Review B 59(8), 5470–5478 (1999)
[32] Butler, W.H., Zhang, X.G., Schulthess, T.C., MacLaren, J.M.: Spin-dependent tunneling conductance of Fe / MgO / Fe sandwiches. Physical Review B 63(5), 054416 (2001)
[33] Mathon, J., Umerski, A.: Theory of tunneling magnetoresistance of an epitaxial Fe/MgO/Fe(001) junction. Physical Review B 63(22), 220403 (2001)
[34] Wunnicke, O., Mavropoulos, P., Zeller, R., Dederichs, P.H., Grundler, D.: Ballistic spin injection from Fe(001) into ZnSe and GaAs. Physical Review B 65(24), 241306 (2002)
[35] Butler, W.H.: Tunneling magnetoresistance from a symmetry filtering effect. Science and Technology of Advanced Materials 9(1), 014106 (2008)
[36] Parkin, S.S.P., Kaiser, C., Panchula, A., Rice, P.M., Hughes, B., Samant, M., Yang, S.H.: Giant tunnelling magnetoresistance at room temperature with MgO (100) tunnel barriers. Nature Materials 3(12), 862–867 (2004)
[37] Yuasa, S., Nagahama, T., Fukushima, A., Suzuki, Y., Ando, K.: Giant room-temperature magnetoresistance in single-crystal Fe/MgO/Fe magnetic tunnel junctions. Nature Materials 3(12), 868–871 (2004)
[38] Djayaprawira, D.D., Tsunekawa, K., Nagai, M., Maehara, H., Yamagata, S., Watanabe, N., Yuasa, S., Suzuki, Y., Ando, K.: 230% room-temperature magnetoresistance in CoFeB/MgO/CoFeB magnetic tunnel junctions. Applied Physics Letters 86(9), 092502 (2005)
[39] Ikeda, S., Hayakawa, J., Ashizawa, Y., Lee, Y.M., Miura, K., Hasegawa, H., Tsunoda, M., Matsukura, F., Ohno, H.: Tunnel magnetoresistance of 604% at 300 K by suppression of Ta diffusion in CoFeB/MgO/CoFeB pseudo-spin-valves annealed at high temperature. Applied Physics Letters 93(8), 082508 (2008)
[40] Slonczewski, J.C.: Current-driven excitation of magnetic multilayers. Journal of Magnetism and Magnetic Materials 159(1-2), L1–L7 (1996)
[41] Berger, L.: Emission of spin waves by a magnetic multilayer traversed by a current. Physical Review B 54(13), 9353–9358 (1996)
[42] Waintal, X., Myers, E.B., Brouwer, P.W., Ralph, D.C.: Role of spin-dependent interface scattering in generating current-induced torques in magnetic multilayers. Physical Review B 62(18), 12317–12327 (2000)
[43] Heide, C., Zilberman, P.E., Elliott, R.J.: Current-driven switching of magnetic layers. Physical Review B 63(6), 064424 (2001)
[44] Stiles, M.D.: Anatomy of spin-transfer torque. Physical Review B 66(1), 0114407 (2002)
[45] Zhang, S., Levy, P.M., Fert, A.: Mechanisms of spin-polarized current-driven magnetization switching. Physical Review Letters 88(23), 236601 (2002)

[46] Myers, E.B., Ralph, D.C., Katine, J.A., Louie, R.N., Buhrman, R.A.: Current-induced switching of domains in magnetic multilayer devices. Science 285(5429), 867–870 (1999)
[47] Tsoi, M., Jansen, A.G.M., Bass, J., Chiang, W.C., Tsoi, V., Wyder, P.: Generation and detection of phase-coherent current-driven magnons in magnetic multilayers. Nature 406(6791), 46–48 (2000)
[48] Katine, J.A., Albert, F.J., Buhrman, R.A.: Current-induced realignment of magnetic domains in nanostructured Cu/Co multilayer pillars. Applied Physics Letters 76(3), 354–356 (2000)
[49] Katine, J.A., Albert, F.J., Buhrman, R.A., Myers, E.B., Ralph, D.C.: Current-driven magnetization reversal and spin-wave excitations in Co/Cu/Co pillars. Physical Review Letters 84(14), 3149–3152 (2000)
[50] Grollier, J., Cros, V., Hamzic, A., George, J.M., Jaffres, H., Fert, A., Faini, G., Ben Youssef, J., Legall, H.: Spin-polarized current induced switching in Co/Cu/Co pillars. Applied Physics Letters 78(23), 3663–3665 (2001)
[51] Sun, J.Z., Monsma, D.J., Abraham, D.W., Rooks, M.J., Koch, R.H.: Batch-fabricated spin-injection magnetic switches. Applied Physics Letters 81(12), 2202–2204 (2002)
[52] Pufall, M.R., Rippard, W.H., Silva, T.J.: Materials dependence of the spin-momentum transfer efficiency and critical current in ferromagnetic metal/Cu multilayers. Applied Physics Letters 83(2), 323–325 (2003)
[53] Rippard, W.H., Pufall, M.R., Silva, T.J.: Quantitative studies of spin-momentum-transfer-induced excitations in Co/Cu multilayer films using point-contact spectroscopy. Applied Physics Letters 82(8), 1260–1262 (2003)
[54] Urazhdin, S., Birge, N.O., Pratt, W.P., Bass, J.: Current-driven magnetic excitations in permalloy-based multilayer nanopillars. Physical Review Letters 91(14), 146803 (2003)
[55] Petit, S., Baraduc, C., Thirion, C., Ebels, U., Liu, Y., Li, M., Wang, P., Dieny, B.: Spin-torque influence on the high-frequency magnetization fluctuations in magnetic tunnel junctions. Physical Review Letters 98(7), 077203 (2007)
[56] Fullerton, E.E., Conover, M.J., Mattson, J.E., Sowers, C.H., Bader, S.D.: 150-Percent Magnetoresistance In Sputtered Fe/Cr(100) Superlattices. Applied Physics Letters 63(12), 1699–1701 (1993)
[57] Mosca, D.H., Petroff, F., Fert, A., Schroeder, P.A., Pratt, W.P., Laloee, R.: Oscillatory Interlayer Coupling and Giant Magnetoresistance In Co/Cu Multilayers. Journal of Magnetism and Magnetic Materials 94(1-2), L1–L5 (1991)
[58] Parkin, S.S.P., Bhadra, R., Roche, K.P.: Oscillatory Magnetic Exchange Coupling Through Thin Copper Layers. Physical Review Letters 66(16), 2152–2155 (1991)
[59] Shinjo, T., Yamamoto, H.: Large Magnetoresistance of Field-Induced Giant Ferrimagnetic Multilayers. Journal of The Physical Society of Japan 59(9), 3061–3064 (1990)
[60] Sakakima, H., Satomi, M.: Oscillations In Low-Field Giant Magnetoresistance In Ni-Fe-Co/Cu(/Co) Superlattices. Journal of Magnetism and Magnetic Materials 121(1-3), 374–377 (1993)
[61] Dieny, B., Speriosu, V.S., Metin, S., Parkin, S.S.P., Gurney, B.A., Baumgart, P., Wilhoit, D.R.: Magnetotransport Properties of Magnetically Soft Spin-Valve Structures. Journal of Applied Physics 69(8, Part 2a), 4774–4779 (1991)
[62] Parkin, S.S.P., More, N., Roche, K.P.: Oscillations In Exchange Coupling and Magnetoresistance In Metallic Superlattice Structures - Co/Ru, Co/Cr, and Fe/Cr. Physical Review Letters 64(19), 2304–2307 (1990)

[63] Heim, D.E., Fontana, R.E., Tsang, C., Speriosu, V.S., Gurney, B.A., Williams, M.L.: Design and Operation of Spin-Valve Sensors. IEEE Transactions on Magnetics 30(2, Part 1), 316–321 (1994)
[64] Parkin, S.S.P.: Origin of Enhanced Magnetoresistance of Magnetic Multilayers - Spin-Dependent Scattering From Magnetic Interface States. Physical Review Letters 71(10), 1641–1644 (1993)
[65] Tedrow, P.M., Meservey, R.: Spin-Dependent Tunneling Into Ferromagnetic Nickel. Physical Review Letters 26(4), 192–195 (1971)
[66] Stearns, M.B.: Simple Explanation of Tunneling Spin-Polarization of Fe, Co, Ni and Its Alloys. Journal of Magnetism and Magnetic Materials 5(2), 167–171 (1977)
[67] Valet, T., Fert, A.: Theory of The Perpendicular Magnetoresistance In Magnetic Multilayers. Physical Review B 48(10), 7099–7113 (1993)
[68] Pratt, W.P., Lee, S.F., Slaughter, J.M., Loloee, R., Schroeder, P.A., Bass, J.: Perpendicular Giant Magnetoresistances of Ag/Co Multilayers. Physical Review Letters 66(23), 3060–3063 (1991)
[69] Lee, S.F., Pratt, W.P., Loloee, R., Schroeder, P.A., Bass, J.: Field-Dependent Interface Resistance of Ag/Co Multilayers. Physical Review B 46(1), 548–551 (1992)
[70] Lee, S.F., Pratt, W.P., Yang, Q., Holody, P., Loloee, R., Schroeder, P.A., Bass, J.: 2-Channel Analysis of CPP-MR Data For Ag/Co and AgSn/Co Multilayers. Journal of Magnetism and Magnetic Materials 118(1-2), L1–L5 (1993)
[71] Lee, S.F., Yang, Q., Holody, P., Loloee, R., Hetherington, J.H., Mahmood, S., Ikegami, B., Vigen, K., Henry, L.L., Schroeder, P.A., Pratt, W.P., Bass, J.: Current-perpendicular and current-parallel giant magnetoresistances in Co/Ag multilayers. Physical Review B 52(21), 15426–15441 (1995)
[72] List, N.J., Pratt, W.P., Howson, M.A., Xu, J., Walker, M.J., Greig, D.: Perpendicular Resistance of Co/Cu Multilayers Prepared By Molecular-Beam Epitaxy. Journal of Magnetism and Magnetic Materials 148(1-2), 342–343 (1995)
[73] Piraux, L., Dubois, S., Fert, A.: Perpendicular giant magnetoresistance in magnetic multilayered nanowires. Journal of Magnetism and Magnetic Materials 159(3), L287–L292 (1996)
[74] Doudin, B., Blondel, A., Ansermet, J.P.: Arrays of multilayered nanowires. Journal of Applied Physics 79(8, Part 2b), 6090–6094 (1996)
[75] Bass, J., Pratt, W.P.: Current-perpendicular (CPP) magnetoresistance in magnetic metallic multilayers. Journal of Magnetism and Magnetic Materials 200(1-3), 274–289 (1999)
[76] Fert, A., Piraux, L.: Magnetic nanowires. Journal of Magnetism and Magnetic Materials 200(1-3), 338–358 (1999)
[77] Bozec, D., Howson, M.A., Hickey, B.J., Shatz, S., Wiser, N., Tsymbal, E.Y., Pettifor, D.G.: Mean free path effects on the current perpendicular to the plane magnetoresistance of magnetic multilayers. Physical Review Letters 85(6), 1314–1317 (2000)
[78] Tedrow, P.M., Meservey, R.: Spin Polarization of Electrons Tunneling From Films of Fe, Co, Ni, and Gd. Physical Review B 7(1), 318–326 (1973)
[79] Julliere, M.: Tunneling Between Ferromagnetic-Films. Physics Letters A 54(3), 225–226 (1975)
[80] LeClair, P., Hoex, B., Wieldraaijer, H., Kohlhepp, J.T., Swagten, H.J.M., de Jonge, W.J.M.: Sign reversal of spin polarization in Co/Ru/Al2O3/Co magnetic tunnel junctions. Physical Review B 64(10), 100406 (2001)

[81] Slonczewski, J.C.: Conductance and Exchange Coupling of 2 Ferromagnets Separated By A Tunneling Barrier. Physical Review B 39(10, Part b), 6995–7002 (1989)
[82] Belashchenko, K.D., Tsymbal, E.Y., van Schilfgaarde, M., Stewart, D.A., Oleynik II,, Jaswal, S.S.: Oleynik, S.S.Jaswal. Effect of interface bonding on spin-dependent tunneling from the oxidized Co surface. Physical Review B 69(17), 174408 (2004)
[83] Zhang, X.G., Butler, W.H.: Band structure, evanescent states, and transport in spin tunnel junctions. Journal of Physics-Condensed Matter 15(41), R1603–R1639 (2003)
[84] Tsymbal, E.Y.: Theory of Spin-Dependent Tunneling: Role of Evanescent and Resonant States. In: Krohnmuller, H., Parkin, S.S.P. (eds.) Handbook of Magnetism and Advanced Magnetic Materials, vol. 5, pp. 2963–2978 (2006)
[85] Velev, J.P., Dowben, P.A., Tsymbal, E.Y., Jenkins, S.J., Caruso, A.N.: Interface effects in spin-polarized metal/insulator layered structures. Surface Science Reports 63(9), 400–425 (2008)
[86] Mavropoulos, P., Papanikolaou, N., Dederichs, P.H.: Complex band structure and tunneling through ferromagnet/insulator/ferromagnet junctions. Physical Review Letters 85(5), 1088–1091 (2000)
[87] Dederichs, P.H., Mavropoulos, P., Wunnicke, O., Papanikolaou, N., Bellini, V., Zeller, R., Drchal, V., Kudrnovsky, J.: Importance of complex band structure and resonant states for tunneling. Journal of Magnetism and Magnetic Materials 240(1-3), 108–113 (2002)
[88] Butler, W.H., Zhang, X.G., Vutukuri, S., Chshiev, M., Schulthess, T.C.: Theory of tunneling magnetoresistance for epitaxial systems. IEEE Transactions on Magnetics 41(10), 2645–2648 (2005)
[89] Vutukuri, S., Chshiev, M., Butler, W.H.: Spin-dependent tunneling in FM / semiconductor / FM structures. Journal of Applied Physics 99(8), K302 (2006)
[90] Faure-Vincent, J., Tiusan, C., Jouguelet, E., Canet, F., Sajieddine, M., Bellouard, C., Popova, E., Hehn, M., Montaigne, F., Schuhl, A.: High tunnel magnetoresistance in epitaxial Fe/MgO/Fe tunnel junctions. Applied Physics Letters 82(25), 4507–4509 (2003)
[91] Yuasa, S., Fukushima, A., Kubota, H., Suzuki, Y., Ando, K.: Giant tunneling magnetoresistance up to 410% at room temperature in fully epitaxial Co/MgO/Co magnetic tunnel junctions with bcc Co(001) electrodes. Applied Physics Letters 89(4), 042505 (2006)
[92] Zermatten, P.J., Gaudin, G., Maris, G., Miron, M., Schuhl, A., Tiusan, C., Greullet, F., Hehn, M.: Experimental evidence of interface resonance states in single-crystal magnetic tunnel junctions. Physical Review B 78(3), 033301 (2008)
[93] Zermatten, P.-J., Bonell, F., Andrieu, S., Chshiev, M., Tiusan, C., Schuhl, A., Gaudin, G.: Influence of Oxygen Monolayer at Fe/MgO Interface on Transport Properties in Fe/MgO/Fe(001) Magnetic Tunnel Junctions. Applied Physics Express 5(2), 023001 (2012)
[94] Vedyayev, A., Dieny, B., Ryzhanova, N., Genin, J.B., Cowache, C.: Angular-Dependence of Giant Magnetoresistance In Magnetic Multilayered Structures. Europhysics Letters 25(6), 465–470 (1994)
[95] Vedyayev, A., Ryzhanova, N., Dieny, B., Dauguet, P., Gandit, P., Chaussy, J.: Angular variation of giant magnetoresistance for current perpendicular to the plane of the layers. Physical Review B 55(6), 3728–3733 (1997)

[96] Dieny, B., Speriosu, V.S., Parkin, S.S.P., Gurney, B.A., Wilhoit, D.R., Mauri, D.: Giant Magnetoresistance In Soft Ferromagnetic Multilayers. Physical Review B 43(1, Part b), 1297–1300 (1991)

[97] Dieny, B., Cowache, C., Nossov, A., Dauguet, P., Chaussy, J., Gandit, P.: Anisotropy and angular variation of the giant magnetoresistance in magnetic multilayers. Journal of Applied Physics 79(8, Part 2b), 6370–6375 (1996)

[98] Slonczewski, J.C.: Currents and torques in metallic magnetic multilayers. Journal of Magnetism and Magnetic Materials 247(3), 324–338 (2002)

[99] Johnson, M.: Analysis of Anomalous Multilayer Magnetoresistance Within The Thermomagnetoelectric System. Physical Review Letters 67(25), 3594–3597 (1991)

[100] Slonczewski, J.C.: Currents, torques, and polarization factors in magnetic tunnel junctions. Physical Review B 71(2), 24411 (2005)

[101] Meiklejohn, W.H.: Exchange Anisotropy - A Review. Journal of Applied Physics 33(3), 1328–1335 (1962)

[102] Malozemoff, A.P.: Random-Field Model of Exchange-Anisotropy At Rough Ferromagnetic-Antiferromagnetic Interfaces. Physical Review B 35(7), 3679–3682 (1987)

[103] Berkowitz, A.E., Takano, K.: Exchange anisotropy - a review. Journal of Magnetism and Magnetic Materials 200(1-3), 552–570 (1999)

[104] Boulle, O., Cros, V., Grollier, J., Pereira, L.G., Deranlot, C., Petroff, F., Faini, G., Barnas, J., Fert, A.: Shaped angular dependence of the spin-transfer torque and microwave generation without magnetic field. Nature Physics 3(7), 492–497 (2007)

[105] Rychkov, V.S., Borlenghi, S., Jaffres, H., Fert, A., Waintal, X.: Spin Torque and Waviness in Magnetic Multilayers: A Bridge between Valet-Fert Theory and Quantum Approaches. Physical Review Letters 103(6), 066602 (2009)

[106] Brataas, A., Nazarov, Y.V., Bauer, G.E.W.: Finite-element theory of transport in ferromagnet-normal metal systems. Physical Review Letters 84(11), 2481–2484 (2000)

[107] Stiles, M.D., Zangwill, A.: Noncollinear spin transfer in Co/Cu/Co multilayers (invited). Journal of Applied Physics 91(10, Part 2), 6812–6817 (2002)

[108] Shpiro, A., Levy, P.M., Zhang, S.F.: Self-consistent treatment of nonequilibrium spin torques in magnetic multilayers. Physical Review B 67(10), 104430 (2003)

[109] Fert, A., Cros, V., George, J.M., Grollier, J., Jaffres, H., Hamzic, A., Vaures, A., Faini, G., Ben Youssef, J., Le Gall, H.: Magnetization reversal by injection and transfer of spin: experiments and theory. Journal of Magnetism and Magnetic Materials 272(Part 3, SI), 1706–1711 (2004)

[110] Manchon, A., Ryzhanova, N., Strelkov, N., Vedyayev, A., Dieny, B.: Modelling spin transfer torque and magnetoresistance in magnetic multilayers. Journal of Physics-Condensed Matter 19(16), 165212 (2007)

[111] Edwards, D.M., Federici, F., Mathon, J., Umerski, A.: Self-consistent theory of current-induced switching of magnetization. Physical Review B 71(5), 054407 (2005)

[112] Zwierzycki, M., Tserkovnyak, Y., Kelly, P.J., Brataas, A., Bauer, G.E.W.: First-principles study of magnetization relaxation enhancement and spin transfer in thin magnetic films. Physical Review B 71(6), 064420 (2005)

[113] Haney, P.M., Waldron, D., Duine, R.A., Nunez, A.S., Guo, H., MacDonald, A.H.: Current-induced order parameter dynamics: Microscopic theory applied to Co/Cu/Co spin valves. Physical Review B 76(2), 024404 (2007)

[114] Stiles, M.D.: Exchange Coupling In Magnetic Heterostructures. Physical Review B 48(10), 7238–7259 (1993)

[115] Slonczewski, J.C.: Overview of Interlayer Exchange Theory. Journal of Magnetism and Magnetic Materials 150(1), 13–24 (1995)

[116] Ralph, D.C., Stiles, M.D.: Spin transfer torques. Journal of Magnetism and Magnetic Materials 320(7), 1190–1216 (2008)

[117] Baraduc, C., Chshiev, M., Ebels, U.: Introduction to spin transfer torque. In: Nasirpouri, F., Nogaret, A. (eds.) Nanomagnetism and Spintronics, pp. 173–192 (2011)

[118] Xia, K., Kelly, P.J., Bauer, G.E.W., Brataas, A., Turek, I.: Spin torques in ferromagnetic/normal-metal structures. Physical Review B 65(22), 220401 (2002)

[119] Kalitsov, A., Theodonis, I., Kioussis, N., Chshiev, M., Butler, W.H., Vedyayev, A.: Spin-polarized current-induced torque in magnetic tunnel junctions. Journal of Applied Physics 99(8), G501 (2006)

[120] Theodonis, I., Kioussis, N., Kalitsov, A., Chshiev, M., Butler, W.H.: Anomalous bias dependence of spin torque in magnetic tunnel junctions. Physical Review Letters 97(23), 237205 (2006)

[121] Tulapurkar, A.A., Suzuki, Y., Fukushima, A., Kubota, H., Maehara, H., Tsunekawa, K., Djayaprawira, D.D., Watanabe, N., Yuasa, S.: Spin-torque diode effect in magnetic tunnel junctions. Nature 438(7066), 339–342 (2005)

[122] Petit, S., de Mestier, N., Baraduc, C., Thirion, C., Liu, Y., Li, M., Wang, P., Dieny, B.: Influence of spin-transfer torque on thermally activated ferromagnetic resonance excitations in magnetic tunnel junctions. Physical Review B 78(18), 184420 (2008)

[123] Kubota, H., Fukushima, A., Yakushiji, K., Nagahama, T., Yuasa, S., Ando, K., Maehara, H., Nagamine, Y., Tsunekawa, K., Djayaprawira, D.D., Watanabe, N., Suzuki, Y.: Quantitative measurement of voltage dependence of spin-transfer torque in MgO-based magnetic tunnel junctions. Nature Physics 4(1), 37–41 (2008)

[124] Sankey, J.C., Cui, Y.-T., Sun, J.Z., Slonczewski, J.C., Buhrman, R.A., Ralph, D.C.: Measurement of the spin-transfer-torque vector in magnetic tunnel junctions. Nature Physics 4(1), 67–71 (2008)

[125] Oh, S.-C., Park, S.-Y., Manchon, A., Chshiev, M., Han, J.-H., Lee, H.-W., Lee, J.-E., Nam, K.-T., Jo, Y., Kong, Y.-C., Dieny, B., Lee, K.-J.: Bias-voltage dependence of perpendicular spin-transfer torque in asymmetric MgO-based magnetic tunnel junctions. Nature Physics 5(12), 898–902 (2009)

[126] Kiselev, S.I., Sankey, J.C., Krivorotov, I.N., Emley, N.C., Schoelkopf, R.J., Buhrman, R.A., Ralph, D.C.: Microwave oscillations of a nanomagnet driven by a spin-polarized current. Nature 425(6956), 380–383 (2003)

[127] Devolder, T., Crozat, P., Chappert, C., Miltat, J., Tulapurkar, A., Suzuki, Y., Yagami, K.: Instability threshold versus switching threshold in spin-transfer-induced magnetization switching. Physical Review B 71(18), 184401 (2005)

[128] Stiles, M.D., Miltat, J.: Spin-polarized current-induced torque in magnetic tunnel junctions. In: Hillebrands, B., Thiaville, A. (eds.) Spin Dynamics in Confined Magnetic Structures III, pp. 255–308 (2006)

[129] Xiao, J., Zangwill, A., Stiles, M.D.: Macrospin models of spin transfer dynamics. Physical Review B 72(1), 14446 (2005)

[130] Sun, J.Z.: Spin-current interaction with a monodomain magnetic body: A model study. Physical Review B 62(1), 570–578 (2000)

Microfabrication Techniques

Diana C. Leitão[1], José Pedro Amaral[1], Susana Cardoso[1], and Càndid Reig[2]

[1] INESC-Microsistemas e Nanotecnologias,
Lisbon, Portugal
[2] University of Valencia, Valencia, Spain
scardoso@inesc-mn.pt

Abstract. GMR sensors are fabricated by following a sort of techniques in a similar fashion to those related to standard CMOS processes. Deposition, patterning and encapsulation steps are found in both parts. However, characteristics related to the specific materials involved in the GMR technology recommend the use of some particular techniques. In this chapter, we focus on these specific methods, while keeping in mind the interest in merging standard CMOS with GMR technologies.

1 Introduction

The fabrication of giant magnetoresistance (GMR) sensors involves deposition, patterning and encapsulation steps different than those typically used for semiconductor industry. In contrast to semiconductor processing, GMR device fabrication is based on low temperature techniques; therefore doping related processes such as diffusion or implantation are usually not required.

Nevertheless, there are additional differences between the procedure for the fabrication of a GMR device and a commercial process for producing integrated circuits (IC) such as those offered by foundries consortia (e.g. Europractice [1] or MOSIS [2]). These differences are mainly related to the mass production concept. In the case of GMR sensors (except for hard disk magnetic read heads), several options exist for costume-made sensors, where the material properties, geometry, sensor layout can be easily modified upon request. In this scenario, single-wafer processes are often considered, and several groups worldwide have easy and fast access to GMR sensors. In contrast, a typical Bi-CMOS fabrication process can involve the use of 20 lithographic masks, which strongly limits the number of suppliers worldwide. The complexity in process motivated a global effort on developing technological solutions for high yield production in large wafers (12 inch diameter), therefore less flexibility in design modification. In the case of GMR sensors, typically less than 5 lithographic levels are required in order to get functional devices [3], and can be integrated into pre-paterned wafers based on other technologies.

Materials associated to the fabrication of GMR sensors and devices slightly differ from those used in standard Bi-CMOS processes [4]. Regarding the devices holder, GMR can be deposited on silicon wafers but glass, sapphire or flexible substrates can also be used. On Bi-CMOS processes, silicon, silicon oxide and aluminum are the

C. Reig et al.: *Giant Magnetoresistance (GMR) Sensors*, SSMI 6, pp. 31–45.
DOI: 10.1007/978-3-642-37172-1_2

basis materials, as well as the dopants (Boron, Phosphorus, Arsenic, Antimony and related compounds). In the case of GMR devices, the fabrication of magnetic layers requires the use of additional magnetic materials (Iron, Cobalt, Niquel, Manganese, and their alloys), different metals (e.g. Copper, Ruthenium) and additional oxides (Al_2O_3, MgO …), not usually found in conventional semiconductor facilities. Each of these materials has particular requirements in terms of deposition technology and conditions or system contamination that need to be specifically considered and optimized. As a highlighting example we mention the deposition of layers with preferentially aligned magnetic moment which requires the use of a polarizing magnet placed inside the deposition system, therefore not easily compatible with hot deposition tools.

This chapter provides an overall description of several microfabrication techniques, emphasizing those particularly interesting for the fabrication of GMR based devices.

2 Deposition Techniques

GMR structures are composed by multilayered engineered structures [5] based on nanometric to sub-nanometric thick layers of ferromagnetic materials (eg: Co, CoFe, NiFe) separated by a non-magnetic spacer (Cu). Figure 1 shows a typical structure of GMR sensor. These materials have been studied theoretically and experimentally for many years, with large impact for the device application [6]. Such small size structures led to new physical effects, which are the basis of the GMR technology [7, 8, 9]. Therefore, adequate deposition techniques namely those using ultra-high vacuum systems and providing a thorough control of the thickness of the deposited layers are essential for the proper functionality of so obtained devices. In this section, we will briefly describe some of the techniques commonly used for depositing GMR structures.

Fig. 1. Typical structure of a GMR stack

2.1 Physical Vapor Deposition

Sputtering

One of the more common technique used for depositing thin films onto substrates is cathodic sputtering [10]. The sputtering process occurs when an accelerated ion collides with a solid target material. When the ion kinetic energy is sufficient ejection of an atom from the matrix takes place due to the momentum transfer. Figure 2 displays a sketch of sputtering's basic principle.

The thin film deposition requires pumping the reaction chamber to a low enough pressure (usually lower than 10^{-7} Torr), so that the water and oxygen adsorbed at the chamber walls is reduced. The deposition requires inert gas (Argon or Xenon) in order to produce the plasma, typically at few mTorr. A high voltage is then applied to the target holder producing an electrical discharge that allows the ionization of the gas and hence leads to the plasma. The produced ions are then attracted toward the cathode hitting the target.

The ions with energy above the threshold can extract atoms from the target material. These atoms are deposited onto the substrate, usually facing the target, and thus forming a layer of material. During the collision process, some secondary electrons are also produced, promoting a sustainable plasma at lower pressures. Furthermore, a magnetron can also be placed near the target to increase the ionization yield (magnetron sputtering). In particular cases, the ionization process can also be assisted by means of thermo emitted electrons from a filament.

Overall, sputtering is a very efficient process since the incident ions and sputtered atoms are able to retain most of their energy and directionality due to their high mean free paths inside the chamber. The latter is achieved by the use of low base pressure ($<10^{-7}$ Torr) and working pressures (10^{-3}~10^{-4} Torr). Also, the incident ions require energies of ~ hundreds of eV, which allied to their atomic dimension (mass) enables a good sputter yield [11]. The excess thermal energy is used by the deposited atom to diffuse to a more stable position on the substrate or, at later stage, on the growing film, being the remainder (heat generated in successive collisions) dissipated into the substrate, which is usually cooled. Cooling of the substrate holder provides a good control on the temperature of the deposition process.

One advantage of this method when compared with conventional evaporation is the possibility to deposit from a target composed of different materials (alloy or mosaic target) [11]. Due to this, sputtering is one of the preferred techniques to deposit metallic and magnetic layers in GMR devices. It is also commonly used for the deposition of metallic nonmagnetic contacts and also insulating oxides.

Ion Beam Sputter Deposition

Although not as widely used as magnetron sputtering, the Ion Beam Deposition (IBD) system provides a good film thickness uniformity and higher deposition control due to the low deposition rates employed, enabling also epitaxial growth under particular conditions [12, 13] and higher deposition textures [14, 15].

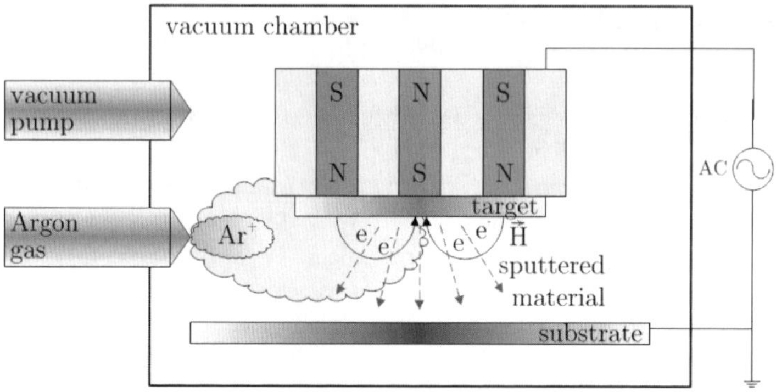

Fig. 2. Basic schematics of a magnetron sputtering process

Fig. 3 shows an example of a basic IBD schematics in 'Z' configuration [16]. Although the physical principle is similar to sputtering, in this case, the plasma is created and confined in an ion gun being then accelerated towards the target through voltage applied into a grid set (graphite or tungsten). RF or DC Kauffman ion sources are usually used, where the ionization process is confined (Ar or Xe), allowing also reactive depositions and in-situ reactions (Nitrogen or Oxygen). The size of commercial sources typically varies from 2.5 to 30 cm and the ion energy is about a few to several hundreds of eV [17]. Furthermore, the basic configuration of a typical IBD system normally includes an assist gun, used either for assisted deposition or ion-milling etching [18, 19]. An automatically interchangeable target holder (4-8 targets) can be used in GMR multilayer deposition without vacuum break, with deposition rates below 1 nm/s.

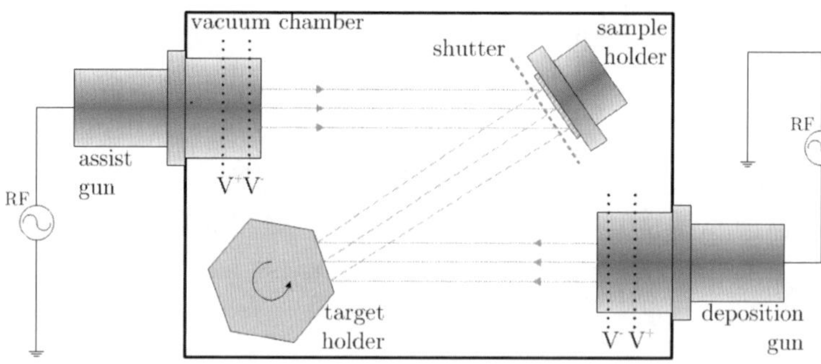

Fig. 3. Ion Beam Deposition basic schematics

One important advantage of this technique resides on the deposition parameters (ion flux, energy and sputtered species, as well as the angle of incidence) which can be more independently controlled and optimized than in other sputtering systems [20-22]. Notice that the angle between substrate/target can be altered according to preferences, as deposition with an angle is known to induce texture during growth [23-25].

Obtaining a good GMR sensor relies on first optimizing the material stack in terms of magnetic properties (e.g: setting a reference layer and a soft, low coercivity free layer) and thermal stability, and secondly, optimize the sensor response under external magnetic fields. Here, the microfabrication process has strong impact on the final sensor properties, as shape and dimensions of the patterned GMR films will determine the sensor sensitivity. Figure 4 shows an example of a sensor transfer curve obtaining upon microfabrication of a GMR film deposited by Ion Beam deposition. The linear response and high sensitivity to detect small magnetic fields at room temperature motivates the use of these type of sensors to many applications nowadays.

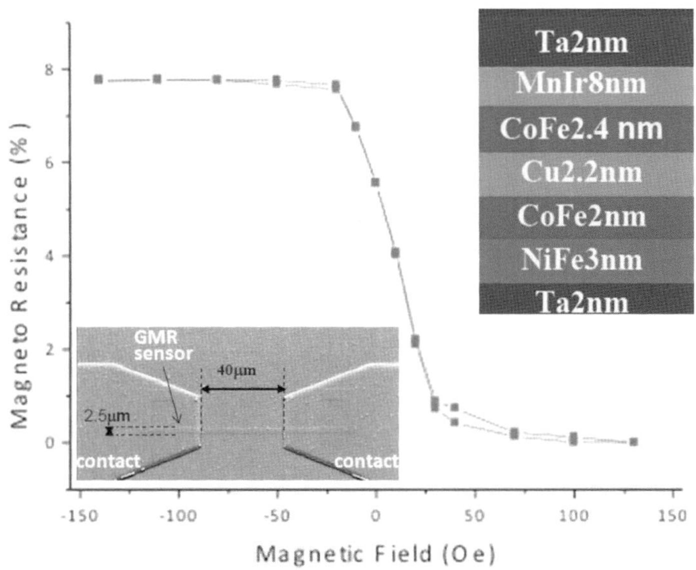

Fig. 4. GMR (spin valve) stack optimized by Ion Beam Deposition and output curve measured with a microfabricated (active area 2.5 x 40 μm^2) sensor

2.2 Chemical Vapor Deposition (CVD)

The deposition of thin films by means of Chemical Vapor Deposition (CVD) is based on the decomposition and/or reaction of different gaseous compounds. With this technique, the considered material is directly deposited onto the substrate surface from a gas phase [26]. High fusion temperature materials such as poly-silicon, silicon oxide or even heavy metals can be deposited by CVD, with excellent stoichiometry, when compared with other deposition techniques.

Regarding GMR sensors fabrication, CVD is mainly used in the deposition of insulating layers (silicon oxide or silicon nitride) leading to good quality layers with moderate cost equipment. This method can be shared with semiconductor processing.

A schematic of a basic CVD system is depicted in Fig. 5. It is composed by a heating system and a reactor. The reactor includes the substrates holders and the required gas inlets, depending of the particular material to be deposited. A gas extraction outlet, together with a vacuum pump, is also included.

Deposition usually occurs at high temperatures (> 300ºC), therefore not compatible with magnetic multilayers. However, since the deposition rates can be very large (therefore fast deposition) and it is a conformal deposition (thus, excellent step coverage), this method is suitable for final GMR device passivation with SiO_2 or Si_3N_4 deposition.

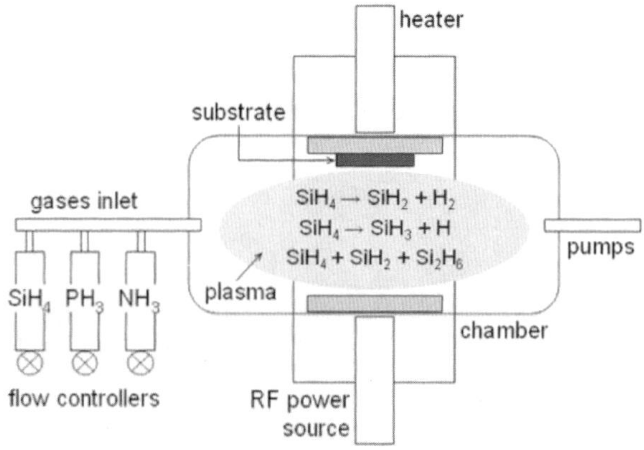

Fig. 5. Chemical Vapor Deposition basic reactor

2.3 Electrodeposition

Electrodeposition usually refers to deposition of a metal or an alloy from an electrolyte by passing a charge between the two electrodes located in the electrolyte [27-29]. This is a widely implemented method, either in industry or research, and enables a control of the length of the nanostructure by the duration of the electrochemical process. Besides, this technique is also not expensive, versatile and does not require vacuum equipment.

Several parameters influence the kinetics of the electrodeposition reaction, namely: (i) electrolyte temperature affects the ions diffusion velocity and the diffusion of already reduced atoms on the substrate surface; (ii) stirring favors the electron diffusion, enables the removal of H_2 gas which can inhibit the process, keeping the electrolyte concentration and the electrolyte/substrate interface pH constant; (iii) the electrodeposition potential determines the species and the corresponding quantity that will be deposited onto the substrate; (iv) finally the electrolyte composition. Usually, for metallic and magnetic single element nanostructures a standard Watts bath can be used [27], however, for multilayers deposition from a single bath, all species need to be present in the electrolyte.

The properties of the deposits such as hardness, stress or resistivity among others, depend on the solution's characteristics and temperature of deposition process [27]. Notice that, the adsorption of H_2 and other species formed at the cathode will influence the microstructural features of the deposits. In general, electrodeposited films present an extremely rough surface, unless additives are used [27].

Moreover, the possibility of depositing GMR was demonstrated early [30-32]. To deposit a multilayer stack using an electrodeposition technique, one starts from a single salt bath containing the metal ions of the elements/alloys to be deposited. A simple scheme of an electrodeposition cell is depicted in Fig. 6. Then a multilayer is formed by periodically varying either the deposition voltage (potentiostatic control) or deposition current (galvanostatic control) between two suitable values [31, 32]. However, GMR multilayer deposition form a single bath can lead to cross contamination between layers, which can become significant for the Cu spacer. Nevertheless, Co/Cu multilayers exhibiting GMR ~ 15 % [31, 32] at 300 K and NiFe/Cu stacks values of GMR~70 % [33, 34] were reported. In addition, NiFe/Cu/Co/Cu pseudo-spin-valves structures with GMR ranging from 0.5 to 1.7 %, depending on Cu thickness [35] have also been prepared. Finally, MnFe-pinned spin-valves with GMR~ 4 - 7 % and a constant pinning field of 650 Oe were demonstrated [36, 37].

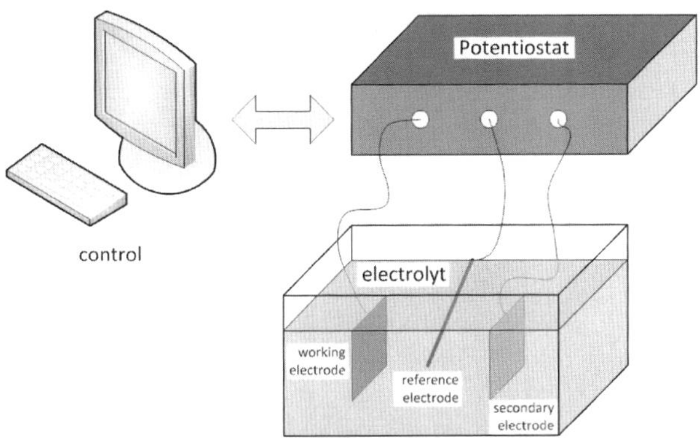

Fig. 6. A basic scheme of a simple potentiostatic electrodeposition cell

3 Patterning

The silicon based semiconductor industry nowadays relies mainly in ultraviolet lithography of hard or software-designed masks, combined with physical/chemical etching processes. Usually one single Si wafer for a complex circuit presents several levels of lithography with intricate patterns. Nonetheless, the planar process allows several wafers or devices to be fabricated in parallel with high reliability and yield, strongly reducing its cost and production time. Current optical lithography methods [38] are the most widely implemented techniques for exposure, meeting the requirements of large number, low-cost and high reliability, to define features down to ~1 μm size. These particularities are still not met by other alternatives methods such as nano imprint, electron-beam and ion-beam lithography, x-ray, nano-indentation or interference lithography methods.

The patterning process of a GMR sensor consists of sequential steps of pattern design and transfer as illustrated in Figure 7 and detailed described in the following. In summary, the standard procedure to define the sensor element implies a lithographic step in order to imprint the photosensitive polymer (resist) with a certain pattern (mask) and a following step where the pattern is transferred to the GMR thin films.

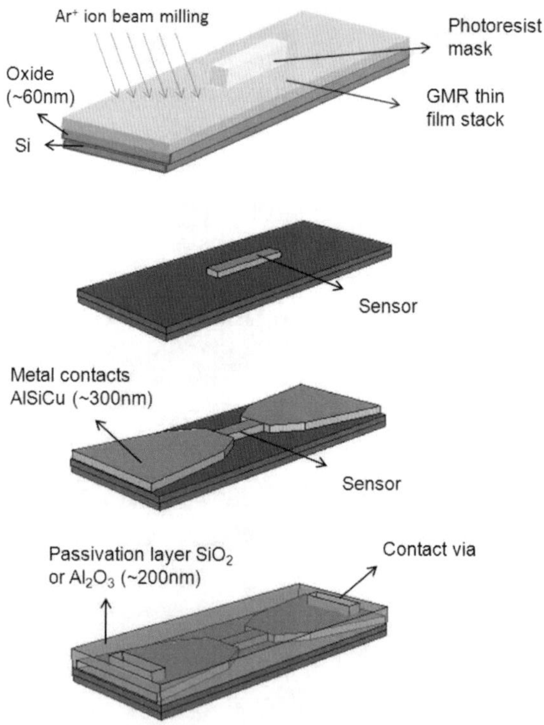

Fig. 7. Basic steps for GMR sensor microfabrication

3.1 Photolithography

The photolithography process (Figure 8) involves three major steps: (i) coating of the sample with a radiation sensitive polymer solution, called photoresist; (ii) exposure of the resist, patterning a certain design (mask), previously prepared; (iii) development of the transferred pattern.

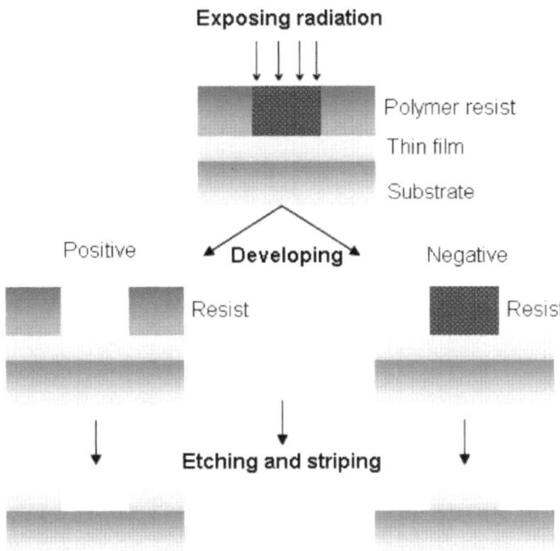

Fig. 8. Pattern definition by lithography: positive –tone (left figure) or negative-tone (right figure) resist can originate complementary features – holes or pillars

Coating

The resist is spun coated on the surface of the sample, where particular conditions such as coating speed, time and resist quantity are optimized for the desired thickness of the sensitive layer [39]. The latter is a crucial parameter which influences the lithography resolution [40]. Prior to coating a surface pre-treatment is usually required to promote the resist adhesion: a monolayer of hexamethyldisilazane (HMDS) is typically used. After being coated, the resist is soft-baked (typical temperature range of 80ºC-100ºC) to remove solvents and stress while improving adherence.

Lithography

The most widely employed standard photolithography systems use a focused laser beam (direct writing systems) or lamps (hard mask aligners) of UV radiation with wavelength typically ranging from 0.5-0.1 μm and resolutions below 1 μm are obtained this way [41].

Using hard masks, lithography can be done with the mask as close as possible to the sample (contact lithography) or through an optical system (projection lithography). In this case hard masks (usually made out of Cr films patterned on quartz) are used with pre-designed pattern which is then transferred to the sample.

In direct writing systems the mask is previously elaborated with the aid of CAD tools and then transferred to the wafer using a collimated beam (usually an UV laser). The information from the pattern is in this case codified to an X-Y displacement system, together with an optical turn on/off mechanism. The spot of the light beam moves through the surface in those zones that need to be illuminated, so directly drawing the pattern. This is a much slower system (full exposure of 150mm wafer area can take 8-16 hours), but still provides a relatively low-cost way for developing prototypes at low scale production, as no hard masks are needed.

On the other hand, if higher resolutions (< 500nm) are required, X-ray, electron or ion beams systems can be used to transfer the pattern to the resist layer. Electron beam lithography is also particularly used for hard mask design. A good overview of the lithography processes can be found in [42].

Development

After lithography, the exposed patterned is developed: first the sample undergoes a post-bake (typically 80-110ºC) to stop uncompleted resist reactions and remove stress. Then resist developer is sprayed or poured over the substrate. If the resist is positive, exposed regions are rendered soluble and will therefore be removed during development (see Figure 8). In contrast, if the resist is negative, exposed regions will harden and remain after developing. In either case, upon developing, the sample is washed to stop the development process and dried, and the pattern is finally printed into the resist layer.

3.2 Pattern Transfer Techniques

After designing the pattern into the resist, one has to transfer it into the underlying film, using either additive techniques such as lift-off, or subtractive methods as etching. Figure 9 compares side-by-side the steps involved for both methods.

Patterning Using Lift-off

Being an additive step, the liftoff process starts by defining the resist mask on a substrate and only then depositing the thin film on top (Figure 9, right). Afterwards, the sample is placed in a resist stripping solution that will remove the resist layer and all the material placed on top of it, leaving the patterned material in the areas previously unoccupied by the mask. This step is widely used for electrical contacts metallization (Figure 10).

Lift-off patterning has the advantage of being independent of the material underneath the photoresist (contrary to etching, where over etch and/or surface oxidation and corrosion can occur), so preventing the substrate from unexpected corrosion. However, it is not possible to lift-off thick films (usually, films with more than half the thickness of the resist layer), nor films deposited with good step coverage (e.g: deposition by CVD).

Etching

Etching concerns a process capable of selective removal of undesired portions of a deposited layer. The selective characteristic is provided by the patterned resist mask and also by the properties of the involved layers. The starting point is usually the film to be patterned deposited on a substrate with the desired pattern defined in the top resist mask.

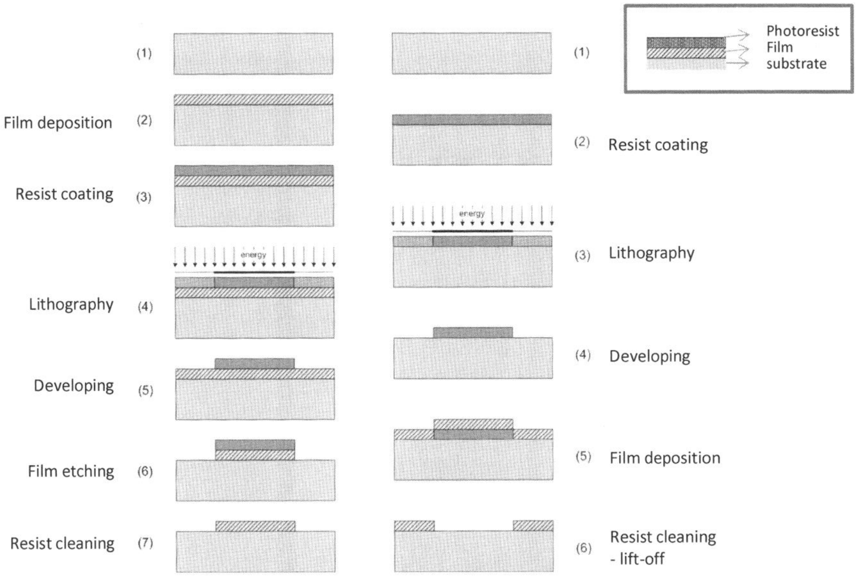

Fig. 9. Basic steps in two typical pattern transfer techniques: (Left) Using etching for film removing and (right) using liftoff

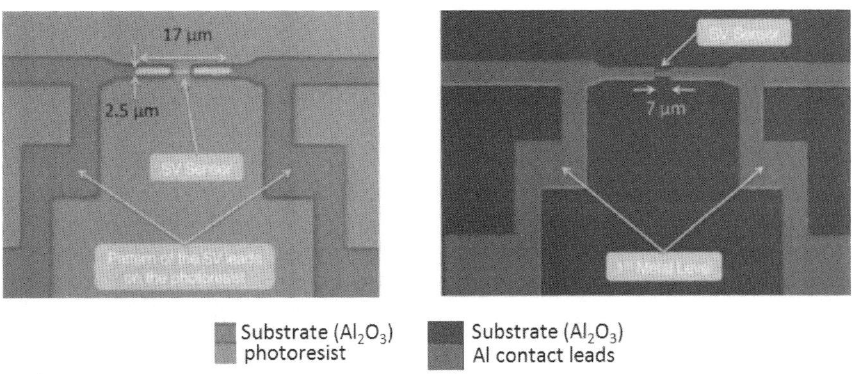

Fig. 10. Top metal contacts to a GMR sensor defined by a liftoff process: first the photoresist mask is defined to cover all areas except on the contact leads (left) and then the metal film is deposited over the surface. Upon photoresist removal (resist strip, acetone), the metal film remains on the contact leads areas (right).

Physical dry etching is usually achieved by a controlled removal of material using plasma etching (reactive etching or an ion beam system). In particular Ion Beam etching (ion milling) offers slow (typically below 0.2nm/s), but very controlled and stable etching ratios and is commonly used for the patterning of GMR devices [16].

Ion-milling etching is an anisotropic process with an etch rate depending on the material type. Moreover, the etched feature profile depends on the angle between the beam and sample, which can be used to control the magnetic properties of patterned materials in sensors or magnetic flux guides [43]. Figure 11 illustrates the impact of the ion beam incidence angle on the patterned element vertical profile.

Wet etching concerns a process taking advantage of the corrosive properties of some substances, usually acids. Because inorganic materials such as polymer based resists are resistant to the inorganic acids action, the wet etching can be performed. The used chemical strongly depends on the material of the layer to be etched (chemically selective), being easily found in the literature with detailed specifications of etchers, time/temperature, according to each material. Due to its aggressive nature and highly isotropic nature, wet etching is not very popular for patterning GMR thin film structures with dimensions ranging below 100 μm.

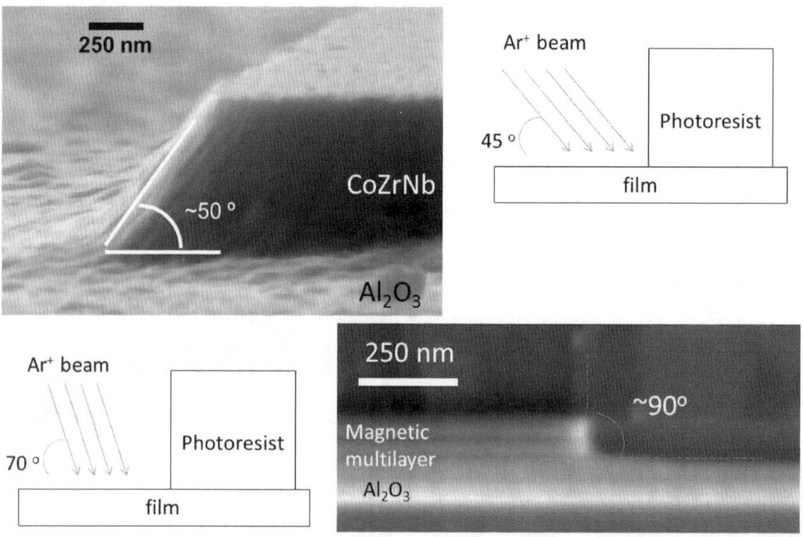

Fig. 11. Top: Sidewall profile at ~50°, resulting of an ion beam etching with an Argon beam incidence angle of 45° from the CoZrNb film surface. Bottom: vertical profile ~90°, promoted by pattering with a beam incidence angle of 70° from the magnetic film surface.

However, combining physical and chemical etching (Reactive Ion etching RIE) is a very good option for thick film patterning, when micrometric dimensions are involved. RIE can be used, for example, to etch silicon substrates, using only conventional microfabrication techniques compatible with CMOS fabrication. Figure 12 shows a successful example of silicon microneedle arrays defined after GMR sensor microfabrication, to be used in neurological applications. [44].

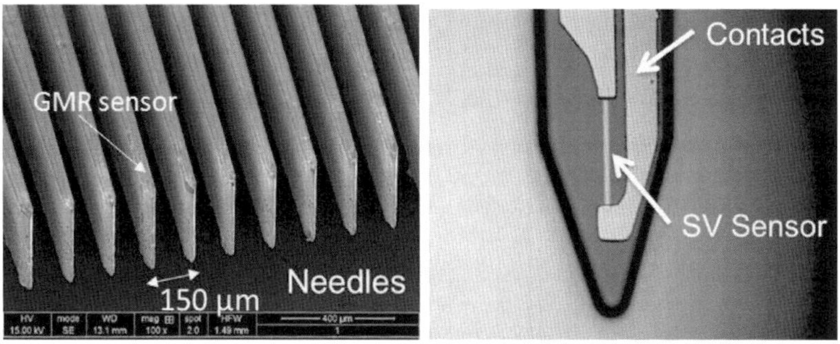

Fig. 12. (left) SEM image of Si needle array fabricated using C_4F_8, SF_6 and O_2 –based RIE; (right) SEM image of the tip of the needle with a GMR sensor

4 Conclusions

This chapter provides a general overview of the microfabrication techniques used for a basic GMR sensor fabrication. Magnetic thin film deposition of GMR stacks are generally done by magnetron sputtering and ion beam tools, and in some cases by electrodeposition. Integration of thin film stacks in a sensor element requires at least 3 processing levels, each one requiring a lithography step: one for sensor area definition, a second for metal contacts definition and finally via opening through an insulating oxide layer. After each lithography, the patterned feature can be transferred using either additive techniques such as lift-off, or subtractive methods as etching. Ion Beam etching is widely used for thin film and small dimensions (below 100 µm, down to nm areas) patterning, while reactive ion etching or wet etch are often preferred for thick films (and micrometric areas) patterning. Finally, arrays of Si needles incorporating GMR sensors are shown, to illustrate how several microfabrication techniques can be used for an integrated device.

References

[1] Europractice. Europractice web site, http://www.europractice-ic.com/
[2] Service MOSIS. The MOSIS Service, http://www.mosis.com/
[3] Fermon, C.: Micro- and Nanofabrication Techniques. In: Thornton, M.J., Ziese, M. (eds.) Spin Electronics, pp. 379–395. Springer (2000)
[4] Hartmann, U.: Magnetic Multilayers and Giant Magnetoresistance. Springer (2000)
[5] Hirota, E., Sakakima, H., Inomata, K.: Giant Magneto-Resistance Devices. Springer (2001)
[6] Grünberg, P., Fert, A.: Received the Nobel Prize in Physics (2007), http://www.nobelprize.org/nobel_prizes/physics
[7] Grunberg, P., Schreiber, R., Pang, Y., Brodsky, M.B., Sowers, H.: Layered magnetic-structures - Evidence for antiferromagnetic coupling of Fe layers across Cr interlayers. Phys. Rev. Lett. 57, 2442–2445 (1986)

[8] Baibich, M.N., Broto, J.M., Fert, A., Nguyen van Dau, F., Petrof, F., Etienne, P., Creuzet, G., Friederich, A., Chazelas, J.: Giant magnetoresistance of (001)Fe/(001)Cr magnetic superlattices. Phys. Rev. Lett. 61, 2472–2475 (1988)
[9] Dieny, B., Speriosu, V.S., Metin, S., Parkin, S.S.P., Gurney, B.A., Baumgart, P., Wilhoit, D.R.: Magnetotransport properties of magnetically soft spin-valve structures. J. Applied Physics 69, 4774–4779 (1991)
[10] Jaeger, R.C.: Introduction to Microelectronic Fabrication. Addison-Wesley (1993)
[11] Ohring, M.: The Materials Science of Thin Films. Academic Press Inc. (1992)
[12] Cheng, N., Ahn, J.P., Kannan, Krishnan, M.: Epitaxial growth and exchange biasing of PdMn/Fe bilayers grown by ion-beam sputtering. J. Appl. Phys. 89, 6597 (2001)
[13] Schwebel, C., Gautherin, G.: Deposition of thin films by ion beam sputtering: Mechanisms and epitaxial growth. In: AIP Conf. Proc., vol. 167, pp. 237–249 (1988)
[14] Gehanno, V., Freitas, P.P., Veloso, A., Ferreira, J., Almeida, B., Sousa, B., Kling, A., Da Silva, M.F.: Ion beam deposition of Mn-Ir spin valves. IEEE Trans. Magn. 35, 4361–4367 (1999)
[15] Dong, L., Srolovitz, D.J.: Texture development mechanisms in ion beam assisted deposition. J. Appl. Phys. 84, 5261 (1998)
[16] Nordiko Technical Services, UK, `http://www.nordiko-tech.com/products.html`
[17] Veeco, US, `http://www.veeco.com/products/ion-beam.aspx`
[18] Ferreira, R., Cardoso, S., Freitas, P.P., Petrova, R., McVitie, S.: Influence of ion beam assisted deposition parameters on the growth of MgO and CoFeB. J. Appl. Phys. 117, 07C117 (2012)
[19] Cardoso, S., Gehanno, V., Ferreira, R., Freitas, P.P.: Ion beam deposition and oxidation of spin-dependent tunnel junctions. IEEE Trans. Magn. 35(5), 2952–2954 (1999)
[20] Schmeusser, S., Hubert, A., Rupp, G.: Ion beam sputtering of magnetoresistive multilayers:Co/Cu and Co-modified $Ni_{81}Fe_{19}$ systems. IEEE Trans. Magn. 32(5), 4722–4724 (1996)
[21] Quan, J.J., Wolf, S.A., Wadley, H.N.G.: Low energy ion beam assisted deposition of a spin valve. J. Appl. Phys. 101, 074302 (2007)
[22] Harper, J.M.E., Gambino, R.J.: Combined ion beam deposition and etching for thin film studies. J. Vac. Sc. Tech. 16(6), 1901–1905 (1979)
[23] Tang, F., Karabacak, T., Morrow, P., Gaire, C., Wang, G.C., Lu, T.M.: Texture evolution during shadowing growth of isolated Ru columns. Phys. Rev. B 72(16), 1–6 (2005)
[24] Fermento, R., Leitao, D.C., Teixeira, J.M., Pereira, A.M., Carpinteiro, F., Ventura, J., Araujo, J.P., Sousa, J.B.: Structural, magnetic and transport properties of ion beam deposited Co thin films. J. Non-Cryst. Sol. 354(47-51), 5279–5281 (2008)
[25] Rijks, T.G.S.M., Sour, R.L.H., Neerinck, D.G., De Veirman, A.E.M., Coehoorn, R., Kools, J.C.S., Gillies, M.F., de Jonge, W.J.M.: Influence of grain size on the transport properties of $Ni_{80}Fe_{20}$ and Cu thin films. IEEE Trans. Magn. 31(6), 3865–3867 (1995)
[26] Pierson, H.O.: Handbook of Chemical Vapor Deposition (CVD) - Principles, Technology and Applications, 2nd edn. William Andrew Publishing Notes (1999)
[27] Schlesinger, M., Paunovic, M.: Modern Electroplating. Wiley (2000)
[28] Whitney, T.M., Jiang, J.S., Searson, P.C., Chien, C.L.: Fabrication and magnetic-properties of arrays of metallic nanowires. Science 261(5126), 1316–1319 (1993)
[29] Evans, P.R., Yi, G., Schwarzacher, W.: Current perpendicular to plane giant magnetoresistance of multilayered nanowires electrodeposited in anodic aluminum oxide membranes. Appl. Phys. Lett. 76, 481 (2000)

[30] Schwarzacher, W., Lashmore, D.S.: Giant Magnetoresistance in Electrodeposited Films. IEEE Trans. Magnetics 32, 3133–3153 (1996)

[31] Blondel, A., Meier, J.P., Doudin, B., Ansermet, J.P.: Giant magnetoresistance of nanowiresof multilayers. Appl. Phys. Lett. 65(23), 3019–3021 (1994)

[32] Piraux, L., George, J.M., Despres, J.F., Leroy, C., Ferain, E., Legras, R., Ounadjela, K., Fert, A.: Giant magnetoresistance in magnetic multilayered nanowires. Appl. Phys. Lett. 65(19), 2484–2486 (1994)

[33] Piraux, L., Dubois, S., Duvail, J.L., Ounadjela, K., Fert, A.: Arrays of nanowires of magnetic metals and multilayers: Perpendicular GMR and magnetic properties. J. Mag. Magn. Mater. 175(1-2), 127–136 (1997)

[34] Tang, X.T., Wang, G.C., Shima, M.: Layer thickness dependence of CPP giant magnetoresistance in individual CoNi/Cu multilayer nanowires grown by electrodeposition. Phys. Rev. B 75(13), 134404 (2007)

[35] Wang, H., Wu, Y., Wang, M., Zhang, Y., Li, G., Zhang, L.: Fabrication and magnetotransport properties of ordered sub-100nm pseudo-spin-valve element arrays. Nanotech 17(6), 1651 (2006)

[36] Wang, H.-X., Wu, Y.-C., Zhang, L., Li, J., Zhang, G., Wang, M., Li, Z.: Synthesis and magnetotransport properties of nanometric spin valve arrays fabricated by electrodeposition. J. Phys. D: Appl. Phys. 38(21), 3841 (2005)

[37] Wang, H., Wu, Y., Li, Q., Wang, M., Li, G., Zhang, L.: Synthesis and characterization of FeMn-pinned spin valve arrays. Appl. Phys. Lett. 89(5), 052107–3 (2006)

[38] Jaeger, R.C.: Lithography. In: Introduction to Microelectronic Fabrication, 2nd edn. Prentice Hall, pper Saddle River (2002)

[39] Middleman, S., Hochberg, A.K.: Process Engineering Analysis in Semiconductor Device Fabrication, p. 313. McGraw-Hill (1993)

[40] Leitao, D.C., Macedo, R.J., Silva, A.V., Hoang, D.Q., MacLaren, D.A., McVitie, S., Cardoso, S., Freitas, P.P.: Optimization of exposure parameters for lift-off process of sub-100 features using a negative tone electron beam resist. In: 2012 12th IEEE Conference on Nanotechnology (2012)

[41] Nabiyouni, G.R.: Design and fabrication of nanomagnetic sensors based on electrodeposited GMR materials. Metrology and Measurements Systems XVI(3), 519–529 (2009)

[42] Okazaki, S., Moers, J.: Lithography. In: Waser, R. (ed.) Nanoelectronics and Information Technology, 2nd edn. Wiley-VCH

[43] Marinho, Z., Cardoso, S., Chaves, R., Ferreira, R., Melo, L.V., Freitas, P.P.: Improving Magnetic Flux Guide Concentrators for MR Sensors. J. Appl. Phys. 109, 07E521 (2011)

[44] Amaral, J., Gaspar, J., Pinto, V., Costa, T., Sousa, N., Cardoso, S., Freitas, P.P.: Measuring brain activity with Magnetoresistive Sensors integrated in Micromachined Probe Needles. Applied Physics A (in press)

Noise in GMR and TMR Sensors

C. Fermon and M. Pannetier-Lecoeur

DSM/IRAMIS/SPEC, CEA Saclay and CNRS URA 2464,
91191 Gif sur Yvette, France
Claude.Fermon@cea.fr

Abstract. Giant Magnetoresistances (GMR) and Tunnel Magnetoresistances (TMR) take an increasing part in many applications like current sensing, magnetometry or position sensing, thanks to their high magnetoresistance at room temperature, which leads to a large output signal variation. But the real performances of such sensors can only be estimated with respect to the sources of noise. In this chapter, we give first some bases on noise theory and data treatment. Fluctuations, ergodicity and volume considerations will be discussed. A second part will detail noise measurement techniques and data analysis of typical noise power spectra. Sources of noise will be discussed in a third part. In the end of the chapter, specific cases of GMR and TMR magnetic noise and non magnetic noise will be discussed with their physical origin and their analytical or phenomenological expression. We will then present ways to design GMR and TMR sensors for noise reduction, depending on the applications targeted.

Introduction

Any macroscopic quantity in a system such as voltage, current, resistance is subject to fluctuations around its mean value. These fluctuations are created by the random contributions to the transport or by internal displacements of atoms. The first point of view is to consider that these fluctuations are strongly related to the properties of the material and their study gives a new approach to understand processes in condensed matter. A second point of view more related to applications is to call these fluctuations noise which is related to the quality of the material investigated. Reduction of this noise becomes a target for magnetic sensor applications.

1 Noise Formalism

In this part we give some bases necessary for the analysis of noise measurements in GMRs with a highlight on some common traps in noise treatment. We will not address here extensively the quantum approach of fluctuations as magnetoresistive sensors have size and working temperature which can be handled with a classical treatment.

C. Reig et al.: *Giant Magnetoresistance (GMR) Sensors*, SSMI 6, pp. 47–70.
DOI: 10.1007/978-3-642-37172-1_3

1.1 Fluctuations, Average and Distribution

We consider a fluctuating quantity $V(t)$. The voltage V is the most common quantity measured in magnetic sensors but the following treatment can be applied to any relevant quantity like the current, the resistance or the charge. In an experiment, $V(t)$ is often measured at discrete times $t_1,\ldots t_n$ by some acquisition system but it can also be treated in an analog way with some integration, derivation, analog multiplication or other mathematical operation. Any non linear operation has to be carefully handled in order to avoid spurious deformation of the signals.

Then a number of new quantities can be derived from that measurement. The first one is the average

$$\bar{V} = \frac{1}{n}\sum_{i=1}^{n} V(t_i) \tag{1}$$

A formal definition of the average, easier to manipulate is:

$$\bar{V} = \lim_{T\to\infty}\frac{1}{T}\int_0^T V(t)dt \tag{2}$$

where T is the duration of the measurement. In the following, we will use the second formalism.

It should be noticed that this average is also a fluctuating quantity on a time scale T corresponding to the total acquisition time. In common systems, if T is long enough this average is representative of V but in some cases like in presence of slow magnetic relaxation, $\bar{V}$ can be very dependent of the history of the system.

The second quantity is the variance of V. It is defined as

$$\sigma^2 = \frac{1}{n}\sum_{i=1}^{n}(V(t_i) - \bar{V})^2 \text{ or}$$

$$\sigma^2 = \lim_{T\to\infty}\frac{1}{T}\int_0^T (V(t) - \bar{V})^2 dt \tag{3}$$

It measures the exploration in amplitude of V. σ is an easy comparison of different signals but sometimes a spurious frequency may appear in the signal like the line frequency (50 or 60 Hz) which dominates and sets the value of σ.

For large systems, the knowledge of σ and $\bar{V}$ is sufficient to know the distribution function ρ of V. This is due to a very useful theorem, the central limit theorem, which demonstrates that a sum of n random identical quantities tends very rapidly to a normal (or Gaussian) distribution law when n grows.

$$\rho(V) = \frac{1}{\sqrt{2\pi\sigma^2}}\exp\left(-\frac{(V-\bar{V})^2}{2\sigma^2}\right) \tag{4}$$

For example, the voltage across a magnetoresistance with a fixed sensing current can be divided in smaller identical magnetoresistances which individually fluctuates and hence one can easily demonstrate that the distribution of the voltage follows a normal distribution. We will see later that in small systems or in presence of an individual strong defect, this law is modified.

Another very common fluctuation is the random jump between two discrete levels. This type of fluctuation is for example the Random Telegraph Noise (RTN) described later. The jumps are arising from a level 1 to a level 2 with a w_1 probability to stay at level 1 and w_2 probability to stay at level 2. The mean period of these jumps is τ. The distribution is then showing peaks centered on the values of the levels with a width related to the additional noise. Hence, analysis of the distribution may help to distinguish the presence of discrete levels.

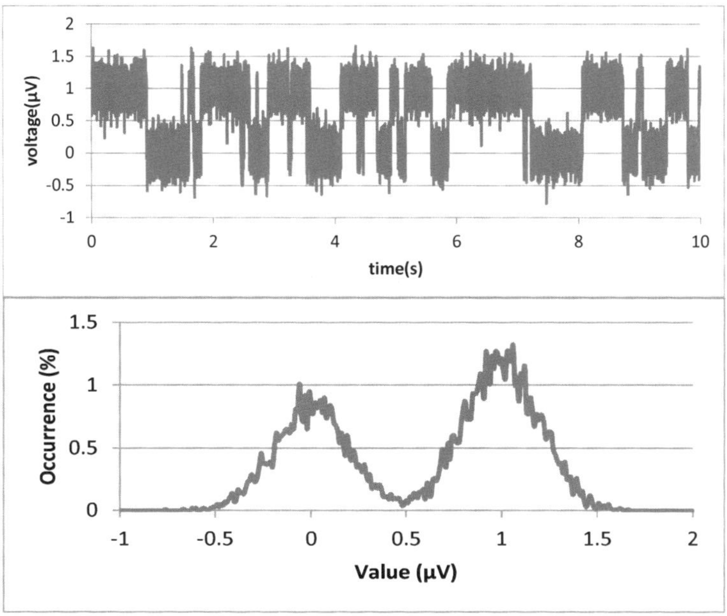

Fig. 1. Example of a two level noise with nearly equal probabilities in states up and down. Time domain (up) and associated distribution function (bottom).

1.2 Correlations

In time domain, correlations are very important because they give a measure of the similarities between two functions (cross-correlations) or for the same function but at two different times (auto-correlation). The auto-correlation is a measure of the memory of the system.

The general form of the correlation of two functions X and Y is defined as the average of the functions at a time difference of τ.

$$C_{xy}(\tau) = \lim_{T\to\infty} \frac{1}{T}\frac{1}{2\pi}\int_0^T X(t)Y^*(t-\tau)dt \tag{5}$$

The auto-correlation function is then

$$g_x(\tau) = \lim_{T\to\infty} \frac{1}{T}\frac{1}{2\pi}\int_0^T X(t)X^*(t-\tau)dt \tag{6}$$

The auto-correlation at τ=0 is simply the mean square value of the fluctuations.

For example, a thermal noise has an auto-correlation function which is zero except for τ=0. This is the signature of a totally random process. However, if you measure a thermal noise through a filter, you may find a non zero auto-correlation due to the memory injected by the filtering.

It should be noticed that the experimental correlations are depending on the initial time chosen for the acquisition. Hence, there is a strong hypothesis behind, usually fulfilled by GMR and TMR systems, the stationarity of the system which assumes that the system is at equilibrium and these quantities are independent on the initial measurement time. In the following parts, we will consider that this hypothesis is fulfilled.

1.3 Frequency Space and Spectral Density

Amplitude and amplitude distribution of fluctuations are analyzed by the tools described previously but spectral analysis gives the frequency dependence of the noise which is essential for separating and understanding the noise sources. The way to switch from time domain to frequency domain is the Fourier transform.

$$V(\omega) = \frac{1}{2\pi}\int_{-\infty}^{\infty} V(t)e^{i\omega t}dt \tag{7}$$

We introduce also a more experimental Fourier transform which considers that V(t) is zero outside of the measuring time.

$$V_T(\omega) = \frac{1}{2\pi}\int_0^T V(t)e^{i\omega t}dt \tag{8}$$

This well-known form of Fourier transform is however source of lot of errors in experiments. Firstly, if the total acquisition time is T, the lowest achievable frequency is $2\pi/T$ and the highest is given by $2\pi/t_{acq}$ where t_{acq} is the acquisition time interval. Secondly, if signals are present in the fluctuations at higher frequencies they will appear by folding in the frequency range. This is the reason why low pass filtering is necessary for noise measurements. Thirdly, Fast Fourier Transform (FFT) algorithms are generally used due to the gain in time of data treatment. As the data are finite in time, a window function is applied on the data. The simplest one is just a rectangular window. The data are then considered as 0 outside the window and multiplied by 1 in the window. However, the jump at the edges created spurious oscillations in the Fourier transform. More sophisticated windows are hence used. The most common are Hann, Hamming and Gaussian windows. In ref [1], you can find a large number of examples. The Hann function consists of multiplying the data by a cosine function which vanishes for the first and last points. So the quantity calculated becomes:

$$V_{Hann}(\omega) = \frac{1}{2\pi}\int_0^T V(t)e^{i\omega t}(1 - \left|\cos\left(\frac{\pi t}{T}\right)\right|)dt \tag{9}$$

Or in the case of Gaussian window,

$$V_{Gauss}(\omega) = \frac{1}{2\pi}\int_0^T V(t)e^{i\omega t}e^{-\frac{(t-\frac{T}{2})^2}{2(\sigma T/2)^2}})dt \tag{10}$$

On rather flat signals like noise, the windowing has no impact but in case of a strong line signal, it allows suppressing artefacts.

The total energy of the signal is given by

$$E = \int_0^T |V(t)|^2 dt \tag{11}$$

which can be expressed through the Parseval theorem by

$$E = 2\pi\int_{-\infty}^{\infty} |V(\omega)|^2 d\omega \tag{12}$$

Hence the average power associated to the fluctuations can be defined by:

$$P = \lim_{T\to\infty}\frac{1}{T}\int_0^T |V(t)|^2 dt = \lim_{T\to\infty} 2\pi\int_{-\infty}^{\infty}\frac{|V_T(\omega)|^2}{T}d\omega \tag{13}$$

This allows us to define the power spectral density (PSD) as:

$$S_V(\omega) = \lim_{T\to\infty} 2\pi\frac{|V_T(\omega)|^2}{T} \tag{14}$$

The PSD is given for voltage fluctuations in V^2/ Hz.

A very important point is the relation between the power spectral density and the auto-correlation function which is the Wiener-Khintchine theorem:

$$S_V(\omega) = 2\int_0^{\infty} g(\tau)\cos(\omega\tau)d\tau. \tag{15}$$

A major example is an exponentially decreasing auto-correlation function. It is the case for a relaxation process but also for the current in a simple R-L circuit. The autocorrelation decreases with a characteristic time τ_c and is written as:

$$g(\tau) = g(0)\exp(-\frac{\tau}{\tau_C}) \tag{16}$$

The corresponding PSD is

$$S_V(\omega) = 4g(0)\frac{\tau_C}{1+\tau_C^2\omega^2} \tag{17}$$

This spectrum called Debye-Lorentzian spectrum presents a flat response at low frequencies and a $1/f^2$ decrease at high frequencies with a corner at $1/\tau_c$. This exponential decrease of the autocorrelation function is also valid for RTN noise and hence the RTN signature in the PSD is this Debye-Lorentzian spectrum. Another important remark is related to very slow decrease of the DC level in noise measurements determining the formal spectral density from a noise mechanism. Very often noise measurements setups present a very low frequency high pass filter (0.1Hz or lower) to avoid amplifier saturation by DC offsets. If there is an external perturbation like a short pulse on the DC line or a jump in the DC signal this induces a decrease of the output signal with a very long characteristic time and hence a Debye-Lorentzian spectrum with a corner frequency below the measurement range. For that reason, in case of measurement of a $1/f^2$ decrease at very low frequencies, investigation of the DC level fluctuations should be done.

1.4 Sensitivity, Signal to Noise Ratio and Detectivity

Noise PSD is given in V^2/Hz but it is usually more convenient to compare a signal given in V to the square root of the PSD which is in V/sqrt(Hz).

In order to evaluate a signal-to-noise ratio, a reference signal at a known frequency generated by a coil is often used. If a signal $V_0 \cos(\omega t)$ is seen in the acquisition system, its PSD is the power associated to this signal taken on a bandwidth of 1Hz so 1s and it corresponds to $V_0^2/2$. This allows direct calibration of a sensor.

The sensitivity for a magnetoresistive signal, β, is usually given in V/V/T. Typical values for GMR sensors are 20-40V/V/T or 2-4%/mT.

The output voltage of a magnetoresistive sensor can be written as:

$$V_{out} = (R_0 + \delta R(H))I \tag{18}$$

If the sensor is linearized and well centered, the output can be written as:

$$V_{out} = (R_0 + \frac{\delta R}{\delta H}.H + \cdots)I \tag{19}$$

where the terms with higher H power are small. We will use that notation in this chapter. $\frac{\delta R}{\delta H}$ is given in Ohms/Tesla.

The sensitivity is given by $\frac{\delta R}{\delta H}/R_0$.

In order to compare different sensors, it is very convenient to use the field equivalent noise power spectral density called sometimes detectivity. It corresponds to the PSD divided by the sensitivity.

For example if a sensor exhibits a thermal noise of 1nV/sqrt(Hz) and a sensitivity of 20V/V/T, the corresponding detectivity will be 50pT for 1V of bias voltage.

2 The Different Sources of Noise

In this part, we will describe the various sources of noise existing in magnetoresistive sensors. Some, like shot noise, are very specific of tunnel junctions but nearly all are common to all kind of magnetoresistive sensors.

2.1 Separation of Magnetic and Non Magnetic Noise

It should be noticed that noise in magnetic sensors can be or not magnetic-field dependent. A magnetic field dependent noise appears or disappears with the application of an external field. This is for example the case of a part of the 1/f noise in magnetoresistive junctions. One way to separate it is to measure the noise of the sensors in the whole range of operation.

2.2 Frequency Independent Noise (Thermal or Johnson-Nyquist Noise), Shot Noise

Thermal Noise

Frequency independent noise is called white noise and corresponds to processes without any auto-correlation except at zero time. It should be noticed that noise appears flat in the range of measurement because its correlation characteristic time is faster that the minimal sampling time.

The most important noise is the thermal noise which is directly related to the resistance of the sensor. The first observation of this noise has been done by Johnson [2] and interpreted by Nyquist[3]. The associated PSD is written as:

$$S_V(\omega) = 4Rk_BT \tag{20}$$

where $k_B = 1.3806\ 10^{-23}\mathrm{J.K^{-1}}$ is the Boltzmann constant. There are several ways to demonstrate this relation but the base is just to say that the energy available for a dipole of resistance R is $k_BT/2$. This formula is valid for frequencies much lower than $k_BT/\hbar$ where $\hbar$ is the Planck constant. This thermal noise cannot be eliminated or reduced except by changing the resistance or the temperature but it has the advantage to be independent on the voltage applied on the sensor.

It should also be noticed that the impact of this noise on the signal-to-noise ratio is directly related to the working bandwidth. The integrated noise will increase as the square root of the bandwidth.

Shot Noise

This noise is due to the fact that the electrical current is not continuous due to the discrete nature of the electrons. This noise is detectable only if there is a barrier to cross where the quantum nature of electrons is revealed. In metals, the electrons inelastic scattering length is very small, typically a few nm at room temperature and a metallic sensor could be described as a very large number of individual elements of

Table 1. Some values of integrated thermal noise for various resistances and temperatures

Resistance	Temperature	Bandwidth	$\int S_V^{\frac{1}{2}}(\omega)d\omega$
50 Ohms	300K	1Hz	0.9nV
50 Ohms	300K	1kHz	29nV
1kOhms	300K	1Hz	4nV
1kOhms	300K	1kHz	129nV
1kOhms	4K	1kHz	15nV

few nm in series and hence the shot noise is divided by the square root of this number [4]. For that reason, shot noise is not present in GMR sensors.

In magnetic tunnel junctions a number of theoretical and experimental works have been recently published. We will describe them later. The electrons are following a Poisson law when they pass through a barrier (without quantum or correlation corrections) and hence it is possible to calculate the PSD of the shot noise. At T=0, it is simply given by

$$S_I(\omega) = 2eI \tag{21}$$

In mesoscopic systems, deviations of the Poisson statistics are observed and modify eq (21). A review of noise in mesoscopic systems can be found in ref [5]. In particular it has been shown that the statistics can be slightly different from a pure Poisson statistics inducing an enhancement or a reduction of the shot noise [6,7]. This enhancement or reduction is characterised by the Fano factor F.

Crossover between Shot Noise and Thermal Noise

When the temperature is increasing, the shot noise expression must take into account thermal fluctuations. A more general formula should be applied [4]:

$$S_V(\omega) = 2eIcoth\left(\frac{eV}{2k_BT}\right)R^2 \tag{22}$$

In the two temperature limits (T >>0 and T → 0) we find the expression of noise given in (20) and (21) respectively.

If a modified Poisson distribution is involved, eq. 22 becomes

$$S_V(\omega) = 4(1-F).k_BRT + 2FeIcoth\left(\frac{eV}{2k_BT}\right)R^2 \tag{23}$$

2.3 Low Frequency Noise

1/f Noise

1/f noise is a general term referring to a frequency decreasing noise with a power law frequency $\frac{1}{f^\beta}$ where β is an exponent typically of the order of 1. This noise is observed in nearly all fluctuating systems including for example biological and geological fluctuations. In GMR and TMRs this low frequency noise is dominant and is often a drawback in performances of magnetoresistive sensors.

The origin of the 1/f noise is resistance fluctuations and not voltage fluctuations like the thermal noise. Hence, these fluctuations can only be revealed by applying a current in the sensor. Voss and Clarke have demonstrated in 1976 [8] that the variance of Johnson noise exhibits a 1/f power spectrum demonstrating the resistance fluctuation nature of the 1/f noise.

This resistance fluctuations behavior implies that the PSD is varying as V^2 or I^2. This is very important because this allows us discriminating between white noise and low *1/f* noise. This general law can be however modified if the current induces itself modifications of the resistance or local heating. We have observed in particular in small GMR devices an increase of *1/f* noise much faster than I^2.

GMR sensors have an isotropic dependence of the resistance unlike AMR where the signal is depending on the angle between field and current or Hall sensors where the voltage appears in a direction perpendicular to the current. Hence it is impossible to play on current direction to separate resistance variation and external field variation. This is a major issue with GMRs where spinning techniques used in Hall sensors or flipping of the current used in AMR sensors cannot be applied.

The second important point is the size effect on 1/f noise. Typically PSD of the 1/f noise is decreasing as the volume of the sensor increases. This can be understood easily by an averaging effect. If we suppose that 1/f fluctuations are coming from small individual sources, you can consider that a resistance is the sum of N small resistances r so the total resistance $R = N.r$ but the fluctuations of R are $\sqrt{N}$ larger than the individual *r* fluctuations. Hence for a given R, the fluctuations are decreasing as $\sqrt{N}$, i.e. as the square root of the volume. This has been observed in several systems for large enough sizes. At very small sizes an individual fluctuator may dominates and this rule is broken.

There is a general phenomenological formula proposed by Hooge [9] which allows us comparing various sensors

$$S_V(\omega) = 2\pi \frac{\gamma_H V^2}{N_c \omega} = \frac{\gamma_H V^2}{N_c f^\beta} \tag{24}$$

where γ_H is a dimensionless constant. This formula is well adapted to semiconductors where the averaging is more on the number of carriers N_c than on the effective volume. For TMRs, the formula becomes:

$$S_V(\omega) = 2\pi \frac{\alpha V^2}{A\omega} = \frac{\alpha V^2}{Af} \tag{25}$$

where A is the active surface of the device and α is then a parameter with the dimension of a surface. This last formula describes rather correctly the evolution of the noise with the size of the sensors.

We note that *1/f* noise can exhibit a non magnetic and a magnetic component with sometimes different slopes.

The evolution of the noise as function of magnetic field can help to separate the two contributions. Often, the noise recorded under a strong field which saturates the different layers of a magnetic sensor is mainly non magnetic and an additional noise due to magnetic fluctuations in the layers appears in the sensitive regions. As described later, the shape of the sensors has a strong impact on the magnetic 1/f noise.

In case of TMRs, the 1/f noise depends also whether the junction is in the parallel or antiparallel state. In the parallel state, the number of channels opened through the barrier is larger than in the antiparallel state and hence the noise is smaller. This is a direct consequence of the reduction of the effective size of a junction.

Random Telegraph Noise

Random Telegraph Noise (RTN) or "popcorn" noise is the noise arising from the fluctuations of a specific source between two different levels. For magnetoresistive sensors, this noise generally appears in devices with a size small enough to let individual defects becoming dominant. However, it can be sometimes observed in reasonably large GMRs at high currents levels.

RTN is, as explained before, a fluctuation between two levels with comparable energies and a barrier height able to give a typical characteristic time in the measurement range. Hence, RTN is very dependent on the temperature, field and applied bias current. Figure 2 gives an example of RTN noise in a micron size TMR sensor as function of temperature.

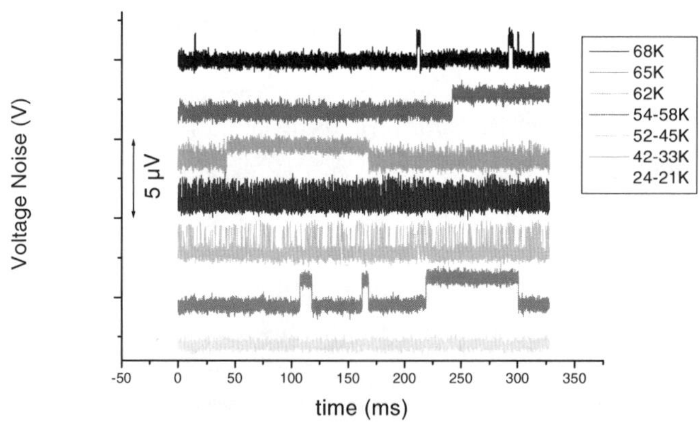

Fig. 2. Example of RTN noise in a TMR sensor as function of temperature. We can see different individual fluctuators with different characteristic times [10].

RTN is difficult to handle and a sensor with RTN noise is in general very difficult to use even if it is theoretically possible to suppress partially this noise by data treatment. That suppression requires a RTN with a low fluctuation frequency and two states well separated. The treatment is then based on the recognition of each transition level from low state to high state and suppression of the step.

2.4 High Frequency Noise and Ferromagnetic Resonance

At high frequencies, the noise is usually dominated by thermal or shot noise. However, noise peaks can be detected in magnetoresistive sensors due to the fluctuations of magnetic layers. In the GHz regime, thin magnetic films present ferromagnetic resonances with frequencies dependent on the material and on the shape of the sensors. In small elements, quantization of the spin waves induces a large number of resonances. The noise detected is coming from two different sources. The first one is due to the GMR effect. The free layer is fluctuating with more important amplitude at resonance and if a DC current is sent in the device, a voltage at the resonance frequency appears. The second one is due to thermal and shot noises. This noise is amplified by the quality factor of the resonance and hence this appears as an extra noise even in absence of bias current. The amplitude of this ferromagnetic resonance enhanced noise might depend on the probing DC current due to spin transfer torque which affects the quality factor [11].

2.5 External Noise

With magnetic sensors and in particular very sensitive magnetic sensors it is essential to take care of the magnetic external noise. Inside a laboratory, there are three types of external perturbations which lead to magnetic noise. First a number of discrete frequencies, including the power supply line (50/60Hz) and its harmonics but also higher frequencies typically up to MHz coming from power supplies, low consumption lights etc. All these lines correspond to real AC magnetic fields and hence are proportional to the bias current in the magnetoresistive sensor. The second source is a 1/f magnetic noise which exists everywhere. In a laboratory, this noise has an intensity of about 100nT at 1Hz and decreases slightly faster than 1/f. In a good magnetic shielded room with passive and active shielding, the noise level at 1Hz can be of the order of 100fT.

The third type of noise is less intuitive. It is created by the vibration of the magnetoresistive sensor in an existing DC field. This noise can only be detected with very sensitive sensors, typically mounted with flux concentrators. The vibration induces a flux the variation in a flux concentrator and hence an additional signal. This noise can be easily recognized because it appears usually as bursts at fixed very low frequencies and their amplitude varies strongly when artificial vibrations are created.

3 Electronics and Noise Measurements

Noise is often difficult to measure quantitatively and there are several approaches to reliably estimate this quantity. Combining field and temperature variation can even complicate the measurement. In this part, we will focus on a "standard" setup with some alternatives and with an emphasis on common errors.

3.1 Electronics Design

The electronics for a noise measurement or for interfacing magnetic sensors in an application can be separated roughly in five parts.

- The sensor system which may contain one or several different sensors
- A biasing source (voltage or current)
- A front end electronics which contains a first preamplifier stage
- An amplification, filtering and shaping stage
- An acquisition system.

For noise measurements, the two last parts may be replaced by a spectrum analyzer. Fig. 3 gives an overview of the electronics.

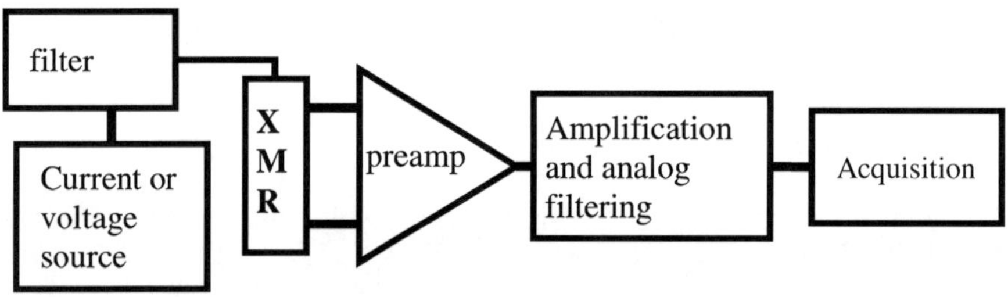

Fig. 3. Schematics of electronics for typical noise measurements

GMR and TMR Configuration
Noise is typically few orders of magnitude smaller than the main voltage applied. For that reason, differential measurements have to be performed. There are several ways for doing that.

- **Full Bridge Configuration**

This configuration is the most usual in applications and very efficient for noise measurements but it requires at least 4 matched elements. For noise measurements, this bridge has to be supplied by a very well stabilized voltage source. Typically, a battery is used with proper filtering.

Depending on the application, magnetoresistances R_A and R_B can be identical and hence the bridge becomes insensitive to applied homogeneous fields but may be sensitive to local fields as generated by current lines for example in current sensors [12] (Fig. 4-right) or may have opposite responses in order to be sensitive to external fields (Fig. 4-left) (magnetometers applications).

For estimating the noise we will consider here the case of four identical magnetoresistances, each one having its own noise. The resistance seen between V_1 and V_2 is $R=R_A=R_B$ and associated noise either due to resistance fluctuations or thermal noise is the average noise of R. If there is some noise in the voltage supply,

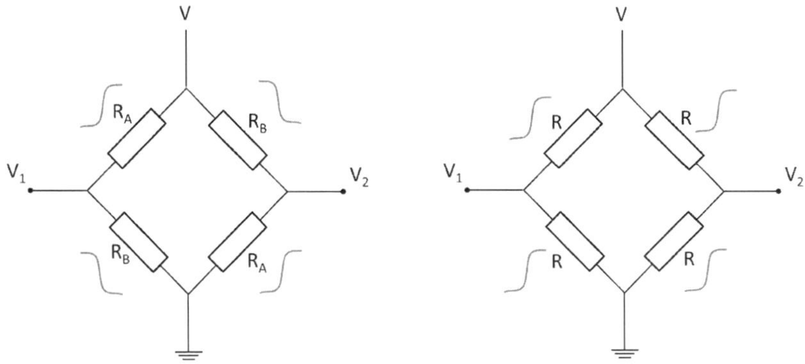

Fig. 4. Full bridge configuration. Left: for local field or current measurements. Right: for external field measurements.

the differential measurement (V_1-V_2) suppresses it at a first order but it should be noticed that in case of application of a field (or current) the bridge is no more balanced and a noise proportional to this non equilibrium may appear.

- **Half Bridge Configuration**

Sometimes, it is difficult to have four matched magnetoresistances or the analysis of a single element is desired. In that case, at least half of the bridge is replaced by external resistors. If a full bridge design is taken and the remaining resistors are replaced the noise of the single element is divided by four and the signal given by this element is also divided by four. For that reason, people often use a half bridge configuration which consists on feeding the one or two elements by a current like source. The schematic is given in figure 5.

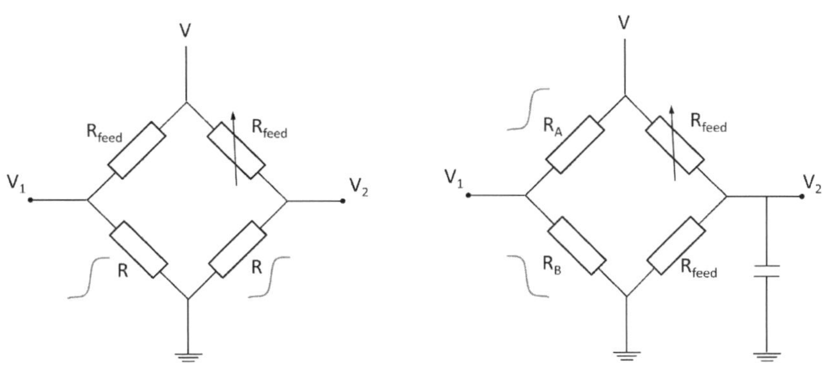

Fig. 5. Half bridge configuration. Left: gives the classical configuration, Right: is an alternative configuration where the noise of feeding resistance (R_{feed}) is suppressed.

In case of Fig. 5-left configuration, the resistances R_{feed} are usually chosen so that $R_{feed}>>R_A$ and R_A and R_B chosen so $R_A = R_B$. The voltages V_1 and V_2 are given by

$$V_1 = V\frac{R_A}{(R_{feed}+R_A)} \text{ and } V_2 = V\frac{R_B}{(R_{feed}+R_B)} \tag{26}$$

This requires a rather large supply voltage but all the variations of R_A and R_B are transferred to V. The noise induced by the resistances can be calculated from the above formulas. If the voltage V is fixed and well filtered, the noise fluctuations on the output voltages is given by

$$\frac{\delta V_1}{V_1} = \frac{\delta R_A}{R_A} - \frac{(\delta R_{feed} + \delta R_A)}{(R_{feed} + R_A)} \tag{27}$$

So the measured noise is the sum of the noise induced by R_{feed} and R_A. If R_{feed} is 10 times larger than R_A, the thermal noise induced by the feeding resistance is only $1/9^{th}$ of the tested resistance thermal resistance but if the tested magnetoresistance is at low temperature, say 4K, then the feeding resistance noise dominates.

For that reason, we prefer the configuration of Fig. 5-left. In that configuration, the noise is exactly the noise of a full bridge but with 2 non noisy resistances because the other branch is shortcut to the supply.

- **Single Element Configuration for AC Measurements**

In case of AC measurements, i.e when signal and noise of interest are in a specific frequency range, AC coupling of the element solves easily the problem of DC bias. It is also possible to work in such configuration for noise measurements but a specific care has to be taken due to low frequency noise induced by large capacitances. The noise determination is very similar to the previous configuration (Fig. 5-left) but, as only V_1 is measured, the noise measured is a factor $\sqrt{2}$ less than the half bridge configuration. The R_{feed} is advantageously replaced by a self (Fig 6) for additional rejection of source noise.

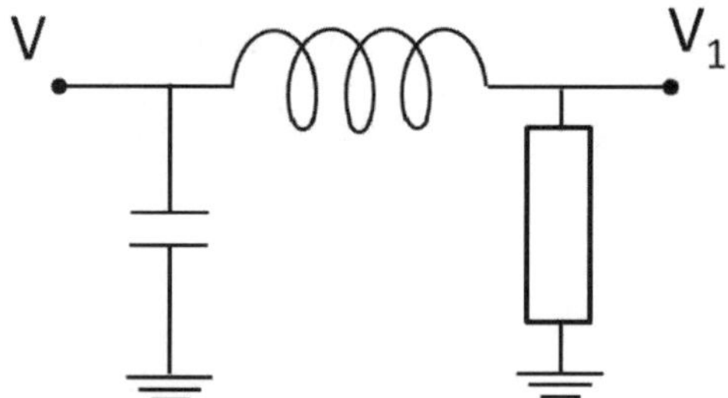

Fig. 6. AC coupling of a single element. The GMR element is fed through a filter and the signal V_1 is AC coupled to the preamplifier.

Biasing

Biasing with a current or a voltage source is often discussed. For the full bridge configuration, voltage biasing is clearly preferable as the working point (V/2) does not change with temperature. For half bridge configuration (Fig. 5-left), the two large resistors make a conversion from voltage-source to current-source so as their variation with temperature will be very different than the magnetoresistances variation, voltage or current source does not change the working point. In the configuration of Fig.5b, voltage biasing is a better option than current biasing.

Front End Electronics

The first stage of the front end electronics is always a very low noise preamplifier. Several options are possible and new low noise preamplifiers are regularly proposed on the market. The choice of the preamplifier has to be driven by the following considerations. Each preamplifier has an input referred voltage noise and a current referred input noise. For a resistance connected to the input, the noise will be given by

$$S_V^{1/2} = \sqrt{4kRT + e_n^2 + R^2 i_n^2} \tag{28}$$

Where e_n is the voltage noise of the preamplifier and i_n the current noise of the preamplifier. Figure 7 gives the noise spectral densities for an INA103KP that can be used for low resistance (<1kOhm) noise measurements.

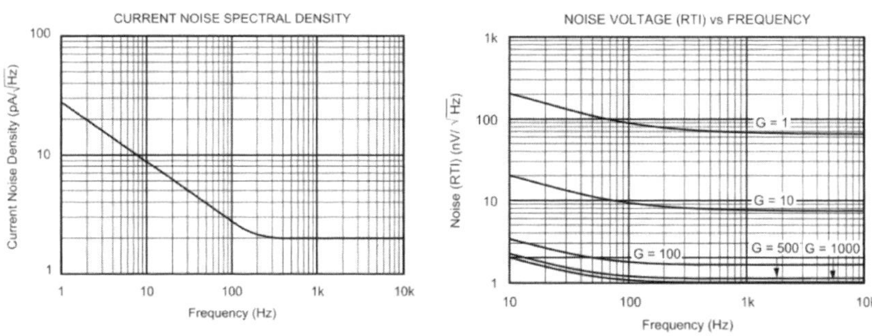

Fig. 7. Current and voltage noise spectral densities as function of frequency for an INA103KP amplifier [13]

For high resistance noise measurements, FET preamplifiers can be preferred since, despite their higher voltage noise, they exhibit a much lower current noise.

After the preamplifier, a typical chain will contain an offset correction, a second amplifying chain, so that the total gain is about 10^4 to 10^5, and a filtering.

A low pass efficient filtering to avoid spectral aliasing is essential for proper noise measurements. Its characteristics are strongly related to the acquisition frequency.

Acquisition and Filtering

The acquisition may be performed either with an acquisition card connected to a computer or with a spectrum analyzer. For noise measurements, it is better to have a 16 bits acquisition for accepting a high dynamic range and a high acquisition frequency even if the spectral range of interest is low. Typically for MR sensors, the range of interest for noise is below 100kHz for low frequency determination or above 1GHz for resonant noise measurements (see Part 2.4). The latter requires a GHz spectrum analyzer and will not be discussed here. For low frequency measurements, we are using typical acquisition rates larger than 1 MHz and an anti-aliasing filter at 200kHz with 8 to 12 poles. Then the 0-100kHz range is really free of contamination of higher frequencies.

3.2 Additional Remarks and Alternatives

Modulation and Demodulation

In all the previous configurations, we have always considered a DC bias source. It is however very interesting to use an AC source (current or voltage) and to perform a demodulation of the signal at the end either digitally after acquisition (the simplest), or through a commercial lock-in. The interest is firstly to suppress the 1/f noise of the preamplifier and secondly to suppress the line contamination in the measured spectrum.

Connections Noise

In noise investigation as for the development of very sensitive magnetic sensors, it is necessary to take care of the connection quality. Often connections are made with wire bounding in Al or Au with an effective small surface contact. Even if the total resistance contact is small, a low frequency noise can appear. Additionally, connector noise along a measurement chain can also be a source of noise which might be not negligible in the total noise if the sensor's noise is low.

Correlation for Preamplification Noise Suppression

When the signal of interest is small and dominated by the preamplification noise, it is difficult to measure it. A possible approach consists of using two preamplifiers which are measuring the same signal. If a cross-correlation as given in Eq.5 is performed, we obtain the spectral density of the signal of interest plus three terms which are not correlated; the cross correlation of the signal and each amplifier noise and the correlation between the noise of the two preamplifiers. By averaging, these 3 terms are reduced as the square root of the averaging number, but the spectral density is not reduced. Hence with 100 averages the preamplification noise is reduced by 10. This can be applied for noise studies but not for applications.

4 Noise in Magnetoresistive Sensors

4.1 Noise in GMR Devices

GMR devices are usually micro-fabricated in shapes that can range from simple rectangles with two contact pads at the edges, to more sophisticated ones like meanders or yoke with four or more contacts, depending on the application. The noise of the device is strongly linked to its shape and size.

The main components of noise in a GMR device are the thermal noise, the 1/f noise and the magnetic noise.

A typical power spectral density spectrum for a 500 Ohms GMR is given on Fig. 8.

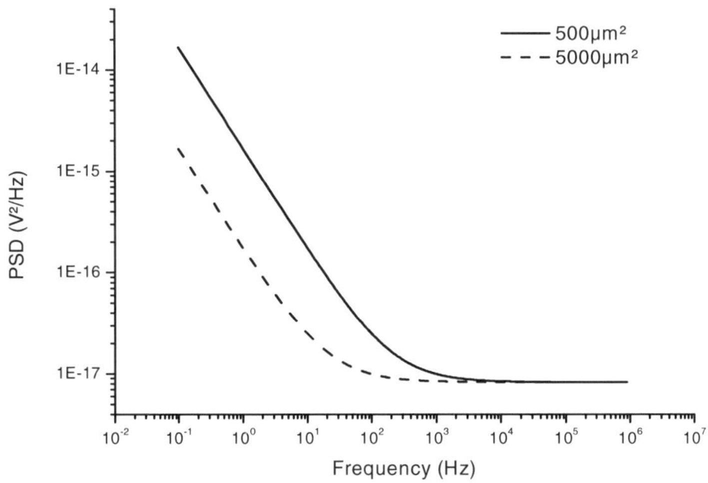

Fig. 8. Noise power spectral density calculated for a 500 Ohms GMR element of two different volumes (500µm² and 5000µm²)

The thermal noise depends on the resistance of the element, which is related to its geometrical dimensions and size. One can design the sensor to target a precise resistance, for instance adapted to the reading electronics. The lowest the resistance the lower the thermal noise is. However, it should be noted that the practical relevant limit is often given by the preamplifier white noise.

The 1/f noise scales as the square root of the volume of the device. For large size devices, the 1/f noise will be smaller, and as a consequence, the 1/f noise knee, which is the crossover between the 1/f noise and the thermal noise, will be displaced to lower frequencies. However, the GMR element size cannot always be increased, for surface occupation reasons on the chip.

γ_H (Hooge parameter) is ranging from 10^{-3} to 10^{-2} for typical GMR spin valves grown on silicon. The number of carriers is always difficult to precisely estimate. In our GMRs, we take $4.10^8/\mu m^2$ which corresponds to the sum of carriers given by each

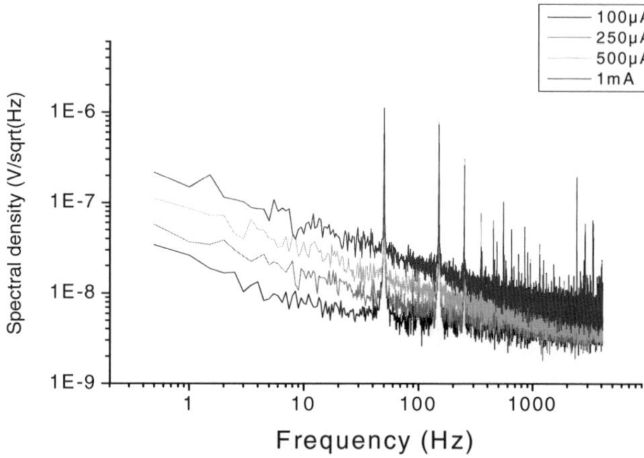

Fig. 9. Noise (square root of the PSD) of a GMR element with a 4μmx150μm yoke shape for various feeding current. The cutoff frequency between 1/f noise and thermal noise varies with the current and is equal to 10kHz for 1mA.

layer. The device given on figure 9 has then a Hooge parameter of 3.10^{-3} which is a rather good value for GMR spin valves.

Another type of noise, specific to magnetoresistive thin films devices, can appear in the spectrum: this is the magnetic noise. This noise is usually due to magnetic domain formation and motion in the soft layer, which can be observed in the V(H) response as small steps in the non-saturated parts of the curve. When the magnetic environment changes, the domain distribution can change, and the domain motion induces large noise.

This specific noise can even be observed as RTN in nanostructures where a small number of magnetic fluctuators are present [14]. In the saturated regime, for instance under a strong applied field, this noise disappears from the spectrum. This is the way to separate magnetic from non-magnetic noise in GMR devices.

This type of noise can dominate strongly the other sources of noise at low frequency, therefore it is important for sensing application to try to reduce or suppress it. As the domain formation depends on the geometry of the device, some specific shapes are more favorable to domain formation than other. Indeed, in a square or rectangular magnetic soft layer, when the applied field is rotated within the plane of the film, the lowest energy configuration leads to formation of opposite domains, which may be displaced during the field change. A strong magnetic noise will then appear in the structure [15].

To avoid this effect, one may use shape where the domain formation is mastered in specific areas outside of the measured part. Two particular configurations can follow this principle. In a meander structure, the domain formation will occur at the corners of the meander. To avoid introducing magnetic noise in this type of device, it is necessary to short cut the edge by a thin metallic layer (like the one used for the contact pads) on the corners of the meander. The current will then flow through this metallic layer instead of the GMR stack at the corners, where the magnetic domain may fluctuate (Fig. 10).

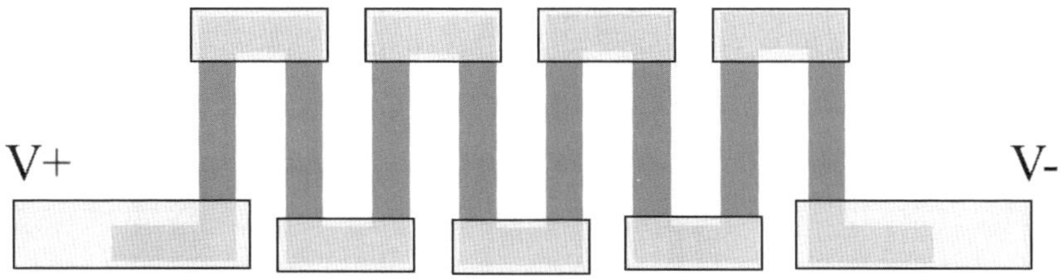

Fig. 10. Example of short-cut meander shape for GMR sensors with low magnetic noise. The short-cut part length should be at least 2 times the width of the GMR element.

Another particularly interesting shape for magnetic noise suppression is the yoke shape. In this case also, the magnetic domains are confined to the edges, and a measurement free of magnetic noise can be performed in four point configuration by injecting the current on the outer contacts while measuring the resultant voltage in the central part, where the magnetization rotation is homogeneous, free of domains (Fig. 11).

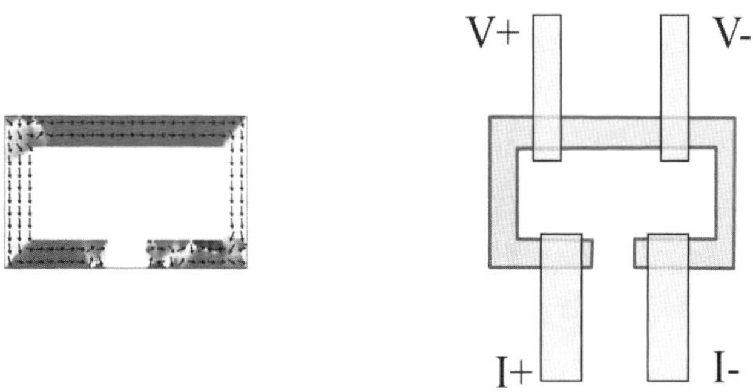

Fig. 11. Example of yoke shape. Left: micromagnetic simulation giving the domain configuration after a magnetic cycle, Right: typical 4 points measurement configuration.

The best design in terms of noise for a GMR-based sensor will therefore depends on several parameters:

- the frequency range targeted: if the sensor operates at frequencies higher than the 1/f knee corner, the most important parameter will be the resistance choice and the available power. If the sensor operates at low frequency, it might be important to increase the size of the sensor to reduce the 1/f noise contribution
- the available size: if surface use is not a limitation, it is interesting to use meander shape and/or large size devices. In case of small dies or multi-element systems, the constraints are higher and often designs like yokes are a better choice.
- the available power. The resistance choice will determine the voltage output depending on the available feeding current and overall power.

The latter plays an important role when detection is done in the thermal noise regime. An increase of the current (or the voltage on the bridge) gives an increase of the signal but does not add noise. In the 1/f regime, an increase of current induces an increase of the signal and of the noise in a same way. A good comparison of performances between sensors and magnetometers is the detectivity or the field equivalent noise which is the field corresponding to a signal-to-noise ratio equal to 1. Figure 12 gives the field equivalent noise for the sensor shown in figure 9.

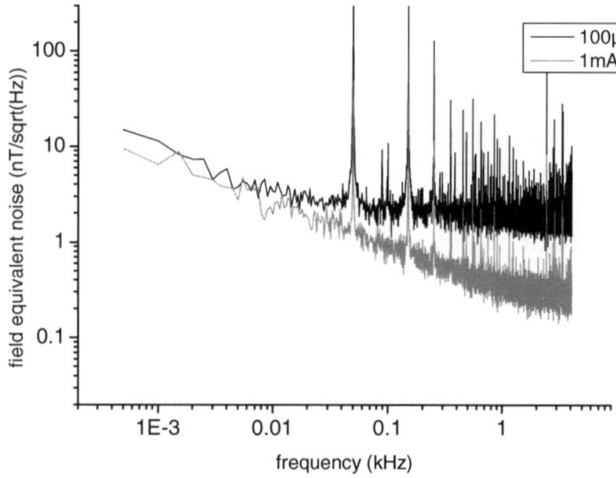

Fig. 12. Field equivalent noise of a GMR element with a 4µmx150µm yoke shape for two values of the feeding current. The increase of current has a strong impact on the field detectivity at high frequencies but no effect in the 1/f regime.

4.2 Noise in Metallic Magnetic Tunnel Junctions

Metallic tunnel junctions (which are made of thin metallic magnetic layers separated by a thin insulating barrier) also exhibit common noise features with GMR elements, which are the thermal noise, magnetic noise and 1/f noise. Additionally, junctions also exhibit shot noise, which may dominate the thermal noise regime. If for GMR sensors, 1/f noise is mainly related to the quality of materials, in MTJs, 1/f noise is large and related to the barrier itself. A recent review [16] done by Z.Q. Lei and co-authors gives a good overview on noise in tunnel magnetic junctions based of Al or MgO barriers.

Thermal Noise and Shot Noise

The white noise in MTJs, is well described by the formula (23). In good quality junctions, the Fano factor is about 1 but if impurities are present in the junction, a Fano factor lower than 1 can be found [6]. It should be noticed that at room temperature, the shot noise is small compared to the thermal noise. Due to the large low frequency noise, accurate measurement of the shot noise is difficult and can only be performed at rather high frequencies.

Low Frequency Noise

The 1/f noise in MTJ is very large compared to GMR sensors. It can be separated in magnetic 1/f noise and electronic 1/f noise.

The electronic noise is the noise measured in a strongly saturated state of the junction. It is well described by the modified Hooge formula

$$S_V(\omega) = 2\pi \frac{\alpha V^2}{A\omega} = \frac{\alpha V^2}{Af} \tag{29}$$

It decreases when the polarization is increasing, which is the signature of a better barrier and more controlled interfaces, and it decreases when the barrier becomes thinner. This variation with the resistance by surface unit (RA) may be related to the roughness of thick barrier. Below a RA of 100kOhms/μm², the α parameter becomes stable. As a reference, good MgO based MTJs exhibit a α parameter of about 10^{-10}μm² (fig. 13).

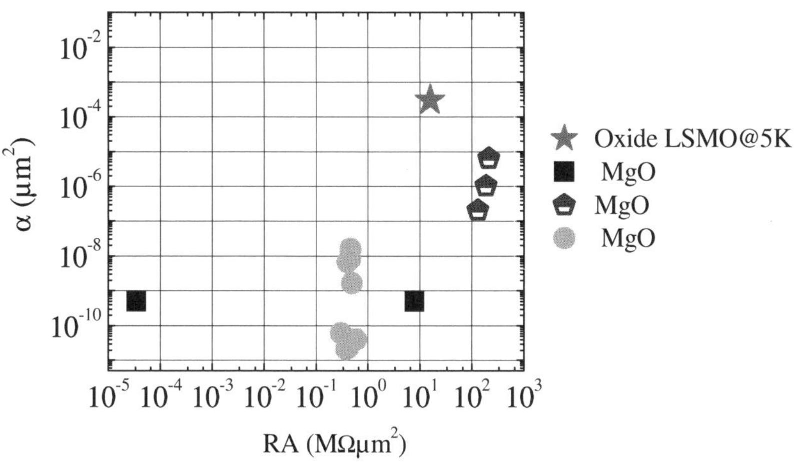

Fig. 13. Hooge parameter for different junctions as function of the resistance square of the tunnel junction. Black squares [17], blue pentagons [18] green circles [19], red stars [20].

When the MTJs is saturated in the antiparallel state, the α parameter is typically increased by a factor 2 or more. This effect can be understood by the fact that a number of conducting channels are closed compared to the parallel state, which is equivalent to a reduction of the effective surface of the junction.

The magnetic 1/f noise is, like for GMRs, mainly due to domains fluctuations. This noise can be reduced by shaping the free layer in a configuration that reduces or eliminates the domain walls. This is however more difficult to realize than on GMR element, due to the specific lithographic process needed to design differently the top and bottom electrode of the barrier. A yoke shape free layer electrode with a rectangular hard layer top electrode MTJs can have a magnetic noise decreased by more than one order of magnitude compared to classical rectangular MTJs.

A way to reduce the 1/f noise is like for GMRs to increase the total surface. This can be done by using a large number of junctions mounted in series and in parallel [21]. A number of junctions in series have the advantage to accept a much larger voltage without a decrease of the TMR factor. This is very important for practical applications where output and bias voltages are in the range of the volt.

4.3 Noise in Oxide Tunnel Junctions

Due to their high spin polarization, magnanite perovskites have been investigated for TMR junction realization. TMR as high as 2000% at low temperature in these devices has been measured [22, 23]. Recent noise studies have been performed on such junction, in order to compare with metallic MTJs. In this study [20], the low frequency noise is one order of magnitude higher than in MTJs with comparable RA product. This is not due to the manganite thin film noise, which has been measured to be two to three orders of magnitude lower than those of the junction, but more probably linked to the noise in the $SrTiO_3$ barrier, where defects like oxygen vacancies or structural transitions can contribute to this noise increase. Studies on other type of barrier in oxide MTJ would be very valuable to identify the noise sources and reduce them in these systems.

5 Conclusion and Perspectives

GMR and MTJ based magnetic sensors are very sensitive and are more and more incorporated in various applications. If the noise impact at high frequencies can be reduced by increasing the power on the sensor, the low frequency noise is still the main limitation of the performances of these devices. By playing on the GMR and MTJ design and increasing the power it is however possible to reduce this noise. Table 2 below gives a summary of what is achievable in terms of noise for GMRs and TMRs sensors at room temperature including the preamplifier noise and without flux concentrators. For TMR, we have considered a free layer shaped in order to eliminate the domain magnetic low frequency noise. In a rectangular element, noise two or three order of magnitude larger can arise.

Table 2. Field equivalent noise for GMR and TMR of various surfaces

	Surface	Noise at 1Hz	White noise	Power consumption
Small GMR	(150x4μm)	10nT	50pT	5mW
Large GMR	$1mm^2$	100pT	20pT	100mW
TMR single	(4x20μm)	4nT	10pT	50μW
TMR in arrays	Surface $1mm^2$	40pT	1pT	5mW

Alternatives presently investigated to reach the pico-Tesla to femto-Tesla range with spin electronics devices are double. The first one consists of amplifying the sensed field by flux concentrators. These flux concentrators can be either superconducting [24] with enhancement of gain of more than 1000 allowing sub-picotesla detection at 1Hz or femtoTesla detection at high frequencies but at lower temperature, either with soft materials [25] with gain up to 100 at room temperature. A more sophisticated approach consists of using a modulated flux concentrator so that the field seen by the sensor is no more at low frequencies but displaced to higher frequencies where the sensors are in the thermal regime. This can be done for example by combining a MEMS oscillator with a flux concentrator [26, 27].

References

[1] http://en.wikipedia.org/wiki/Window_function
[2] Johnson, J.B.: Thermal Agitation of Electricity in Conductors. Nature 119 119, 50 (1927); Phys. Rev. 29, 367 (1927)
[3] Nyquist, H.: Thermal Agitation of Electric Charge in Conductors. Phys. Rev. 32, 110 (1928)
[4] Steinbach, A.H., Martinis, J.M., Devoret, M.H.: Observation of Hot-Electron Shot Noise in a Metallic Resistor. Phys. Rev. Lett. 76, 3806–3809 (1996)
[5] Blanter, Y.M., Büttiker, M.: Shot noise in mesoscopic conductors. Physics Reports 336(1-2), 1–166 (2000)
[6] Guerrero, R., Aliev, F.G., Tserkovnyak, Y., Santos, T.S., Moodera, J.S.: Shot Noise in Magnetic Tunnel Junctions: Evidence for Sequential Tunneling. Phys. Rev. Lett. 97, 266602 (2006)
[7] Garzon, S., Chen, Y., Webb, R.A.: Enhanced spin-dependent shot noise in magnetic tunnel barriers. Physica E 40, 133–140 (2007)
[8] Voss, R.F., Clarke, J.: Flicker (1/f) noise: Equilibrium temperature and resistance fluctuations. Phys. Rev. B 13, 556–573 (1976)
[9] Hooge, F.N., Hoppenbrouwers, A.: 1/f noise in continuous thin gold films. Physica 45, 386–392 (1969); Hooge, F.N.: 1/f noise. Physica B 83, 14-23 (1976)
[10] Scola, J., Polovy, H., Fermon, C., Pannetier-Lecœur, M., Feng, G., Fahy, K., Coey, J.M.D.: Noise in MgO barrier magnetic tunnel junctions with CoFeB electrodes: Influence of annealing temperature. Appl. Phys. Lett. 90, 252501 (2007)
[11] Foros, J., Brataas, A., Bauer, G., Tserkovnyak, Y.: Noise and dissipation in magnetoelectronic nanostructures. Phys. Rev. B 79, 214407 (2009)
[12] Pannetier-Lecœur, M., Fermon, C., de Vismes, A., Kerr, E., Vieux-Rochazet, L.: Low noise magnetoresistive sensors for current measurement and compasses. J. Magn. Magn. Mat. 316, E246–E248 (2007)
[13] http://www.ti.com/lit/ds/symlink/ina103.pdf
[14] Raquet, B., Viret, M., Costes, M., Baibich, M., Pannetier, M., Blanco-Mantecon, M., Rakoto, H., Maignan, A., Lambert, S., Fermon, C.: Electronic noise in magnetic low-dimensional materials and nanostructures. J. Magn. Magn. Mat. 258, 119–124 (2003)
[15] Schaefer, R., Desimone, A.: Hysteresis in soft ferromagnetic films. IEEE Trans. Mag. 38, 2391–2393 (2002)
[16] Lei, Z.Q., Li, G.J., Egelhoff Jr., W.F., Lai, P.T., Pong, P.W.T.: Review of Noise Sources in Magnetic Tunnel Junction Sensors. IEEE Trans. Mag. 47, 602–612 (2011)

[17] Polovy, H., Guerrero, R., Scolab, J., Pannetier-Lecœur, M., Fermon, C., Feng, G., Fahy, K., Cardoso, S., Almeida, J., Freitas, P.P.: Noise of MgO-based magnetic tunnel junctions. J. Magn. Magn. Mat. 322, 1624–1627 (2010)
[18] Gokce, A., Nowak, E.R., Yang, S.H., Parkin, S.S.P.: 1/f noise in magnetic tunnel junctions with MgO tunnel barriers. J. Appl. Phys. 99, 08A906 (2006)
[19] Aliev, F.G., Guerrero, R., Herranz, D., Villar, R., Greullet, F., Tiusan, C., Hehn, M.: Very low 1/f noise at room temperature in fully epitaxial Fe/MgO/Fe magnetic tunnel junctions. Appl. Phys. Lett. 91, 232504 (2007)
[20] Guerrero, R., Solignac, A., Fermon, C., Pannetier-Lecœur, M., Lecoeur, P., Fernandez-Pacheco, R.: Low frequency noise in $La_{0.7}Sr_{0.3}MnO_3$ based magnetic tunnel junctions. Appl. Phys. Lett. 100, 142402 (2012)
[21] Guerrero, R., Pannetier-Lecœur, M., Fermon, C., Cardoso, S., Ferreira, R., Freitas, P.P.: Low frequency noise in arrays of magnetic tunnel junctions connected in series and parallel. J. Appl. Phys. 105(11), 113922 (2009)
[22] Bowen, M., Maurice, J.-L., Barthelemy, A., Bibes, M., Imhoff, D., Bellini, V., Bertacco, R., Wortmann, D., Seneor, P., Jacquet, E., Vaures, A., Humbert, J., Contour, J.-P., Colliex, C., Blügel, S., Dederichs, P.H.: J. Phys.: Using half-metallic manganite interfaces to reveal insights into spintronics. Condens. Matter. 19, 315208 (2007)
[23] Werner, R., Petrov, A.Y., Alvarez, L., Kleiner, M.R., Koelle, D., Davidson, B.A.: Improved tunneling magnetoresistance at low temperature in manganite junctions grown by molecular beam epitaxy. Appl. Phys. Lett. 98, 162505 (2011)
[24] Pannetier-Lecœur, M., Parkkonen, L., Sergeeva-Chollet, N., Polovy, H., Fermon, C., Fowley, C.: Magnetocardiography with sensors based on giant magnetoresistance. Appl. Phys. Lett. 98, 153705 (2011)
[25] Marinho, Z., Cardoso, S., Chaves, R., Ferreira, R., Melo, L.V., Freitas, P.P.: Improving Magnetic Flux Guide Concentrators for MR Sensors. J. Appl. Phys. 109, 07E521 (2011)
[26] Patil, S.B., Guedes, A., Freitas, P.P., Cardoso, S., Chu, V., Conde, J.P.: On-chip magnetoresistive detection of resonance in microcantilevers. Appl. Phys. Lett. 95, 023502 (2009)
[27] Egelhoff, W.F., Pong, P.W.T., Unguris, J., McMichael, R.D., Nowak, E.R., Edelstein, A.S., Burnette, J.E., Fischer, G.A.: Critical challenges for picoTesla magnetic-tunnel-junction sensors. Sens. and Act. A-Physical 155, 217–225 (2009)

Resistive Sensor Interfacing

Andrea De Marcellis, Giuseppe Ferri, and Paolo Mantenuto

University of L'Aquila
Department of Industrial and Information Engineering and Economics
via G. Gronchi 18, 67100 – L'Aquila, Italy
andrea.demarcellis@univaq.it

Abstract. This chapter has the aim to give a complete overview on the first analog front-ends, describing some circuit and system solutions for the design of electronic interfaces suitable for resistive sensors showing different variation ranges: small, as in dedicated-application GMR sensors; wide, especially referred to GMR sensing devices whose baseline is unknown. After a description of the main interface parameters, the authors present several solutions, most of which do not require any calibration. These solutions are different according to the entity of resistive sensor variations, can utilize either AC or DC excitation voltages for the employed sensor and are developed in Voltage-Mode (VM, which considers the use of either the Operational Amplifier (OA) or the Operational Transconductance Amplifier (OTA) as main active block) as well as in Current-Mode (CM) approach (being in this case the Second Generation Current Conveyor (CCII) the active device). The described interfaces can be easily fabricated both as prototype boards, for a fast characterization, and as integrated circuits, also using modern microelectronics design techniques, in a standard CMOS technology with Low Voltage (LV) and Low Power (LP) characteristics, especially when designed for portable applications and instrumentation. Moreover, thanks to their reduced sizes in terms of chip area, the proposed solutions are suitable for being used for sensor arrays applications, where a number of sensors is employed, as in portable systems, to detect different environmental parameters.

1 Sensors and Electronics

Introduction on Signal Conditioning

A measurement sensor system is typically formed by a number of active and/or passive blocks, as shown in Figure 1, able to reveal and quantify physical/chemical phenomenon variations by means of a sensing element named sensor. In particular, the latter has to be processed by a suitable analog signal conditioning circuit, named interface, which allows to readout of the information coming from the sensor so providing a suitable output signal easy to display or to elaborate through an analog-to-digital converter (ADC) processing element. On the other hand, through an electronic interface, it is possible to detect any measurand variation as an electrical quantity which can be furtherly processed by means of a Personal Computer (PC), a microcontroller (μC), a microprocessor (μP), and so on.

C. Reig et al.: *Giant Magnetoresistance (GMR) Sensors*, SSMI 6, pp. 71–102.
DOI: 10.1007/978-3-642-37172-1_4

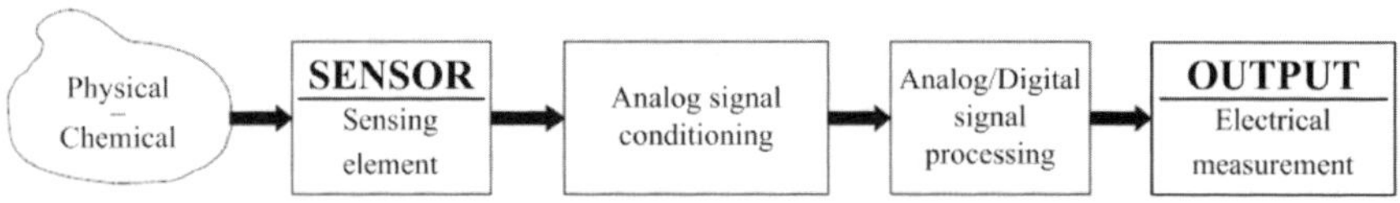

Fig. 1. Block scheme of a measurement system composed by sensor, signal conditioning and processing circuits

The need of novel sensors and related electronic interfaces showing small dimensions and the capability of working with reduced both supply voltages (Low-Voltage, LV) and power consumption (Low-Power, LP), especially in portable applications, is in a continuous growth. In particular, when both sensor and electronic circuitry for its interfacing, which have to be developed in a suitable integrated technology (*e.g.*, a standard CMOS), can be also combined into only one chip, it is possible to implement the so-called "Smart Sensor" [1-4].

Clearly, as stated above, sensors and electronic interfaces are a sub-set of measurement systems and, therefore, their performance should be expressed through suitable parameters, which will be listed and detailed in the next Paragraph. In this sense, the design or the use of an electronic interface is strictly related to the problem of the detection and quantification of the physical/chemical measurand. More in general, the measurement corresponds to a comparison of the measurand with a reference quantity (which, ideally, is a constant value coming from theoretical calculations).

The measurement equipment must be as "ideal" as possible, so to avoid the introduction of errors. This means that the perturbation introduced by the measurement action should be negligible for the desired level of accuracy (otherwise impedance loading effects must always be taken into account and properly evaluated). Therefore, in practice, some preliminary simulations are necessary for a more detailed analysis of the circuit behaviour.

Another important concept is the linear time-invariance of the sensor system, related to its transfer function. The latter, in practical cases, may be only approximately constant within a determined range of frequencies, called bandwidth. All the non-ideal systems have a limited speed and, therefore, have a finite bandwidth. Since that non-ideal systems are slowly time-variant, typically the time-invariance hypothesis is possible.

Let us now give more details about the interfacing of resistive sensors. Generally, sensors that behave as pure resistors as well as those sensing elements which do not bear an alternating voltage (*i.e.*, an AC excitation signal), since they give bad responses and lower lifetimes, can be biased through a constant voltage value (*i.e.*, a DC excitation signal), especially when, for several specific applications, it is also possible to neglect the effect of their parasitic components (*e.g.*, parasitic capacitance). On the contrary, in some application fields, these parasitics have to be known so to have a more complete information about the sensor [3-7].

Nevertheless, when the sensor can be modelled through a resistance, whose baseline is known and/or can be estimate, and that, in particular, varies into a reduced range (generally few percent but, however, not more than two-three decades, see next Paragraphs and Sections) and/or its baseline is known (*e.g.*, previously evaluated), the

Resistance-to-Voltage (*R-V*) conversion can be utilized for the resistive sensing interfacing [7-14]. Typically, this technique applies a constant voltage as sensor excitation so to measure the change of conductivity of the resistive sensing element. In this case, the simpler electronic interfaces, which perform the *R-V* conversion, are both the voltage divider and its "fully differential" version, known as Wheatstone bridge. Drawbacks of this approach are the very reduced signal level (typically comparable to the noise level) and the limitations due to the saturation (limited by supply voltage) of employed devices [7,10,15-18].

Therefore, if larger variations of sensor resistive values occur (see next Paragraphs and Sections) and/or, more in particular, the sensing element baseline is unknown and/or unpredictable, it is preferable to perform the Resistance-to-Time (*R-T*) conversion, where the "time" is often the period of a repetitive (i.e., periodic) waveform [7-18]. In this sense, more in detail, considering the state of the art of the manufacturing, the sensor resistance value may vary also across several decades, being normally the combination of three variable components: the nominal baseline, the deviation from this nominal baseline (due to ageing, working temperature, operating condition, etc.) and the resistive sensing element variation due to the physical/chemical phenomenon to be revealed. Since each contribution can be in the order of one-two decades, wide range sensors have to be considered (*e.g.*, in GMR sensors the starting values can be very different, varying from few Ω up to hundreds of kΩ). Typically, an *R-T* approach is based on an oscillator architecture which exploits the sensor as resistive element to be excited by a switched voltage (i.e., the AC excitation voltage). In this case, the simpler electronic interface which operates an *R-T* conversion is formed by a basic square wave generator, whose output voltage period *T* is directly dependent on the sensor resistance value.

This kind of solution allows both to avoid the use of high-resolution pico-ammeters, scaling factors, switches, etc., and to employ the same output periodic waveform to provide the AC-excitation to the sensor.

Moreover, since this type of wide range sensor signal conditioners covers several magnitude decades, it does not require any calibration procedure and/or manual settings (i.e., the so-called "uncalibrated" system) and its frequency output (i.e., "digitalized" output signal) offers a number of benefits compared to voltage output circuits, such as improved noise immunity (*e.g.*, offsets, frequency disturbs, etc.), easiness in multiplexing, insulation, signal processing, and so on. Unfortunately, sometimes these interfaces can show higher errors in sensor resistance estimation, when compared to *R-V* (*e.g.*, bridge-based) solutions; therefore, active elements must be accurately designed with good performances, especially in terms of time responses (*e.g.*, high Slew-Rate, SR, values) and low voltage and current offsets.

However, in the literature, different solutions for wide range resistive sensor interfaces which perform *R-V* conversion are also available. Some of them utilize amplifiers with scaling factors, but their drawbacks are related to the need of high-resolution ADCs and the difficult calibration procedures which are required when the sensor baseline is unknown. On the contrary, new recent approaches, considering "uncalibrated" interfaces always operating the *R-V* conversion, have been proposed in the literature so that, especially when the resistive sensor baseline or its variation changes for different decades (also up to 5-6 or more), better estimation characteristics are ensured.

Active Blocks Main Basic Concepts: OA and CCII

The name of the "*Operational Amplifier*" was originally adopted for a series of high performance DC amplifiers used in analog computers. These amplifiers were used to perform mathematical operations applicable to analog computation such as summation, scaling, subtraction, integration, etc. [19-21].

In practice, an OA is a DC-coupled high-gained electronic voltage amplifier showing a differential input and a single-ended output. It produces an output voltage that represents the difference between the two input terminals, multiplied by the gain A. Since this active block has been designed for use in a feedback loop, ideally it shows the following characteristics: an infinite input impedance (i.e., no current flows into the input terminals); a zero output impedance (i.e., it can drive any load impedance to any voltage); an infinite open-loop voltage gain A; no bandwidth limitations; a zero output voltage for a null input voltage difference (i.e., zero voltage offset). As a consequence, without any negative feedback, the OA would act like a comparator of its inputs.

The OA can be internally implemented by different cascaded stages, such as a differential input to single output amplifier, a high gain stage with Miller capacitive compensation, a voltage buffer providing a high output current and a low output impedance.

Nowadays, the applications of OAs have become very widely diversified, in both linear and non-linear applications, such as: single-input single-ended voltage amplifier, differential voltage instrumentation amplifier, integrator, differentiator, comparator, voltage follower, ADC, DAC, etc.

Nevertheless, interfacing a sensor system with a voltage amplifier, based on OA, requires the matching between the sensor/signal conditioning output with the amplifier input. In general, sensors can provide either a single-ended or differential output. In the first case, all the inputs are referenced to system ground. Differential signals provide a positive and a negative signal with the positive output referenced to the negative one. In addition, it is important to consider also that a common-mode signal refers to a common voltage, with the same magnitude and phase that appears on both differential inputs of an amplifier. On the contrary, the Common-Mode Rejection Ratio (CMRR), generally defined as the ratio between the differential voltage gain versus the common-mode voltage gain, is specified for fully-differential inputs and describes the amplifier capability to reject a common-mode signal.

Starting from these considerations, it is possible to introduce the following three common input structures related to single-ended or differential output sensors: single-ended, pseudo-differential or fully-differential voltage amplifiers. Obviously, there are trade-offs with each structure that should be considered. In addition, consider that if the analog signal-conditioning circuitry is used between the sensor and next digital processing sub-system (*e.g.*, an ADC), this circuitry can affect, for example, the digital block input structure choice.

Anyway, the simplest method to be considered is to use a single-ended amplifier when measuring single-ended signals. In this case, all the signals are referenced to a common ground and each channel is connected to a specific input pin. It must be highlighted that the analog ground pin is shared between all inputs. Because of its behavior, the single-ended amplifier suffers DC offset and noise in the signal paths; those effects can decrease the dynamic range of the input signal, unless using suitable conditioning circuits, so single-ended input structures are best employed when the signal source and amplifier are close one each other (*e.g.*, on the same PCB/chip, so that signal traces can be kept as short as possible).

Differential input amplifiers can offer a performance improvement because measure the difference between the "positive" and "negative" terminals of a sensor. Obviously, it is still possible to use the differential amplifier to measure single-ended signals by connecting one input terminal to analog ground (*e.g.*, typically the inverting one is preferred so to do not affect signal phase). Fully-differential inputs offer the best performance in rejecting DC and dynamic common-mode voltages. Moreover, another advantage in the use of differential signals is the capability to extend the amplifier dynamic range. In fact, because the two differential inputs can be also 180° phase shifted, differential inputs amplifier have two times the full-scale input voltage level, so they have a superior DC and AC common-mode rejection and a higher Signal-to-Noise Ratio (SNR). As drawback, in noisy environments, it is possible that coupled-noise could cause the differential inputs to exceed the amplifier allowable input voltage range (in this cases it is sufficient to reduce the input signal range to avoid the amplifier input stage saturation). However, it is important to select a fully-differential input amplifier when dynamic time-varying signals occur and dynamic common-mode rejection is mandatory.

In a floating differential system, ideally and supposing that the sensor and the next amplifier stage are isolated one each other, common-mode voltages beyond electrical supplies can exist, provided that the differential voltage does not exceed the amplifier maximum input range. One way to do it is to employ separate voltage supplies with galvanically isolated grounds. As long as isolation between the grounds exists, the amplifier block only detects the differential voltage between its differential inputs and the sensor can be regarded as floating. Thus, pseudo-differential inputs are similar to fully-differential inputs since they separate signal ground from the amplifier ground, allowing the reduction of only the DC common-mode voltages. However, unlike fully-differential inputs, they have a little effect on dynamic common-mode noise (they do not provide AC common-mode rejection). Pseudo-differential inputs are applied when biased (to an arbitrary DC level) sensors are employed.

Concerning integrated applications, the Current-Mode (CM) approach can be also considered as a possible alternative to traditional Voltage-Mode (VM) circuits to obtain high performance architectures, especially for LV LP applications, because the designer deals with current levels for circuit operation instead of voltage signals. In this manner, as well known, CM circuits, which are able to overcome the limitation of the constant Gain-Bandwidth (GBW) product and the trade-off between speed and bandwidth, typical of OA, provide others possible suitable choices. In particular, CM topologies improve integrated circuit performances in terms of SR and Bandwidth (*BW*), through the development and the use of a suitable "*Second Generation Current Conveyors*" (CCII), which represent the main basic building active blocks in the CM design [22,23]. Typically, CCII-based circuit topologies have a low operating supply voltage, related to the drain-source (saturation) voltage required by the biasing transistors, which has to be minimised so to reduce the circuit total supply voltage. However, a basic well-known CM circuit is the Current-Feedback Operational Amplifier (CFOA). This circuit, if compared to the traditional voltage active block OA, shows a constant bandwidth with respect to the closed-loop gain. This makes it of primary importance in the design of modern *ICs*; in addition, the first stage of CFOA is exactly a Current Conveyor. The only commercially available CCII is the AD844 by Analog Devices which, even if it is a CFOA with very a high slew-rate and a wide bandwidth, is heavily utilized in discrete component prototype PCB implementations of CCII-based circuits, among which also sensor interface topologies. On the contrary, several CCII solutions presented in the literature are based on a differential pair followed by a class-AB output stage. This alternative approach can be also considered particularly useful to different GMR sensor-based applications which employ a current biasing instead of that, more traditionally, based on a voltage [7,9,11-14].

More in particular, the CCII is a three terminal active block that operates, simultaneously, as both voltage and current buffer between its terminals. Moreover, it has a low impedance (ideally zero) current input (*X* node, which is, at the same time, also a voltage output). On the contrary, the other voltage input terminal (*Y* node) shows a high impedance (ideally infinite), while the last terminal (*Z* node) shows also a high impedance level (ideally infinite) resulting an output current node. In this way, currents flowing at *X* and *Z* nodes are always equal in magnitude (the current flowing at *X* node is "conveyed" to the current output *Z* node), while if a voltage is applied to *Y* node, the same voltage will appear at *X* node. In particular, for what concerns the current direction, when the current at *Z* node goes in the same direction of that flowing in *X* node, we can refer to the *CCII+*; otherwise, in case of opposite current flow directions, one has to speak about the *CCII–*. Obviously, also for this active block, parasitic impedances are the main drawback that affects the CCII ideal behavior and, sometimes, its utilization in typical analog applications. Their kind and value mainly depend on the CCII internal topology, developed at transistor level. In fact, due to non-ideal behavior of CCII, the *X* node voltage not exactly equal that at *Y* node as well as the current flowing into *Z* terminal can be slightly different from that of the *X* node. Therefore, the two CCII parameters which represent the ideally-unitary voltage and current gains are α and β, respectively.

2 The Main Parameters of Sensor Electronic Interfaces

In the analysis and characterization of circuits and systems for signal conditioning coming from sensors, it is opportune to evaluate the performances given by the sensor and the interface also under different operating conditions [2-7].

In this sense, the following main parameters, typically referred to a sensor, have to be considered for evaluating performances and characteristics of a more complete front-end:

- ***Sensitivity*:** It is the variation of output electrical parameter with respect to the sensor variation, corresponding to a measurand variation. It represents the relationship (i.e., the transfer function) between the output electrical signal and the sensing element. An interface shows a high sensitivity when, for the same sensor variation, to be revealed, corresponds a larger variation of the generated electrical signal. Generally, sensitivity value depends on the operating point and on the electronic system setting.
- ***Resolution*:** Mathematically given by the ratio of the output noise level with respect to the interface sensitivity value, it is the minimum detectable measurand value that can be determined under the condition of unitary SNR, that is smallest variation of the sensor appreciable by the interface which provides a detectable output variation. Resolution is definitively the most important characteristic in sensor applications; numerically speaking, it must be minimized. A system with a very low resolution value is typically mentioned as a "high-resolution system". Sensitivity and resolution must be evaluated in the typical variation range of the sensor parameter where, possibly, have to be constant or linear; in this case their value does not depend on the operating point.

Moreover, other important interface parameters are the following:

- ***Linearity*:** Proportionality between input and output signals, concerning the interface response curve, which correlates the generated output signal with respect to the sensor parameter variations. For small sensor variation, linearity is always ensured.
- ***Repeatability*:** Capability to provide the same performances after repeated utilizations, when applied consecutively and under the same conditions.
- ***Accuracy*:** Agreement of the output values with a standard reference (*e.g.*, ideal characteristic, theoretical calculation, etc.). Accuracy is closely related to precision, also called reproducibility. As a consequence, accuracy is related typically to percentage relative error between ideal (or expected) and generated values, as shown in Figure 2.
- ***Precision*:** Capability to give output signals with similar values, for different and repeated measurements, when the same sensor value is applied (i.e., repeatability in the same measurement conditions).
- ***Reproducibility*:** Repeatability obtained under different measurement conditions (*e.g.*, in different times and/or places).
- ***Stability*:** Capability of a system to provide the same characteristics over a relatively long period of time (time-invariability).
- ***Drift*:** Slow and statistically unpredictable temporal variation of interface characteristics, related to electronic circuits, due to aging, operating temperature and/or other effects.
- ***Hysteresis*:** Difference among the output signal values, generated by the interface in correspondence of the same sensor variation range, achieved a first time for increasing values and a second time for decreasing values of the sensor parameter.

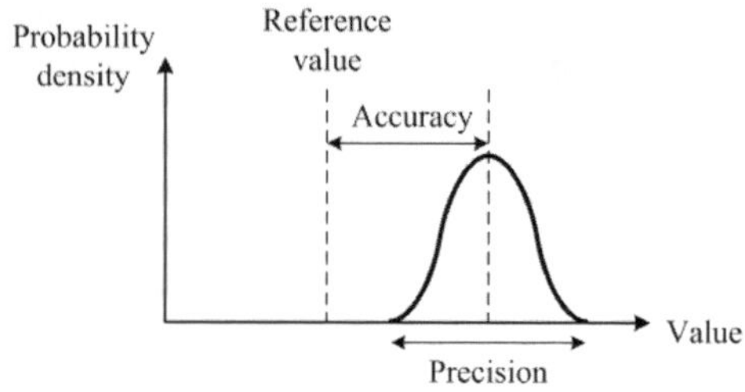

Fig. 2. Accuracy and precision definitions and their relationship

Furthermore, output signals coming from sensors, typically, have the following characteristics: low-level values, relatively slow sensing parameter variations and the need of initial calibration for long-term drift (it means they generally can be considered time-variant) and temperature dependence. For these reasons, in order reduce measuring errors, the design and the use of suitable low-noise low-offset analog electronic interfaces with low parasitic transistors and impedances, and compensation techniques (offset and *1/f* noise reduction by auto-zero circuits, chopper circuits and dynamic element matching) are essential [19-21,24,25]. In this sense, another important feature to be considered is the electrical impedance of the sensor, which determines also the frequency measurement range.

In the following we will present a review of the main resistive sensor analog interfaces, suitable both for the integration on chip in a standard CMOS technology, also with LV LP electronic characteristics, and, in particular, for GMR sensor applications. We have chosen to classify them according to the amount of resistive variations and/or values, and then, considering the kind of sensor excitation (DC, AC), describing both Voltage-Mode and Current-Mode solutions [3].

3 Small Range Resistive Sensor Interfaces

DC Excited Sensor Interface Solutions

Voltage Divider and Wheatstone Bridges: When the resistive sensing element varies into a reduced range (about one-two decades, or less), a simple resistive voltage divider circuit, operating an *R-V* conversion, can be utilized as first and simple analog interface [3,19]. More in detail, considering Figure 3 and, as an example, if a DC supply voltage V_{IN} is applied to drive the sensing element R_{SENS} and utilizing a reference load resistance R_{REF}, the output voltage V_{OUT} can be revealed and processed instantaneously, so to determine the sensor resistance value.

As a consequence, from the voltage divider, changes of the sensor resistance R_{SENS} can be evaluated, once R_{REF} and V_{IN} are known, by measuring the circuit output voltage V_{OUT}, as follows:

$$V_{OUT} = \left(\frac{R_{SENS}}{R_{REF} + R_{SENS}} \right) V_{IN}, \tag{1}$$

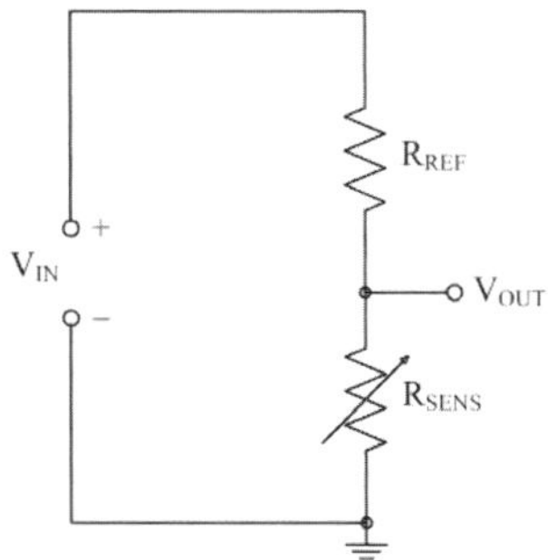

Fig. 3. The voltage divider as resistive sensor interface circuit (V_{IN} = circuit excitation voltage; R_{REF} = reference load resistance; V_{OUT} = circuit output voltage; R_{SENS} = sensing element)

from which:

$$R_{SENS} = R_{REF}\left(\frac{V_{OUT}}{V_{IN} - V_{OUT}}\right). \tag{2}$$

The fully differential version of the voltage divider (for what concerns the output voltage) is the well-known Wheatstone bridge, whose schematic circuit is shown in Figure 4, which still operates an *R-V* conversion, better rejecting the common-mode signal. In particular, it can be used for converting low sensor resistance variations into a differential voltage signal V_{OUT}. It is composed by four resistances and, usually, a resistive sensor is placed into one of the four branches of the bridge whose resistive sensing element varies when an external physical or chemical phenomenon occurs.

Referring to Figure 4, the bridge is balanced when the ratio of resistances of a bridge branch is equal to that of the other: $R_1/R_2=R_3/R_{SENS}$. As a particular case, the bridge is balanced when all the four resistances are the same value: $R_1=R_2=R_3=R_{SENS}$. In this case, the generated differential output voltage V_{OUT} is equal to zero.

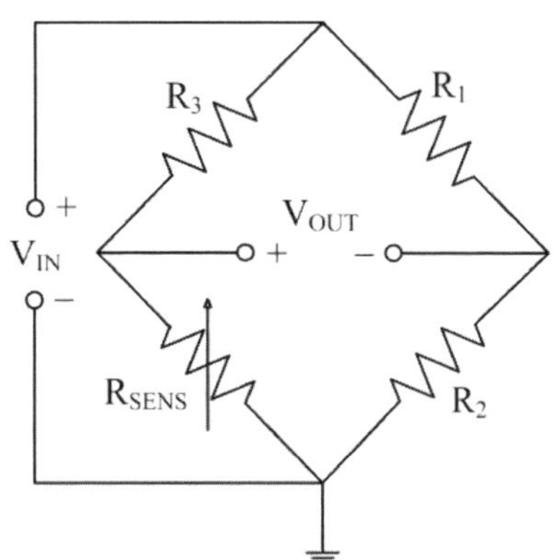

Fig. 4. The Wheatstone bridge as a resistive sensor interface: the *R-V* conversion

On the contrary, starting from equilibrium condition (balanced bridge), when R_{SENS} varies its resistance, a non-zero differential voltage V_{OUT} is provided at the output of the bridge, whose value is proportional to the sensor resistance variation (but only when these variations are small, in particular, referring to Figure 4, if $x<<1$, supposing $R_{SENS}=R_{S0}(1+x)$, being R_0 the baseline sensor value). More in general, the output voltage can be expressed as follows:

$$V_{OUT}=\left[\frac{R_1 R_{SENS}-R_2 R_3}{(R_1+R_2)(R_3+R_{SENS})}\right]V_{IN}, \quad (3)$$

from which:

$$R_{SENS}=\frac{R_2 R_3}{R_1}\left[\frac{1+\frac{V_{OUT}}{V_{IN}}\left(\frac{R_1+R_2}{R_2}\right)}{1-\frac{V_{OUT}}{V_{IN}}\left(\frac{R_1+R_2}{R_1}\right)}\right]. \quad (4)$$

Unfortunately, this kind of resistive sensor interface shows a low and unsettable sensitivity value; in particular, it can be expressed as:

$$S_{R_{SENS}}=\frac{\partial V_{OUT}}{\partial R_{SENS}}=V_{IN}\frac{R_3}{(R_3+R_{SENS})^2} \quad (5)$$

If $R_1=R_2=R_3=R_{SENS}$, in the basic Wheatstone bridge, the sensitivity is constant and equal to $V_{IN}/4$ considering a small variation of only one bridge resistance (R_{SENS}). This value of the sensitivity is exactly the same of the simple voltage divider (see Figure 3). In fact, in the both cases, if the relative variation of the sensor resistance ($x=R_{SENS}/R_0$) is reasonably small (*e.g.*, lower than 5% with respect to the sensor resistance baseline R_0), a quasi-linear relation between the differential output voltage V_{OUT} and the relative variation x exists, as follows:

$$V_{OUT}=V_{IN}\frac{x}{4+2x}\cong V_{IN}\frac{x}{4}. \quad (6)$$

Alternatively, referring to Figure 5, through a suitable null detector (*e.g.*, a simple multimeter or voltmeter), which reveals the balanced condition of the bridge (that is, the output voltage equal to zero), by changing the value of a variable resistor R_{VAR} (one of the resistors in the branch), it is possible to determine the unknown resistance value provided by the resistive sensor R_{SENS}, that changes as a function of an external (physical or chemical) phenomenon to be detected and measured.

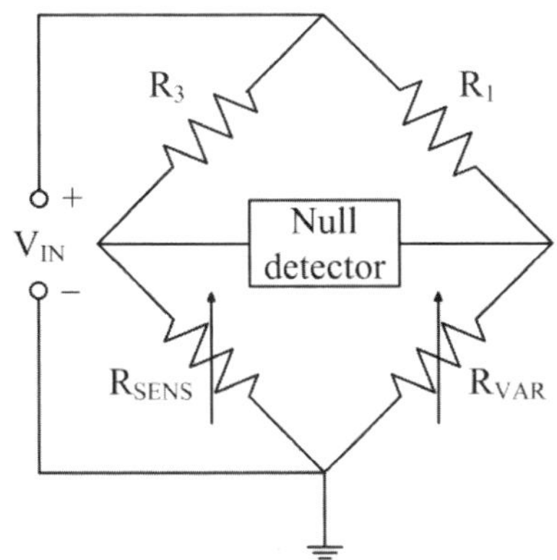

Fig. 5. The null detector in a resistive Wheatstone bridge

The use of a differential input OA-based voltage amplifier, as reported in Figure 6, allows to enhance the front-end circuit sensitivity. This VM circuit, performing also the single-ended conversion, can be placed at the output nodes of the bridge.

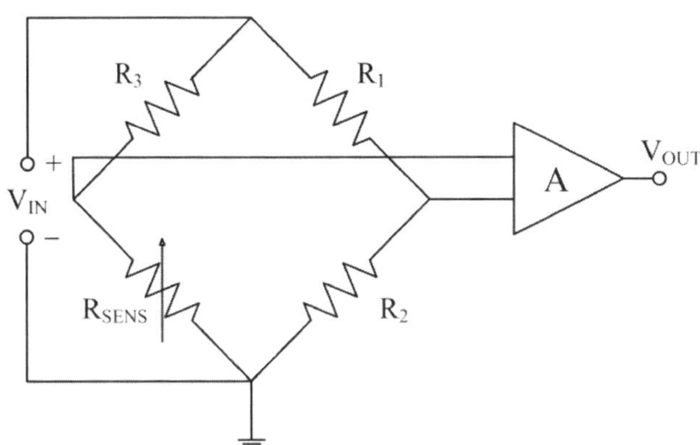

Fig. 6. Differential-to-single ended Wheatstone bridge output by using a voltage differential amplifier

In this case, an instrumentation differential amplifier is the best circuit topology since shows a very high input impedance and, through its internal feedback configuration, gives a well-defined and controlled amplification factor. Another fundamental characteristic of this amplifier must be its low input voltage offset. If A is the OA gain and supposing $R_1=R_2=R_3=R_{SENS}$, we can write, for low-resistive variations (x):

$$V_{OUT} \cong A\left(V_{IN}\frac{x}{4}\right). \tag{7}$$

Current-Biased Solution: A basic interface for resistive sensor, utilizing a current biasing, is shown in Figure 7 [3]. This solution is based on a Resistance-to-Current (*R-I*) conversion, allows to generate an output current I_{OUT} dependent on the sensor resistance value R_{SENS}.

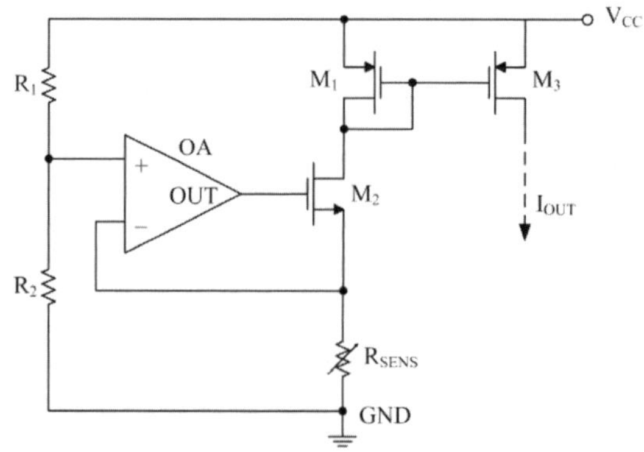

Fig. 7. Resistance-to-Current converter as a resistive sensor interface

Through a simple analysis it is possible to evaluate ideally the generated current, as follows:

$$I_{OUT} \cong V_{CC} \frac{R_2}{R_1 + R_2} \frac{1}{R_{SENS}}, \tag{8}$$

assuming that M_1 and M_3 are matched and equal *p-MOS* transistors. Obviously, the output current I_{OUT}, if required, can be also further converted into a voltage output signal through an additional Current-to-Voltage (*I-V*) conversion.

Current-Mode Resistive Sensor Interface: Figure 8 shows a CCII-based analog interface suitable for DC-excited resistive sensor applications. The advantage of this CM circuit in the sensor interfacing is its capability to perform the offset compensation, in this way the output voltage is linearly proportional to the resistive variation [26]. The only feature to be considered is the design of *CCIIs* having negligible parasitic impedances (see a quasi-ideal configuration reported in Figure 23). In particular, also in this case, it is assumed that the sensor is modelled by the resistance $R_{SENS}=R_0(1+x)$, being R_0 the resistance value referred to the sensor baseline value and x the relative sensor variation.

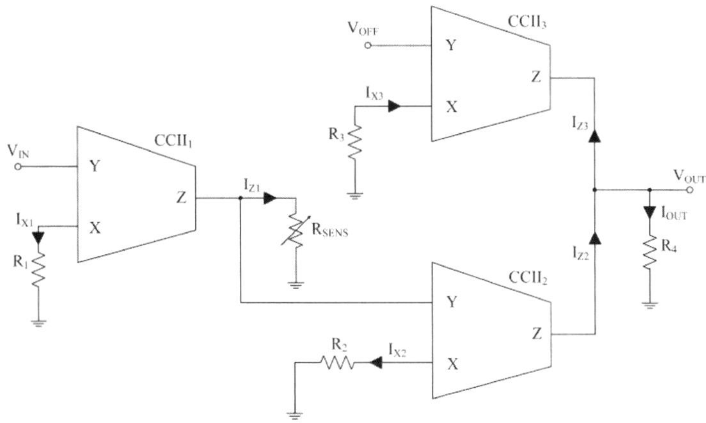

Fig. 8. CCII-based electronic interface for resistive sensors

A straightforward analysis gives the following expression of V_{OUT}:

$$V_{OUT} = \left(\frac{R_4 R_0 V_{IN}}{R_1 R_2} \right) x + \left(\frac{R_4 R_0 V_{IN}}{R_1 R_2} - \frac{R_4 V_{OFF}}{R_3} \right). \tag{9}$$

The first term is linearly proportional to the relative variation x of the sensor resistance, while the second one can be set to zero by a suitable choice of V_{IN}, V_{OFF}, R_1, R_2, R_3 and R_4, so cancelling the voltage offset without reducing the speed of the interface, even if, in this way, time-varying errors, such as drift and *1/f* noise, might not be compensated.

AC Excited Sensor Interface Solutions

Current-Mode Uncalibrated Solution for High-Valued Resistive Sensors: A CM interface for AC-excited high-valued resistive sensors, performing a current differentiation rather than a voltage integration as in typical oscillators, is shown in Figure 9 [27]. In particular, this solution, based on an oscillating circuit (*R-T* conversion), allows to neglect the *Z* and *Y* nodes saturation effects in the square waveform generation. Moreover, it is possible to easily set the interface working range through several passive components which allow also to set the desired sensitivity of the readout circuit. As a consequence, this circuit configuration, which can be employed as a suitable solution for small-range resistive sensor analog front-ends, allows to reveal, with a good accuracy, variations of grounded resistive sensors typically ranging in [MΩ÷GΩ], even if the same circuit is also suitable for wide-range floating capacitive sensors [pF÷μF].

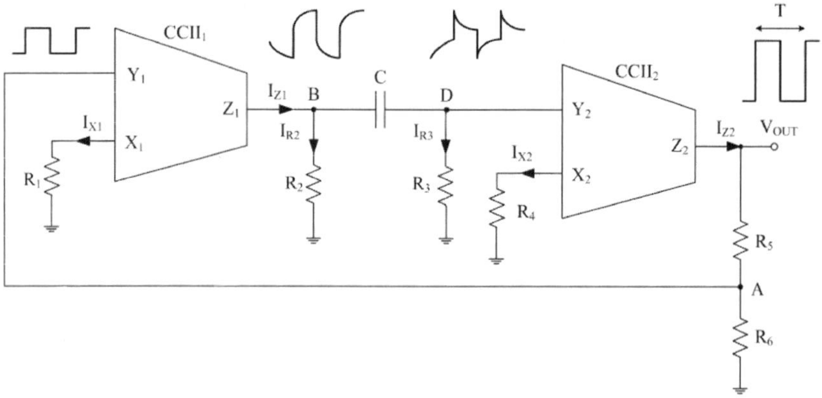

Fig. 9. Block scheme of a CCII-based AC-excited interface

Through a straightforward analysis, considering an ideal CCII behaviour, it is possible to determine the expression for the period T of the generated output square wave signal, revealed at V_{OUT} node, as a function of the sensor resistance (*e.g.*, R_2 or R_3), as follows:

$$T = 2C(R_2 + R_3)\ln\left[\frac{2R_2R_3R_6 - R_1R_4(R_2 + R_3)}{R_1R_4(R_2 + R_3)}\right]. \tag{10}$$

OA–Based Uncalibrated Solution: In Figure 10 we present a square-wave oscillator, based on OA, performing an *R-T* conversion, also in this case based on a voltage differentiation, so ensuring a good immunity to low-frequency disturbs [28].

In the circuit, OA_1 serves as a voltage differentiator, while OA_2 is a hysteresis voltage comparator. Thanks to a suitable closed loop, which avoids any system calibration, resistive sensors can be excited by an AC signal. The block scheme of the circuit shows also the voltage signals at the main circuit nodes, from which the differentiating effect on V_C can be seen.

Through a straightforward circuit analysis, considering ideal OAs, it is possible to achieve the following expression for the period T of the generated square waveform, revealed at V_{OUT} node:

$$T = 2CR_5\ln\left(\frac{\frac{R_5}{R_6}\left(\frac{R_3}{R_3 + R_4} - \frac{R_1}{R_1 + R_2}\right) - 2\frac{R_4}{R_3 + R_4}}{\frac{R_5}{R_6}\left(\frac{R_3}{R_3 + R_4} - \frac{R_1}{R_1 + R_2}\right)}\right). \tag{11}$$

From eq.(11) it is evident the direct proportionality between the output period and capacitance C (useful for capacitance estimation and/or for the sensitivity setting) but, under particular conditions about the resistance values, it is possible to consider R_5 as a resistive sensor achieving a good linear response for a reduced variation range.

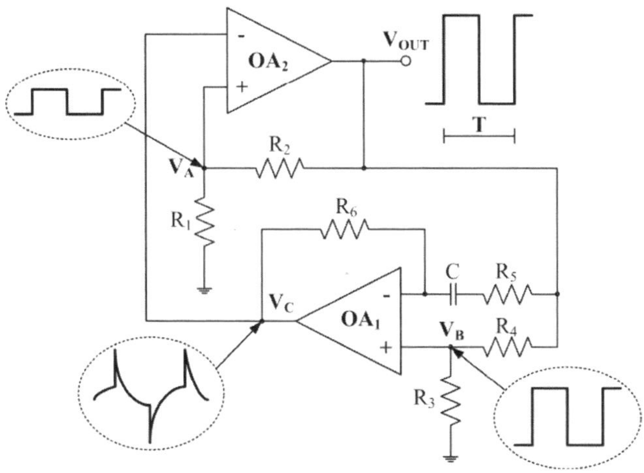

Fig. 10. Block scheme of the proposed capacitive/resistive sensor interface

4 Wide Range Resistive Sensor Interfaces

DC Excited Sensor Interface Solutions

Automatic Uncalibrated Wheatstone Bridges: The Wheatstone bridge configuration can be made "automatic" (so that the circuit does not need initial calibration) through the development of the topology shown in Figure 11 which employs a tuneable resistor implemented through a voltage controlled resistance, based on a novel use of an analog four quadrant multiplier [29], whose variations follow those of the resistive sensor, and a suitable feedback loop.

More in detail, the circuit reported in Figure 11 represents the configuration suitable for a grounded resistive sensor placed in the lower position of the left branch of the bridge. The differential bridge output is connected to an OA-based differential amplifier with a voltage gain A; then, the single-ended output is sent to a voltage inverting integrator whose aim is both to create the stable negative feedback loop and to provide the correct control voltage value (V_{CTRL}) for the tuneable resistor R_{VCR}.

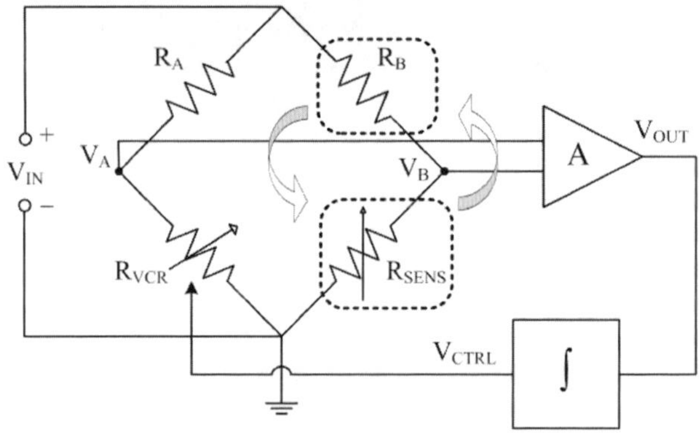

Fig. 11. Block schemes of the proposed bridge-based interfaces (according to the position of the two right branch elements, it is possible to study the "Grounded" and "Floating" sensor configuration)

If a measurand variation occurs into a determined range, the unbalanced output voltage is amplified and the integrator produces a ramp that tunes the active element R_{VCR} until a new equilibrium condition is reached (i.e., the automatic range). The complete expressions for the estimation of the sensor resistance R_{SENS}, as a function of the other three bridge resistances (R_A, R_B and R_{VCR}), the supply voltage V_{CC} and the bridge differential output voltage $\Delta V=V_A-V_B$, is given by:

$$R_{SENS_{GROUNDED}}=\frac{R_{VCR}R_B}{R_A}\cdot\left(\frac{1-\dfrac{\Delta V}{V_{CC}}\dfrac{R_A+R_{VCR}}{R_{VCR}}}{1+\dfrac{\Delta V}{V_{CC}}\dfrac{R_A+R_{VCR}}{R_A}}\right). \tag{12}$$

Similar results can be obtained with another configuration of the automatic bridge, obtained by swapping the two dashed elements in Figure 11, achieving a solution suitable for a floating resistive sensor. In this case, the resistive sensor estimation can be performed through the following equation:

$$R_{SENS_{FLOATING}}=\frac{R_A R_B}{R_{VCR}}\cdot\left(\frac{1+\dfrac{\Delta V}{V_{CC}}\dfrac{R_A+R_{VCR}}{R_A}}{1-\dfrac{\Delta V}{V_{CC}}\dfrac{R_A+R_{VCR}}{R_{VCR}}}\right). \tag{13}$$

This modification maintains the same working principle, but the effect of the variable floating sensor implies an opposite trend, with respect to the first configuration, of the control voltage signal V_{CTRL}, as depicted in Figure 12 where the behaviours of V_{CTRL} and ΔV voltages, for both the configurations, vs. R_{SENS}, have been reported. Due to the electrical limits of the considered analog multiplier inputs, the circuit is not able to follow any bridge unbalancing out of the automatic range, where the control voltage V_{CTRL} reaches the saturation level. However, since the differential output ΔV is not zero and, through eq.s (12) and (13), it is possible to estimate resistive sensor values for more than 5 decades (in a settable range) in a very fast way by only reading the two voltages V_{CTRL} and ΔV.

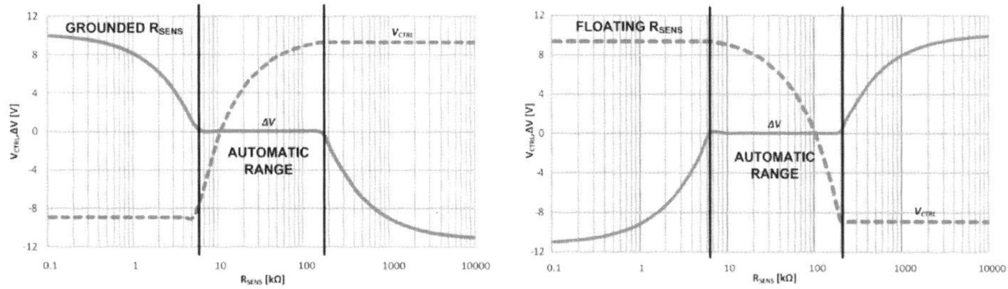

Fig. 12. ΔV and V_{CTRL} vs. sample resistance for "grounded" (left) and "floating" (right) configurations

Voltage-Mode Uncalibrated Solution: In Figure 13 an OA-based interface, performing an *R-T* conversion, is presented. This circuit, based on an oscillator topology and exciting the sensor with a DC voltage V_{EXC}, is able to reveal more than 4 decades of high resistance variations (*e.g.*, 1MΩ÷10GΩ) [30]. It employs three OAs and four switches in order to properly control the voltage signal V_I generated by the first stage, dependent on the sensor resistance value R_{SENS}, while OA_2 operates both as an inverting integrator and as a non-inverting one, through the suitable use of the four switches.

Considering an ideal behaviour for the OAs, through a straightforward analysis, it is possible to evaluate the relationship between the sensor resistance R_{SENS} and the period T of the output square wave signal, as follows:

$$T = 2R_2C_1\left[\left(\frac{R_4}{R_3+R_4}\frac{V_{SAT+}-V_{SAT-}}{R_1V_{EXC}}\right)R_{SENS}-1\right], \tag{14}$$

being V_{SAT+} and V_{SAT-} the OA saturation voltages.

Since the voltage integrator has a double operating function, the presence of the capacitance C_I involves a charge effect, which influences instantaneously the ramp signal when there is the operating function commutation (from inverting to non-inverting and vice versa), through a vertical edge on V_A, as also depicted in Figure 13, whose value depends on the V_I level, thus on R_{SENS} (low values of sensor resistance provide a high voltage gap).

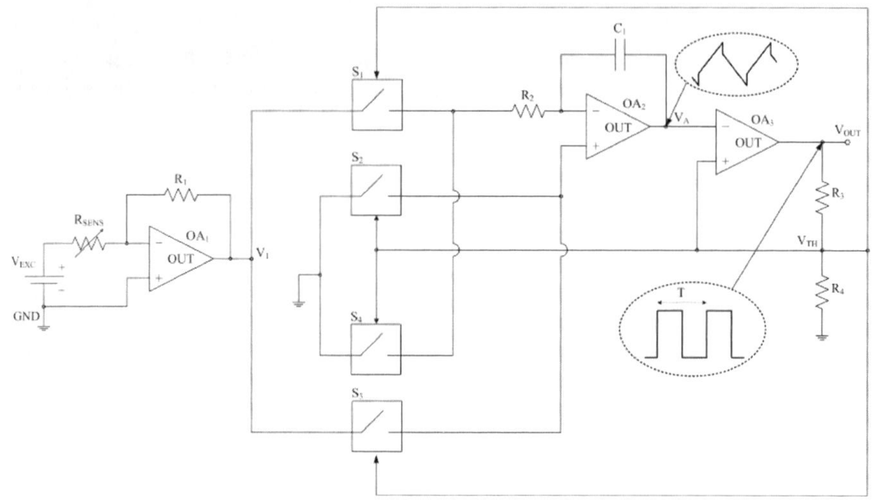

Fig. 13. The proposed OA-based interface with a DC resistive sensor excitation voltage

Current-Mode Uncalibrated Solution: In Figure 14, a similar solution, developed with the CM approach, based on an oscillating circuit, always suitable for resistive sensing elements which do not tolerate an AC excitation voltage, is reported. This interface does not require any preliminary calibration and operates, once again, an *R-T* conversion [31].

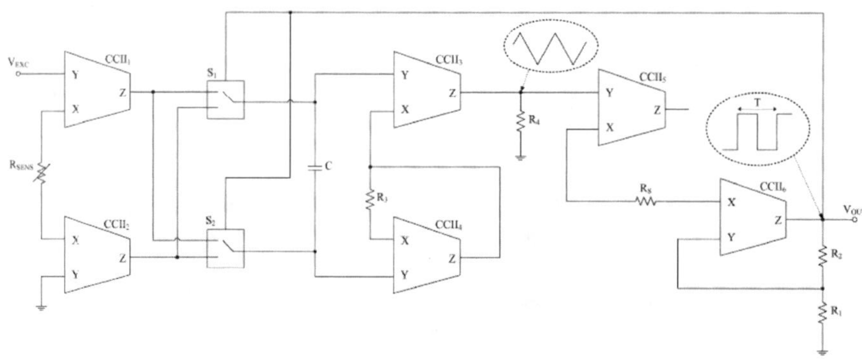

Fig. 14. Scheme of the proposed CCII-based front-end

Through a straightforward analysis, considering ideal *CCIIs* and switch behaviours, it is possible to determine the expression for the period T of the generated output square wave signal as a function of the sensor resistance R_{SENS} as follows:

$$T = 2R_{SENS}C\left[\frac{k(V_{SAT+} - V_{SAT-})}{A \cdot V_{EXC}}\right], \tag{15}$$

where $k = (R_1 - R_S)/(R_1 + R_2)$, A is the voltage gain of instrumentation amplifier ($A = 2R_4/R_3$), V_{EXC} is the DC sensor excitation voltage, while V_{SAT+} and V_{SAT-} are the positive and negative saturation voltages at output terminal V_{OUT}.

AC Excited Sensor Interface Solutions

OA-Based Astable Multivibrator: Generally, when large variations of sensor resistive values occur, the most used strategy is related to an AC-excitation voltage for a (floating) resistive sensor, operating the *R-T* conversion.

The simpler electronic interface which converts a pure resistive variation into a period (or a frequency) can be implemented by an OA in astable multivibrator configuration, as shown in Figure 15 [3,19]. This circuit solution implements a square wave generator, whose output voltage signal period is linearly dependent on the sensor resistance value. Obviously, the same topology is certainly suitable also as a capacitive sensor interface.

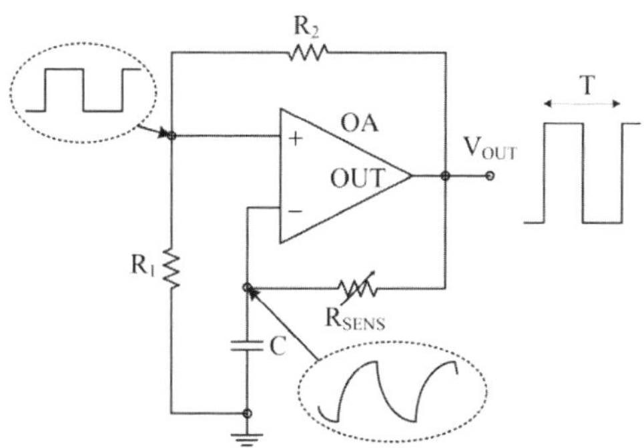

Fig. 15. OA-based astable multivibrator circuit as resistive sensor interface

Through a straightforward circuit analysis, it is possible to evaluate the output period T of the generated square waveform V_{OUT}, dependent on R_{SENS}, according to the following equation:

$$T = 2R_{SENS}C\ln\left(\frac{1+\beta}{1-\beta}\right), \tag{16}$$

being:

$$\beta = \frac{R_1}{R_1 + R_2}. \tag{17}$$

Eq. (16) is valid only for an ideal OA and, more in detail, if we consider R_1=R_2, it simply becomes $T \approx 2.2R_{SENS}C$. However, considering ideal conditions, the circuit has no limitations for high period values (except for the fact that a long measurement time occurs), so it is able to operate, for an example, at least for 6 decades of resistance variations, which correspond to a period span of the same number of decades. The sensitivity, for this kind of resistive sensor interface, is relatively low and, consequently, the main problem related to this front-end concerns the detection of small resistance values or variations. In addition, it is also important to employ accurate values of R_1 and R_2 resistances and non-linear effects (among which the temperature) have to be taken into account and verified so to be eventually considered both in the period measurement and, consequently, in the sensor resistance estimation.

CCII-Based Astable Multivibrator: A CM version of the astable multivibrator, implemented with a single CCII, performing a square wave generation, is reported in Figure 16 [32]. This solution can be used in resistive sensor interfacing, showing a linear relation, between sensor and oscillation period, in an operating frequency range up to about 50MHz.

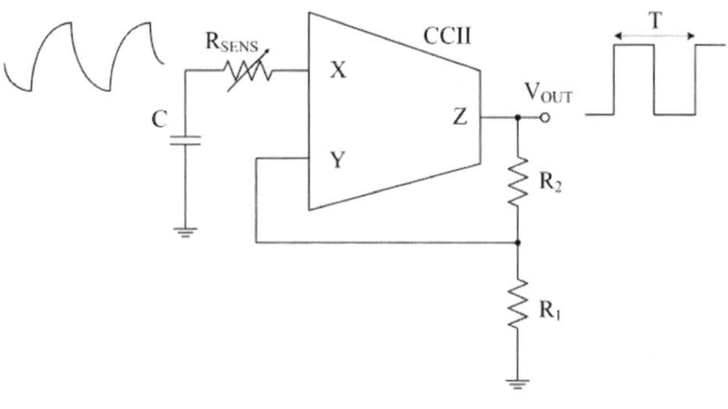

Fig. 16. CCII-based astable multivibrator

The output square-wave signal period is ideally given by:

$$T = R_{SENS} C \ln\left(\frac{V_{TH-} - V_{X+}}{V_{TH+} - V_{X+}} \cdot \frac{V_{TH+} - V_{X-}}{V_{TH-} - V_{X-}} \right) \tag{18}$$

being:

$$V_{TH+} = \frac{R_1 - R_{SENS}}{R_1 + R_2} V_{SAT+}; \quad V_{TH-} = \frac{R_1 - R_{SENS}}{R_1 + R_2} V_{SAT-}, \tag{19}$$

$$V_{X+} = \frac{R_1}{R_1 + R_2} V_{SAT+}; \quad V_{X-} = \frac{R_1}{R_1 + R_2} V_{SAT-}, \tag{20}$$

where V_{SAT+} and V_{SAT-} are the saturation voltages that the CCII is able to reach at its output. Therefore, the period T can be varied by changing R_{SENS}; in this sense, the CCII internal series parasitic resistance at X node must be carefully considered.

CCII-Based Uncalibrated Solution: A CM interface circuit, for AC-excited sensors showing a wide resistive variation is reported in Figure 17 [33]. It is composed by: a voltage integrator ($CCII_1$), a voltage buffer ($CCII_2$) and a CM hysteresis comparator ($CCII_3$). Through a straightforward analysis, considering ideal CCII behaviour, it is possible to determine the expression for the period T of generated output square wave signal, revealed at V_{OUT} node, as a function of the sensor resistance R_{SENS}, as follows:

$$T = 4R_{SENS}CG, \tag{21}$$

being:

$$G = \frac{R_2 - R_1}{R_2 + R_3}. \tag{22}$$

From eq.s (21) and (22), the circuit sensitivity can be opportunely set by choosing C, R_1, R_2 and R_3 values.

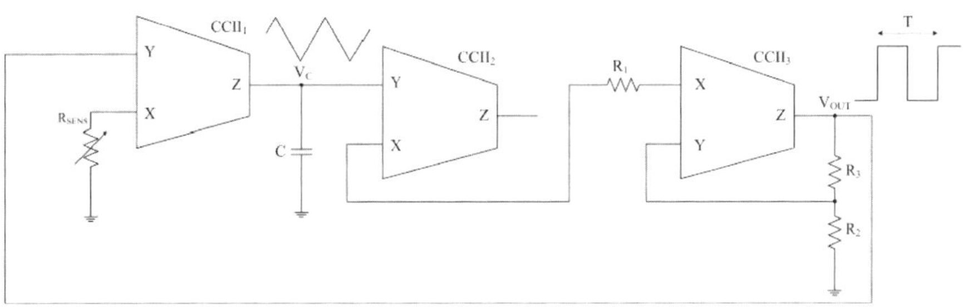

Fig. 17. Block scheme of the proposed uncalibrated CCII-based interface

OA-Based Uncalibrated Solution: Starting from the OA-based astable multivibrator circuit, Figure 18 shows the block scheme and related main node voltage behaviours of an improved VM wide range resistive sensor interface, always based on an oscillator topology, formed by: an inverting amplifier (AMP), a voltage comparator (COMP) and an inverting integrator (INT). This circuit, able to reveal about over 6 decades of sensor resistance variations (*e.g.*, kΩ÷GΩ) without any initial calibration, performs an *R-T* conversion, where the oscillation period T of the generated output square wave signal V_{OUT} is directly proportional to sensor resistance value R_{SENS} [34].

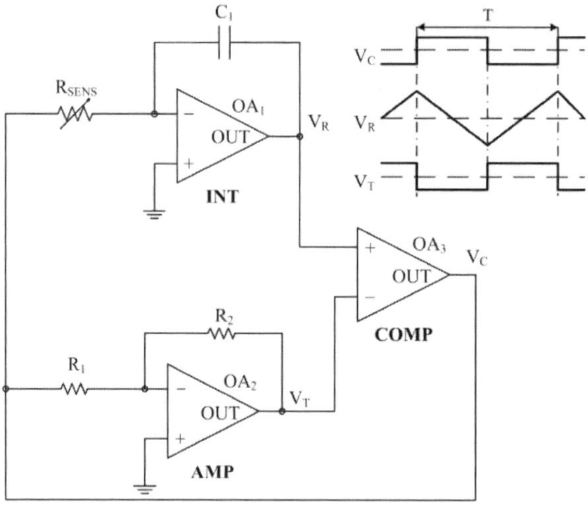

Fig. 18. Block scheme of the resistive sensor interface based on an *R-T* conversion and voltage behaviours at its main nodes

In this case, the ideal relationship between the period T and the sensor resistance R_{SENS} is the following:

$$T = 4GC_1R_{SENS} \tag{23}$$

being G the ratio between R_2 and R_1, typically lower than 1. From eq. (23), it comes that the proposed interface shows two degrees of freedom, in particular C_1 and G (i.e., R_1 and R_2), that helps to choose its sensitivity and so the oscillation frequency range.

OA-Based Time-Controlled Oscillating Uncalibrated Solution: Possible evolutions of previous schemes go towards a main direction: the development of solutions showing a reduced measurement time for high resistive sensor values; some of the latter will be shown in the following.

A suitable interface circuit, capable to overcome the main limit of the solutions based on the *R-T* conversion (i.e., the long measuring time occurring in the evaluation of high-value sensor resistances), is reported in Figure 19. This solution, always performing an *R-T* conversion, is based on a particular oscillating circuit architecture which operates a suitable "compression" of the higher part of the resistive wide range, thus limiting the measuring time, by means of an "ad-hoc" oscillator architecture utilizing suitable feedbacks [35]. This solution provides always an AC excitation voltage for the sensor and results capable to estimate its resistance over a wide range (about 5 decades, *e.g.*, 100kΩ÷10GΩ) with a maximum measuring time lower than hundreds of ms (settable values).

More in detail, referring to Figures 19 and 20, INT_1 is an inverting integrator which generates the sensor ramp V_R, INT_2 is a non-inverting integrator providing the ramp threshold V_Z, COMP is a hysteresis comparator; this last block detects the intersections between the ramp threshold V_Z and the sensor ramp V_R, generating a square-wave signals V_Y. Moreover, V_Y is the main signal which is also employed both to close the circuit loop and to provide the AC sensor excitation voltage (i.e., $\pm V_{CC}$), so performing the oscillating behaviour. Therefore, the output period T, calculated as T_1+T_2, depends on both V_Z (independent from R_{SENS}) and V_X (related to V_R and therefore to R_{SENS}). On the other hands, the measuring time T of the sensor interface is properly regulated by means of both the ramp threshold slope (i.e., the "fixed" time constant of the non-inverting integrator) and the sensor ramp slope (i.e., the "variable" time constant of the inverting integrator depending on sensor resistance value).

Thus, if R_{SENS} is very high, V_R and V_X are almost constant, but the oscillator output period is limited by the presence of the threshold voltage V_Z that, since it is a ramp signal having an opposite slope with respect to both V_R and V_X, "moves" towards the sensor ramp V_R. As a consequence, for very high R_{SENS} values, the output period maximum value is always limited by the ramp threshold V_Z to a finite value, through the implemented "time-compression". On the contrary, for low R_{SENS} values, V_R and V_X show very fast ramps (i.e., having a high slope) when compared to the threshold voltage V_Z which is, in this case, "approximately" constant, since it is a ramp signal with reduced slope.

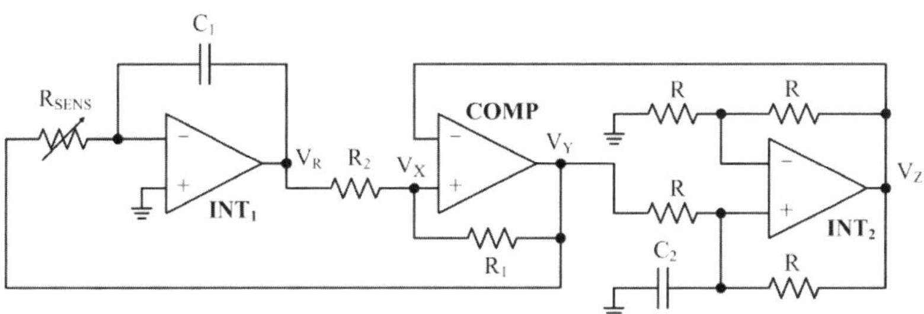

Fig. 19. Electronic implementation of the time-controlled oscillating circuit

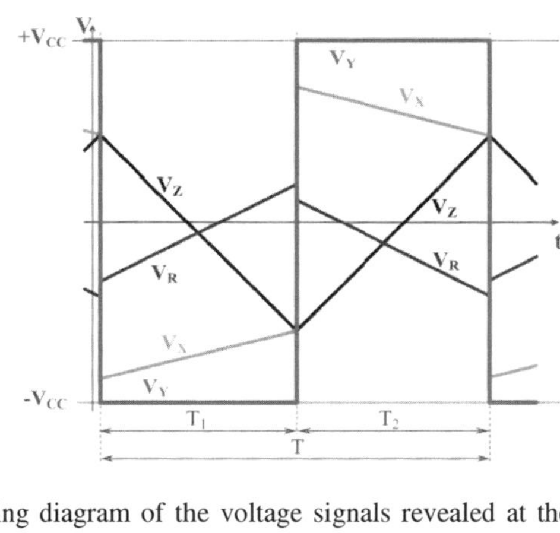

Fig. 20. Typical timing diagram of the voltage signals revealed at the main terminals of the interface solution shown in Figure 19

In this way, the R_{SENS} estimation is performed by measuring the period T of the output V_Y of the comparator COMP, according to the following ideal relationship:

$$T = \frac{4B}{\frac{(1-B)}{R_{SENS}C_1} + \frac{2}{RC_2}}, \tag{24}$$

being $B = R_2/(R_1+R_2)$. Moreover, in this case, the oscillating circuit sensitivity, expressed as $\delta T/\delta R_{SENS}$, shows a quasi-constant value for small resistances, while for high resistor values it decreases for the effect of the "time-compression" of T with respect to R_{SENS}, as shown in Figure 21.

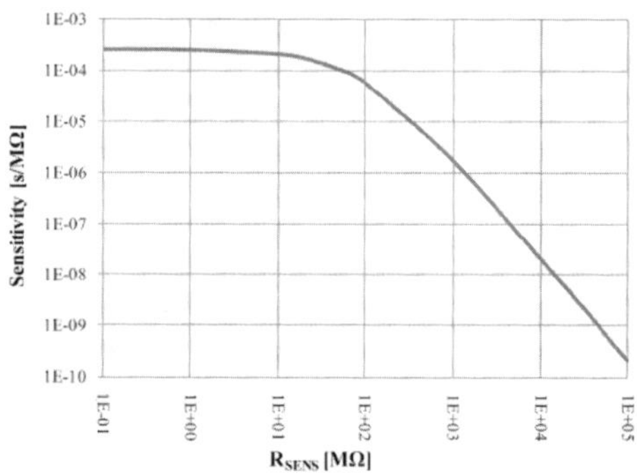

Fig. 21. Example of the time controlled sensitivity response vs. sensor resistance value

5 Integrated Microsystems

Introduction and Basic Main Concepts

The recent evolution in the Very Large Scale Integration (VLSI) industry contributing to the rapid technology changes and leading to an improved interest in analog circuit design (especially for what concerns *ICs*), the tremendous competition among vendors and the demand in the market for *ICs*, represent all together the factors which have led to consider the time-to-market factor with utmost importance. Regarding the electronic/sensor systems, with maximum performance and least turnaround time, an *ASIC* seems to be the best option to meet the ever growing demands for quality chips.

In particular, a huge design effort has been put toward the placement of a larger number of elements and devices on a single chip, together with supply voltage and power dissipation reductions, as much as possible. The suitable analog integrated circuit design, widely utilized in portable single-cell battery operated applications (*e.g.*, biomedicals, cellular phones, etc.), has led to implement new design microelectronic strategies, to be developed in low cost standard CMOS integrated technologies, even if analog blocks, as voltage/current amplifiers, must be carefully designed in integrated sensors interface applications since these processes show transistors with high input offset voltages and a not negligible low-frequency noise levels.

In this sense, as regard integrated microelectronics, the continuous reduction of the threshold voltage in standard CMOS has definitively directed LV design towards CMOS itself, which is also typically characterized by a very low quiescent power consumption. Reducing the supply voltage, CMOS transistor is often biased to work in weak inversion region. In addition, a suitable interfacing of the sensitive element with a proper *ICs* is fundamental, especially when they are fabricated on the same chip. In this sense, CMOS technology is widely used, because it allows to match the reduction of costs of the silicon with the possibility of designing new LV LP interface circuits to be easily dedicated to the portable sensor applications market [21-23,36].

Moreover, referring particularly to the AMS 0.35μm standard CMOS technology, also the passive components have to be taken into account and properly designed. In this sense, in fact, note that there are several problems, especially related to the required silicon area. Therefore, sometimes, analog electronic designers have to evaluate the limitations related to resistors and capacitors integrations. Detailing, an *R-poly* resistance requires 7Ω/ (i.e., square resistance) and a *Poly-Sub Plate* capacitance needs $0.12fF/\mu m^2$; on the contrary, typical areas requested by an active block, such as *OTA* and CCII, are lower than $0.1mm^2$ (*e.g.*, see next Sections).

However, concerning the analog circuit design, the reduction of the supply voltage does not necessarily correspond to a decrease of related power consumption. Therefore, in order to reduce the power dissipation, analog circuits have to be designed as much simple as possible. Moreover, it is important to consider that a trivial decrease of biasing currents, which can reduce circuit dissipation, degrades the circuit performance, first of all bandwidth and dynamic range. As a consequence, chip area cannot be drastically reduced with the lowered feature dimensions. As a result, LP design is characterised by an efficient use of the supply current (*e.g.*, through the utilisation of class-AB output stages) and an efficient frequency compensation strategy.

Finally, for example, referring to GMR sensor integrated technologies, it is important to consider that high density arrays have been integrated and fabricated on a single chip, in a standard CMOS process and with a very small silicon area (*e.g.*, lower than $1mm^2$), for different applications, especially in biomedical fields (*e.g.*, *DNA* hybridization detection) [8,9,15,16]. In fact, integrating GMR resistive sensor arrays and signal conditioning electronics on a single silicon-based chip, fabricated in standard CMOS technologies, yield low-cost complete microsystems (System-on-Chip, *SoC*) that constitutes a promising tool for the future of portable applications. This kind of GMR-based CMOS integrated solutions, compared for example to complex and expensive optical detection approaches or imaging systems, measures electrical signal directly from the sensor and makes a low-cost, highly portable device feasible, especially for applications such as fuel cells monitoring, current sensing in power electronic modules, *ICs* current monitoring and power measurements, etc. [7,10-14,17,37-39].

OTA and CCII CMOS Transistor-Level Solutions

Generally, in microelectronic systems, the active block used as OA is implemented by a suitable *OTA*. A possible *OTA* internal topology, designed at transistor level in a standard CMOS technology, is reported in Figure 22 [3,34].

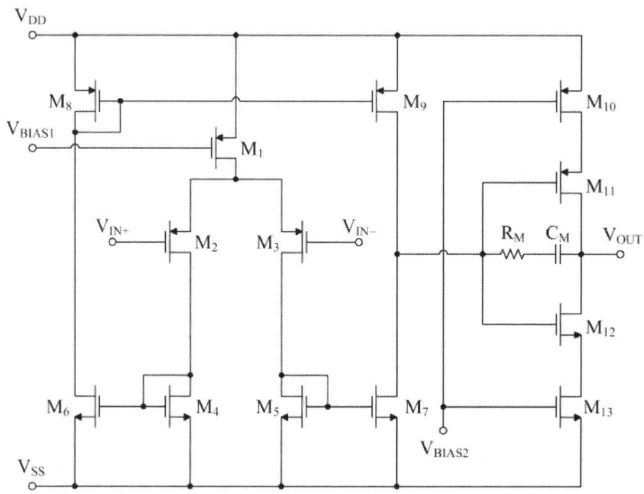

Fig. 22. OTA schematic at transistor level

More in detail, the *OTA* shown in Figure 22 is composed by two stages: the input stage (formed by transistors M_1–M_9), which is a symmetrical *OTA*, and the output stage (formed by transistors M_{10}–M_{13}), that is an AB-class inverter amplifier, based on a push–pull configuration (M_{11},M_{12}), that allows to obtain a full dynamic output range, with a source degeneration (M_{10},M_{13}). In particular, transistors M_{10} and M_{13} allow a better control of the current flowing in the output branch, even if, through the source degeneration, this same current has been done slightly dependent on supply voltage variations. Moreover, this second stage allows to get a high open loop voltage gain, so to make the amplifier more ideal. Frequency stability has been obtained by an *R-C* series Miller compensation (i.e., R_M and C_M components, Figure 22). The choice of a symmetrical configuration as *OTA* input stage and a careful layout implementation have allowed to reduce both the systematic and the random input offset voltages. Furthermore, two *p-MOS* matched transistors, M_2 and M_3, have been utilized as input differential pairs so to have a low input equivalent noise.

This schematic has been developed so to obtain better performances, in terms of very high *SR* and very low input voltage offset, and to operate at reduced supply voltage with low power consumption, as detailed in Table 1. In this sense, it can be employed so to develop suitable integrated versions of previous described VM sensor interface solutions. Moreover, in this way, especially referring to those solutions based on oscillating circuits, the relative error between ideal/theoretical and measured oscillation periods becomes negligible.

Table 1. Main characteristics of the implemented *OTA*, designed in AMS 0.35μm standard CMOS technology

***OTA* parameter**	**Post-layout simulated value**
Voltage supply	3.3 V
Power dissipation	992 μW
GBW	65.8 MHz
Output dynamic range	Full
Open Loop DC Voltage Gain	66 dB
Slew-Rate	40 V/μs
Input voltage offset	100 μV
Input equivalent noise	169 nV/√(Hz) @ 1 kHz
Silicon area	0.05mm^2

At least, in this paragraph, we will show also a possible transistor level integrated solution of a CCII [26], developed in AMS 0.35μm standard CMOS technology. It can be employed so to develop suitable integrated versions of previous described CM sensor interface solutions [3]. This CCII internal topology, reported in Figure 23, shows negligible parasitic impedances and unitary voltage and current gains for a very large bandwidth (quasi-ideal characteristics), as detailed in Table 2.

More in detail, the circuit shown in Figure 23 is formed by a differential input stage (M_1–M_7; R_3), an AB-class output stage (M_8–M_{11}; R_1,R_2; M_{16}–M_{17}) and a LV cascode Wilson current mirror (M_{12}–M_{15}; M_{18}–M_{21}). The AB-class output stage allows to decrease the *X* parasitic impedance, whereas the cascode current mirror increases the *Z* impedance.

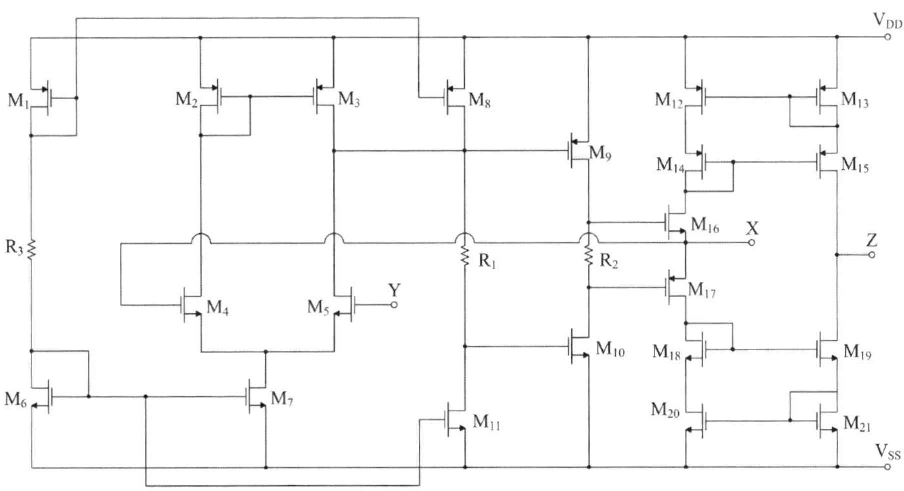

Fig. 23. Quasi-ideal CCII schematic at transistor level

Table 2. Main characteristics of the implemented quasi-ideal CCII

CCII parameter	Post-layout simulated value
Supply voltage	± 0.75V
Power consumption	118μW
3dB Bandwidth	10.5MHz
Biasing Currents	6μA
Voltage Gain (α)	1.00
Current Gain (β)	1.00 (R_{load_X}=R_{load_Z}=10kΩ)
X Parasitic Resistance R_X	13Ω
X Parasitic Inductance L_X	0.4μH
X Parasitic Capacitance C_X	0.1pF
Z Parasitic Resistance R_Z	2.6MΩ
Z Parasitic Capacitance C_Z	0.03pF
Y Parasitic Capacitance C_Y	0.1pF
Silicon area	0.09mm^2

Sensor Arrays Management

As stated before, the modern advances in CMOS and VLSI technologies have enlarged the ever-increasing demand of high resolution sensors by integrating a large number of identical microsensors (i.e., arrays) on a single chip [8,18,40-44]. The multiple resistive sensing elements, typically having identical behaviours, when combined, usually in two-dimensional, *N*×*M*, array configurations, generate patterns. Obviously, the quality or the resolution of this information is enhanced by increasing the array size. Unfortunately, accessing all the elements for information collection and signal processing puts limitations on the array size [45].

Moreover, these sensors are interfaced using a suitable integrated readout circuit. In this way, the ultimate performance of the sensor arrays, employing the proper front-end electronics, depends not only on the sensor chip, but also on the same interface and its matching with the sensors. Typically, the development of a matched readout circuit itself requires a careful evaluation of sensors before its design, resulting in the iterative improvement process and making the implementation of the high resolution multi-sensor arrays complicated, costly and time consuming. Therefore, in general, the sensor array technology development needs the initial fabrication of small/moderate sized arrays as test structures that do not require the use of a particular electronic interface.

Starting from these considerations, one of the main keynodes in integrated system is the interconnection complexity of sensor arrays. In general, the access of all individual sensors requires two physical connections from each sensor resulting in a total of 2[*N*×*M*] connections. This number becomes large for even a small sized array, considering the required on-chip interconnecting metal lines and the number of bonding/probing pads.

In order to solve this problem, different scheme solutions, having reduced interconnections for the readout of all the sensors in a *N*×*M* resistive sensor array, have been proposed in the literature. One of them is reported in Figure 24 where, by means of a suitable control switching circuit, the internal connections related to the matrix rows and columns have been reduced and simplified. Moreover, Figure 24 shows also the basic signal conditioning circuit, considered to the readout of the array elements, based on the simple inverting voltage amplifier [46,47].

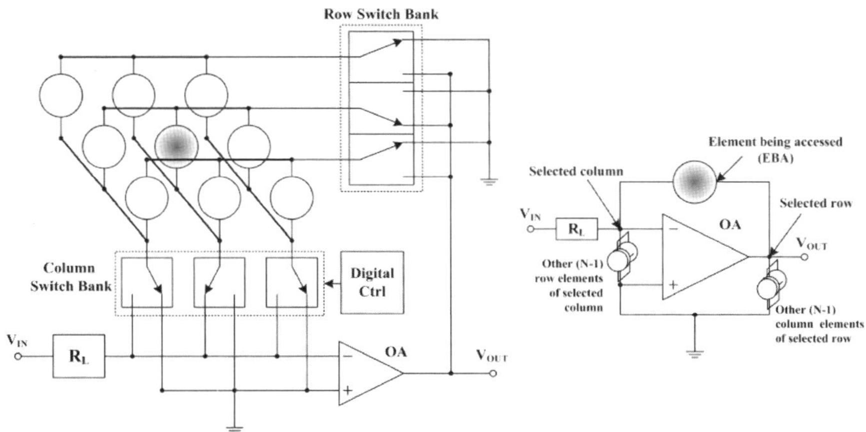

Fig. 24. A solution with reduced interconnections for a sensor array (left) and its simplified equivalent connection scheme at the conditioning circuit (right)

However, in some solutions, it is important to consider also a further significant matter resulting in the undesired information, spreading in the array, due to the interconnect overloading and the crosstalk among the sensor elements. This effect has to be necessarily avoided or reduced in integrated microsystems [44-47].

Finally, it is important to mention that in array-based sensor microsystems, composed by several sensors sensing various measurands with different sensitivities and selectivities, feature extraction techniques can be used to pull out information also from the transient sensor response [48].

References

[1] Huijsing, J.-H.: Integrated smart sensors. Sensors and Actuators A 30, 167–174 (1992)
[2] Meijer, G.C.M.: Smart sensor systems. Wiley (2008)
[3] De Marcellis, A., Ferri, G.: Analog Circuits and systems for Voltage-Mode and Current-Mode sensor interfacing applications. Springer (2011)
[4] Falconi, C., Martinelli, E., Di Natale, C., D'Amico, A., Maloberti, F., Malcovati, P., Baschirotto, A., Stornelli, V., Ferri, G.: Electronic Interfaces. Sensors and Actuators B 121, 295–329 (2007)

[5] Middelhock, S., Audet, S.A., French, P.: Silicon sensors. Academic Press, London (2000)
[6] Fraden, J.: Handbook of modern sensors: physics, design and applications, 3rd edn. Springer (2003)
[7] Hirota, E., Sakakima, H., Inomata, K.: Giant magneto-resistance devices. Springer (2002)
[8] Han, S.J., Xu, L., Yu, H., Wilson, R.J., White, R.L., Pourmand, N., Wang, S.X.: CMOS integrated DNA microarray based on GMR sensors. In: International Electron Devices Meeting (IEDM 2006), pp. 719–723 (2006)
[9] Piedade, M.S., Sousa, L., Almeida, T.M., Germano, J., d'Andrade da Costa, B., Lemos, J.M., Freitas, P.P., Ferreira, H.A., Cardoso, F.A.: A new hand-held microsystem architecture for biological analysis. IEEE Transactions on Circuits And Systems I: regular papers 53, 2384–2395 (2006)
[10] Hartmann, U.: Magnetic Multilayers and Giant Magnetoresistance: Fundamentals and Industrial. Applications Springer Series in Surface Sciences, vol. 37 (2000)
[11] Roldán, A., Reig, C., Cubells-Beltrán, M.D., Roldán, J.B., Ramírez, D., Cardoso, S., Freitas, P.P.: Analytical compact modeling of GMR based current sensors: Application to power measurement at the IC level. Solid-State Electronics 54, 1606–1612 (2010)
[12] Cubells-Beltrán, M.D., Reig, C., Martos, J., Torres, J., Soret, J.: Limitations of magnetoresistive current sensors in industrial electronics applications. International Review of Electrical Engineering 6, 423–429 (2011)
[13] Freitas, P.P., Ferreira, R., Cardoso, S., Cardoso, F.: Magnetoresistive sensors. J. Phys.: Condens. Matter 19, 165221 (2007)
[14] Reig, C., Cubells-Beltrán, M.D., Muñoz, D.R.: Magnetic field sensors based on giant magnetoresistance (GMR) Technology: Applications in Electrical Current Sensing. Sensors 9, 7919–7942 (2009)
[15] Heinrich, B., Bland, J.A.C.: Ultrathin magnetic structures. Application of Nanomagnetism IV. Springer (2005)
[16] Germano, J., Martins, V.C., Cardoso, F.A., Almeida, T.M., Sousa, L., Freitas, P.P., Piedade, M.S.: A portable and autonomous magnetic detection platform for biosensing. Sensors 9, 4119–4137 (2009)
[17] Blagojevic, M., Kayal, M., Gervais, M., De Venuto, D.: SOI Hall-sensor front end for energy measurement. IEEE Sensors Journal 6, 1016–1021 (2006)
[18] Edelstein, A.S., Burnette, J., Fischer, G.A., Cheng, S.F., Egelhoff, J.W.F., Pong, P.W.T., McMichael, R.D., Nowak, E.R.: Advances in magnetometry through miniaturization. Journal of Vacuum Sci & Tech A: Vacuum, Surfaces, and Films 26, 757–762 (2008)
[19] Sedra, A., Smith, K.C.: Microelectronic Circuits, 5th edn. Oxford University Press (2007)
[20] Hogervorst, R., Huijsing, J.H.: Design of low-voltage low-power operational amplifier cells. Kluwer Academic Publishers, Boston (1996)
[21] Maloberti, F.: Analog design for CMOS VLSI systems. Kluwer Academic Publishers (2001)
[22] Palumbo, G., Palmisano, S., Pennisi, S.: CMOS current amplifiers. Kluwer Academic Publishers, Boston (1999)
[23] Ferri, G., Guerrini, N.: Low Voltage, Low Power CMOS Current Conveyors. Kluwer (2003)
[24] Pertijs, M.A.P., Aita, A.L., Makinwa, K.A.A., Huijsing, J.H.: Voltage calibration of smart temperature sensors. In: Proc. of IEEE Sensors, Lecce, Italy, pp. 756–759 (2008)

[25] Muñoz, D.R., Sánchez, J., Casans, S., Castro, E., Reig, C., Navarro, A.E.: Temperature compensation of Wheatstone bridge magneto-resistive sensors based on generalized impedance converter with input reference current. Revi. Sci. Instrum. 77, 034980 (2006)
[26] Ferri, G., Stornelli, V., Fragnoli, M.: An integrated improved CCII topology for resistive sensor application. Analog Integrated Circuits and Signal Processing 48, 247–250 (2006)
[27] De Marcellis, A., Di Carlo, C., Ferri, G., Stornelli, V.: A CCII-based wide frequency range square waveform generator. International Journal of Circuit theory and Applications 41, 1–13 (2013)
[28] De Marcellis, A., Ferri, G., Mantenuto, P.: A Novel OA–based oscillating circuit for uncalibrated capacitive and resistive sensor interfacing applications. In: Proc. of IMCS 2012, Nuremberg, pp. 1707–1709 (2012)
[29] Mantenuto, P., De Marcellis, A., Ferri, G.: Uncalibrated analog bridge-based interface for wide-range resistive sensor estimation. IEEE Sensor Journal 12, 1413–1414 (2012)
[30] De Marcellis, A., Ferri, G., Stornelli, V., Depari, A., Flammini, A., Marioli, A.: A novel Op-Amp based front-end for high valued resistive sensors. In: Proc. of 13th AISEM Conference, Rome, Italy, pp. 540–544 (2008)
[31] Ferri, G., De Marcellis, A., Di Carlo, C., Stornelli, V., Flammini, A., Depari, A., Marioli, D., Sisinni, E.: A CCII-based low-voltage low-power readout circuit for DC-excited resistive gas sensors. IEEE Sensors Journal 9, 2035–2041 (2009)
[32] Del Re, S., De Marcellis, A., Ferri, G., Stornelli, V.: Low voltage integrated astable multivibrator based on a single CCII. In: Proc. of PRIME, Bordeaux, pp. 177–180 (2007)
[33] Ferri, G., Stornelli, V., De Marcellis, A., Di Carlo, C., Flammini, A., Depari, A., Marioli, D.: Uncalibrated Current-Mode oscillator for resistive gas sensor integrable applications. In: Proceedings of ISOEN, Brescia, Italy, pp. 297–300 (2009)
[34] Ferri, G., De Marcellis, A., Di Carlo, C., Stornelli, V., Flammini, A., Depari, A., Nader, J.: A single-chip integrated interfacing circuit for wide-range resistive gas sensor arrays. Sensors and Actuators B 143, 218–225 (2009)
[35] De Marcellis, A., Depari, A., Ferri, G., Flammini, A., Sisinni, E.: A CMOS integrated low-voltage low-power time-controlled interface for chemical resistive sensors. Sensors and Actuators B 179, 313–318 (2013)
[36] Barrettino, D., Graf, M., Taschini, S., Hafizovic, S., Hagleitner, C., Hierlemann, A.: CMOS monolithic metal–oxide gas sensor microsystems. IEEE Sensors Journal 6, 276–286 (2006)
[37] Sebastia, J.P., Lluch, J.A., Vizcaino, J.R.L.: Signal conditioning for GMR magnetic sensors applied to traffic speed monitoring GMR sensors. Sensors and Actuators A 137, 230–235 (2007)
[38] Pannetier-Lecoeur, M., Fermon, C., Giraud, A.: GMR based integrated non-contact voltage sensor for fuel cell monitoring. In: Proc. of IEEE International Conference on Sensing Technology, pp. 101–105 (2011)
[39] Cubells-Beltran, M.D., Reig, C., Munoz, D.R., De Freitas, S.I.P.C., De Freitas, P.J.P.: Full Wheatstone bridge spin-valve based sensors for IC currents monitoring. IEEE Sensors Journal 9, 1756–1762 (2009)
[40] Gardner, J., Yinon, J.: Electronic Noses and Sensors for the Detection of Explosives. Springer (2004)
[41] Yang, W.: Sensor Array. InTech (2012)
[42] Naidu, P.S.: Sensor array signal processing. CRC Press (2009)

[43] Robogiannakis, P., Chatzandroulis, S., Tsamis, C.: Integrated circuit interface for metal oxide chemical sensor arrays. Sensors and Actuators A 132, 252–257 (2007)

[44] Guo, B., Bermak, A., Chan, P.C.H., Yan, G.Z.: A monolithic integrated 4x4 tin oxide gas sensor array with on-chip multiplexing and differential readout circuits. Solid-State Electronics 51, 69–76 (2007)

[45] Liu, H., Zhang, Y.F., Liu, Y.W., Jin, M.H.: Measurement errors in the scanning of resistive sensor arrays. Sensors and Actuators A 163, 198–204 (2010)

[46] Saxena, R.S., Bhan, R.K., Aggrawal, A.: Reducing readout Complexity of Large Resistive Sensor Arrays. IEEE Sensors Journal 8, 1862–1863 (2008)

[47] Saxena, R.S., Bhan, R.K., Aggrawal, A.: A new discrete circuit for readout of resistive sensor arrays. Sensors and Actuators A 149, 93–99 (2009)

[48] Martinelli, E., Falconi, C., D'Amico, A., Di Natale, C.: Feature extraction chemical sensors in phase space. Sensors and Actuators B 95, 132–139 (2003)

GMR Based Sensors for IC Current Monitoring

Càndid Reig and M.D. Cubells-Beltrán

University of Valencia
Valencia, Spain
{candid.reig,m.dolores.cubells}@uv.es

Abstract. The Giant MagnetoResistance (GMR) effect is a magnetic coupling mechanism that can be obtained in multilayer structures of few nanometers thick. In these devices, and at room temperature, the resistance is a function of the external magnetic field, at optimal levels for being used as sensors. Since the GMR effect was reported, scientists and engineers have dedicated their effort to this topic. This way, after two decades, a a very good knowledge of the GMR underlying physics together with notable designs of GMR based devices are nowadays available. They were initially used in the read heads of hard drives, but the constant evolution that this technology has experienced has open new fields of application, mainly related to the measurement of small magnetic fields using miniaturized devices, such as biotechnology and microelectronics.

Regarding the microelectronics case, these sensors can be potentially used in those scenarios that require a detection or measurement of non-intrusive power by the indirect measurement of the magnetic field.

In this chapter, an overview to the current research regarding the application of GMR sensors in the measurement of electrical currents at the integrated circuit (IC) level is drawn. In this particular case is important to take account of particular parameters of the GMR devices such as the sensing structures, the geometric arrangement and implementation of the sensors, the considered linear range, undesired couplings, biasing, hysteresis, temperature drifts, ... We have also described some cases of success describin particular applications of these devices. Finally, some aspects related to the monolithic integration of GMR devices onto standard CMOS engineered chips are also considered in this chapter.

1 Introduction

Being a very well established concept, the measurement of the electrical current is still a matter of concern [1, 2, 3, 4]. Because of that, limitations of classical current measuring techniques are well known. Shunt resistances are cheap and easy to use but they produce insertion losses, display bandwidth and thermal limitations, and do not offer galvanic isolation. On the other hand, 'transformer principle' based techniques display good isolation, but they have hard frequency limitations at DC regime and they usually involve big devices. Theoretically, by using solid state magnetic sensors for indirectly sensing electrical current by means of the measurement of the generated magnetic field, the most of these

C. Reig et al.: *Giant Magnetoresistance (GMR) Sensors*, SSMI 6, pp. 103–131.
DOI: 10.1007/978-3-642-37172-1_5

handicaps are overcome. This general scheme can be applied to the measurement of a current driven by a wire or by a conductive strap in a printed circuit board or an integrated circuit. AC/DC currents can be measured in this way with small, cheap and contact-less systems.

Among all the magnetic field solid state sensors, GMR based ones have demonstrated to own intrinsic characteristics overcoming the performance displayed by Hall or AMR based devices. GMR based sensors can display sensitivities up to more than ten times those given by Hall and AMR devices. So high sensitivity are necessary when electric currents at the IC level need to be measured. If this sensitivity needs to be accomplished by additional amplifiers, the useful bandwidth is hardly reduced. Moreover, these technologies become in devices with active sensing areas of about 10 μm^2, so allowing a very high level of integration. This advantage is very useful in our particular case or during the design of sensor arrays for magnetic mapping purposes, as in eddy current measurements [5]. Finally, and as will be explained later, GMR structures are able to do measurement of in-plane fields which, in our case, is fundamental. In fact, GMR sensors have been successfully applied in the current measurement in industrial applications, by using different sensors of solid-state magnetic field mounted in different configurations [6, 1, 2].

It is important to remember that all the aforementioned devices are usually developed directly on substrates (semiconductors such as silicon, insulators such as glass, alumina, or technologies Silicon-On-Insulator, SOI), following conventional micro fabrication techniques (UV lithography, sputtering, physical and chemical etching, ...). The electronics needed for the proper functionality of these devices must be added later, either with hybrid integration technology, using wafer-bonding techniques (Sistem-on-Chip, SoC and similar) or more classical approaches (use of Printed Circuit Boards, PCB's, and traditional wiring). The advantages of the generalizability of the processes necessary to monolithically integrate GMR and CMOS technologies, would give the ultimate accolade to current applications, and the emergence of many more.

The measurement of electrical current at the IC level is, then, a matter of concern. Built-in current sensing mechanisms for I_{DDQ} and ΔI_{DDQ} have been proposed for long [7,8]. Such approaches imply the break of a current line in order to include transistor-based in-line circuits performing the current measurement. With the increasing demand on hybrid monolithic designs and different sorts of system-on-chip, the interest in off-line current measurement at the IC level has also been increased. In this scenario, the use of GMR based devices is imperative.

In this chapter, the basis associated to this proposal are firstly revised, including basic structures, sensing mechanism and sensor configurations. The particular issues related to the measurement of low currents by means of GMR sensors are then analyzed. Some successful applications are also described. Finally, the current trends and challenges related to the monolithic integration of GMR devices onto standard CMOS chips are also commented.

2 Fundamentals

The sensors considered for these applications need to fulfill a series of specifications regarding particular capabilities such as linearity, sensitivity, size, thermal behavior, technological requirements, ...

2.1 Sensing Structures

The GMR was firstly described on Fe/Cr thin multilayers [9, 10], and they attributed this effect to the scattering of the electrons flowing through the conductive media with the surrounding magnetic interfaces. These interfaces are also found in granular alloys [11]. These later structures also have been successfully used in sensing applications [12].

Basis

For our specific application, among the various GMR structures described in the literature, two present a picture of very attractive properties for use as sensors in applications: the spin-valves (SV) and magnetic tunnel junctions (MTJ) [13].

Spin Valves (SV)

The SV are multilayer devices composed of two ferromagnetic layers separated by a nonmagnetic layer, usually copper. One layer of ferromagnetic material (pinned) has a fixed magnetic moment (usually through an additional antiferromagnetic coupling layer) and the other, the moment is free to rotate. The relative orientation between magnetic moments, caused by the scattering of the spin of these electrons, produces a variation of the resistance of these devices as a function of external magnetic field. For linear applications, and by applying a magnetic field during deposition, these magnetic moments usually are deposited at an angle of 90° (crossed axis), which significantly reduces its hysteresis. A graphical explanation of these concepts is included in Fig. 1(a), left. As observed, in these structures, a current-in-plane (CIP) scheme is followed (see Fig. 1(b), left). For this structure, the sensor output voltage is given by [13]:

$$\Delta V = \frac{1}{2}\frac{\Delta R}{R} R_{\square} \frac{iW}{h} \cos\left(\Theta_f - \Theta_p\right) \qquad (1)$$

where $\Delta R/R$ is the maximum MR level (5-20%), $R_{\square}$ is the sensor sheet resistance (15-20 $\Omega/\square$), W is its width, h is the thickness, i is the sensor current, and Θ_f and Θ_p are the angle of the magnetization angle of free and pinned layers, respectively. Assuming uniform magnetization for the free and pinned layers, for a linearized output, $\Theta_p = \pi/2$ and $\Theta_f = 0$.

Magnetic Tunnel Junctions (MTJ)

The MTJ's is distinguished from SV due to the separation layer is insulating (see Fig. 1(a), right), so in this case the current is perpendicular to the planes

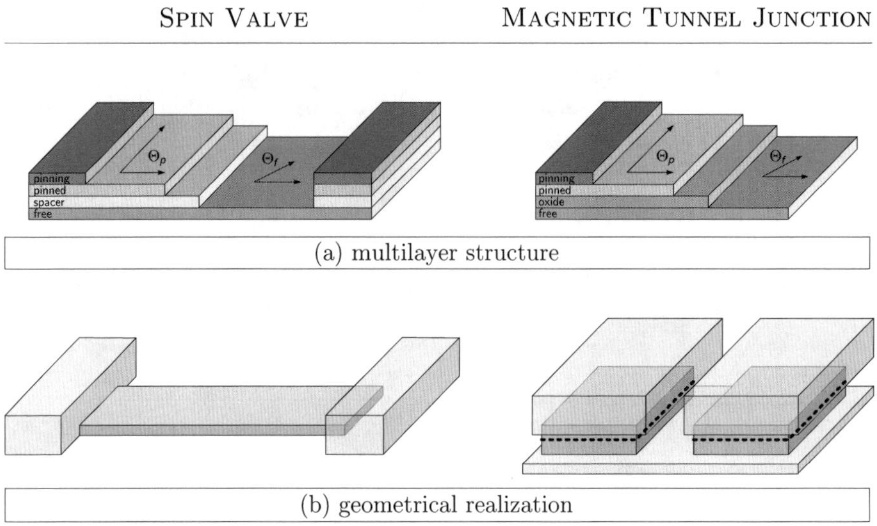

Fig. 1. Linear spin valve and magnetic tunnel junction structures

(CPP) of the layers through the insulating layer by tunneling ((see Fig. 1(b), right)). Each of these devices has the specific characteristics listed below [3].

$$\Delta V = \frac{1}{2} \cdot \mathrm{TMR} \cdot I \frac{R \cdot A}{Wh} \cos\left(\Theta_f - \Theta_p\right) \tag{2}$$

where TMR is the maximum magnetoresistance level, $R \cdot A$ is the resistance by area factor, I is the bias current, Θ_f and Θ_p are the angle of the magnetization angle of free and pinned layers, respectively, and W, h are the geometrical parameters.

Both family of structures have slightly different figures, that are summarized in Tab. 1.

Table 1. Typical features of the SV and MTJ

	R_0	Level (MR)	height (thickness)	saturation field	lineal range	device (area)	
SV	$100\,\Omega$-$50\,k\Omega$	4-20 %	< 50 nm	1-20 Oe	1-10 Oe	50-500 μm^2	~150°C
MTJ	$10\,\Omega$-$10\,k\Omega$	10-500 %	< 50 nm	1-10 Oe	1-5 Oe	20-50 μm^2	~300°C

Deposition

The deposition of these structures can be done with low temperature processes and then patterned by selective physical etching, so avoiding damages in the substrate. In this sense, the deposition of these structures can be accomplished by ion beam deposition (IBD) or by sputtering. In any case, the substrate temperature does not exceed 120°C. Thus, both processes can be directly masked with photoresist without damaging the substrate. In some cases, it is required a final heat treatment between 200°C and 300°C in order to increase the MR ratio [14].

Spin Valves (SV)
A typical SV structure is composed by: NiFe/CoFe/Cu/CoFe/MnIr, with layers' thickness in the order of few nm. Additional layers are commonly used for enhancing conductivity or improving the lithography. For SV, only one lithographic step is required for patterning the structures, and then another to design the contacts, at the ends of the SV strip, as shown in Fig. 2, left. Devices of $200\,\mu\text{m} \times 3\,\mu\text{m}$ give nominal resistances in the order of $1\,\text{k}\Omega$.

Magnetic Tunnel Junctions (MTJ)
Typical MTJ devices are based in structures like: NiFe/CoFe/MgO/CoFe/MnIr. For MTJ, due to their current-perpendicular-to-plane nature, two masks are required for defining an elemental devices. In a first step (see Fig. 2, right), a mesa structure is defined. Then, a second mask is applied in order to define the pillars comprising the active region. Usually, a conductive bottom layer is included in the multilayered structure in order to connect devices in pairs, and then facilitating the electrical contacts on the top of the structures.

2.2 Sensing Mechanism

As well known, the magnetic field induced by an electrical current, I, flowing through an infinite cylindrical conductor can be described by the Ampère Law:

$$\oint \mathbf{H} \cdot \mathbf{dl} = I$$

where $\mathbf{H}$ is the magnetic field intensity, I is the electrical current intensity and $\mathbf{dl}$ is an infinitesimal section of the integral loop. When the integral is solved for a given distance, $r > a$, vectors $\mathbf{H}$ and $\mathbf{dl}$ are parallels, so obtaining:

$$H(r) = \frac{I}{2\pi r}$$

By considering the magnetic field induced by a perfectly conducting metallic strip (Fig. 3), then we have to write:

$$\mathbf{H}(\mathbf{r}) = \frac{1}{4\pi} \int_S \frac{\mathbf{J}(\mathbf{r}') \times \mathbf{R}}{R^2} \mathbf{dr}'$$

where

$$\mathbf{R} = \mathbf{r} - \mathbf{r}'; R = |\mathbf{R}|$$

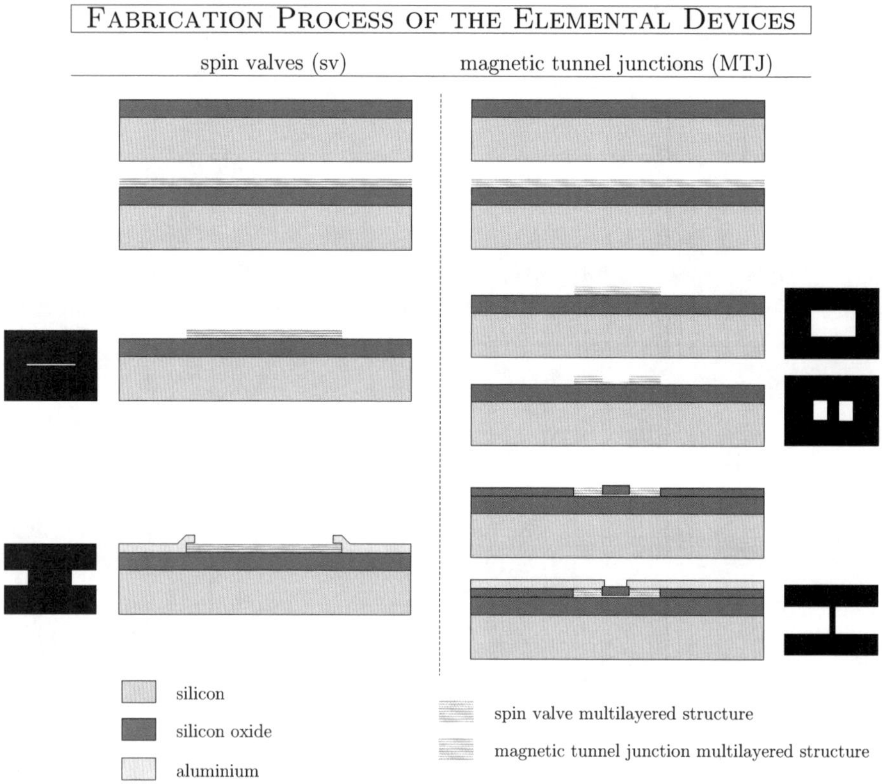

Fig. 2. Fabrication process details

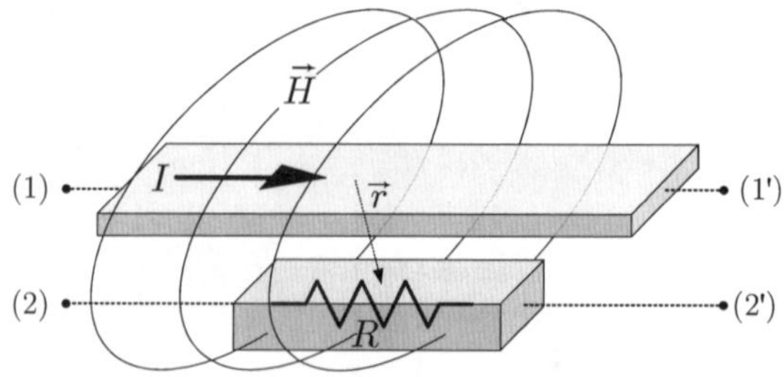

Fig. 3. Fundamentals concept and physics of the considered devices

being **r** the position vector where the field is calculated and **r**′ the 'local' integration vectorial variable.

This way, we got a relationship between the electrical current density, depending on the specific driving shape, and the generated magnetic field at any point in the space. This relationship is hugely useful. From an intuitive point of view, being the distance in the denominator, the magnetic field decreases with the distance. From a mathematical point of view, we have a very powerful tool for modeling purposes.

From an electrical point of view, each of every basic magnetoresistive elements can be understood as a current-dependent-resistance quadripole, as illustrated in Fig. 3.

Out of our experience, in most cases, the function $(f\,(I))$ linking the driven current (I) with the resistance (R) can be accurately expressed as a polynomial, as given below:

$$R = R_0 + \sum_{i=1}^{n} MR_i \cdot I^i \tag{3}$$

where I is the driven current applied between the input terminals (1) and (1'), R is the output resistance between terminals (2) and (2'), R_0 is the quiescent resistance (resistance when $I = 0$) and MR_i are the parameters describing the function. This general form can take into account moderate second order effects such as hysteresis, thermal drifts or unidentified nonlinearities. Even if high order polynomial can be discretionally used, when the range of application is limited, a relatively low order polynomial (3, for example) is often sufficient [15]. In addition, the most of these undesired effects are modulated by the linearizing properties of the Wheatstone bridge.

2.3 Sensing Configurations

Given a resistance based sensing element, a Wheatstone bridge setup is always a good recommendation as the starting step in the conditioning process (See Fig. 4). This case, we will get a differential output as a function of the resistance variation. Depending on the considered case or the particular requirements, we can make use of several bridge configurations.

One active element

As shown in Fig. 4(a), we have a four resistances bridge of identical nominal value, being only one of them active, *i.g.*, $R_2 = R_3 = R_4 = R$; $R_1 = R \pm \Delta R$. This case, the output voltage is given by:

$$V_{\mathrm{out}} = V_{CC} \frac{\frac{\Delta R}{R}}{2\left(2 + \frac{\Delta R}{R}\right)} \tag{4}$$

As easily observed, the relationship is not linear and this configuration is only used in particular cases where small variations of ΔR occur.

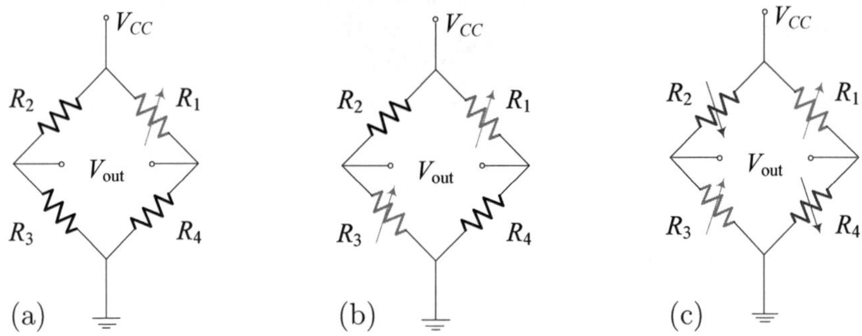

Fig. 4. Different resistive bridge configurations

Two active elements – Half bridge

This case, two opposite resistances in the bridge are forced to be sensitive: $R_2 = R_4 = R; R_1 = R_3 = R \pm \Delta R$ (See Fig. 4(b)). The output voltage is now:

$$V_{\text{out}} = V_{CC}\frac{\frac{\Delta R}{R}}{2 + \frac{\Delta R}{R}} \tag{5}$$

The output signal is again non-linear, but twice the previous value. For small resistance variations ($\Delta R/R << 1$), we have the following linear approximation:

$$V_{\text{out}} = V_{CC}\frac{\frac{\Delta R}{R}}{2} \tag{6}$$

Four active elements – Full bridge

When paired resistances are made active, we can get a full bridge configuration ($R_1 = R_3 = R \pm \Delta R; R_2 = R_4 = R \mp \Delta R$)(See Fig. 4(c)). The output voltage for this case is fully linear and given by:

$$V_o = V_{CC}\frac{\Delta R}{R} \tag{7}$$

This later arrangement is preferable when possible due to this linearity and the higher output signal.

Physical Realization

The election of any of the above mentioned configurations is linked to the specific application and the degree of freedom associated to the design.

If the path driving the current to be measured has been previously designed and it can not modified (for example, in pre-defined current strips in an integrated circuit), we can use a single element sensing scheme or a half bridge configuration as shown in Fig. 5(a). This approach can be improved with a two step definition of the sensing elements, making them oppositely sensitive in pairs, as described for medium current sensing in [16]. Because two deposition (and patterning) steps are required, so obtained resistors can present mismatches.

If we have access to the current paths design, geometrical arrangements like those presented in Fig. 5 (b) and (c) can be considered. In both, current flow from left to right below resistances R_1 and R_3 and from right to left below resistances R_2 and R_4, so giving the full Wheatstone bridge behavior. Such arrangements have been successfully used in medium [6] and low [17] current measurement.

We should note that current strips can be considered to both onto the sensing structures or below them.

A selection of successful realizations are shown in Fig. 6. In Fig. 6(a), a single SV with contact at its ends can be observed under a 100 μm wide current strip. In Fig. 6(b), numerous MTJ elements are connected in series in order to increase the overall resistance of the device. A bridge structure is shown in Fig. 6(c). In Fig. 6(d), contacted SV's are fabricated onto a standard CMOS chip.

3 Particular Issues in IC Current Monitoring

When used in real applications, being conceptually simple devices (they are dependent on magnetic field intensity) it has been shown that by using certain biasing, acquisition and conditioning circuits, the response of these sensors can be significantly improved. In this sense, it is recommended to feed these devices with current [18]. To improve the voltage offset and hysteresis, AC polarizations have also been proposed [19]. For the particular case of MTJ devices, the bias voltage must be below certain values [20]. Whether is a problem, we can use associations of elements in series and in parallel that, as an added value, increase the impedance of the same and reduce partially the noise [21]. Noise in GMR devices limits the detectivity and should also being considered. Although in high-frequency applications (hard disk read heads) only the thermal noise and shot noise (shot noise, in the case of the MTJ) are significant, in applications with sensors, the type of noise is predominantly $1/f$ [13]. This noise can be reduced partly by AC polarization [19] or by placing a diode in series with the GMR element [20]. The correction of thermal drift can be achieved by including a temperature-sensitive element in the compensation circuit [22]. Low-noise differential amplifiers are usually used for reading the output voltage. To minimize the noise in these devices we can also make use of new acquisition schemes, such as converting resistance to frequency, which can be implemented for both resistive elements [23] and bridges [24].

3.1 Low Signal Level Detection

IC level electric currents generate low magnetic field to be measured. The sensitivity of the sensors is conditioned by several parameters that must be considered.

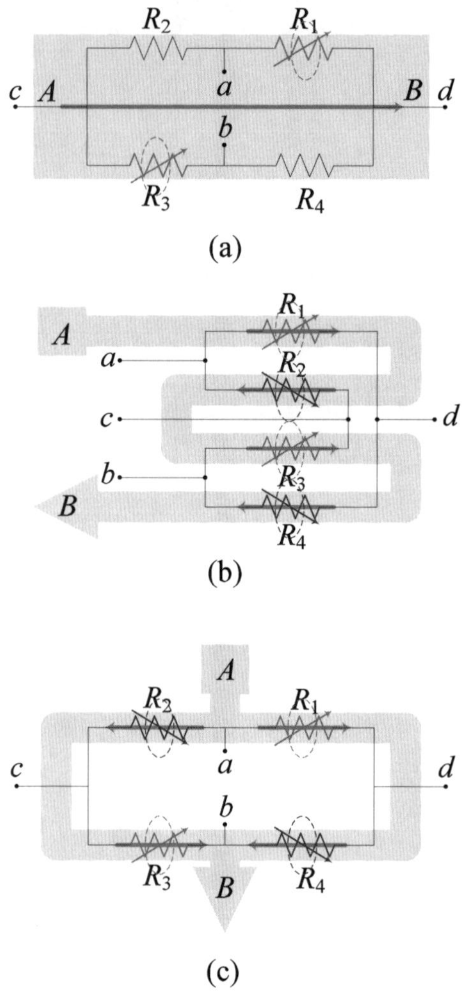

Fig. 5. Different current sensing configurations: (a) single current line, (b) S-type configuration, (c) P-type configuration

Intrinsic Sensitivity of GMR Structures

Initial GMR structures displayed MR ratios of about 4-8% [25, 26, 27]. Then, Al_2O_3 isolation based MTJ were developed with TMR levels around 100% [13]. More recently, MgO isolation based MTJ have been reported with TMR levels up to 500% [28].

For specific IC level current measurement, spin valves [29, 30, 17] and Al_2O_3-MTJ [31, 32, 33] have been successfully used. MgO-MTJ devices are currently under development.

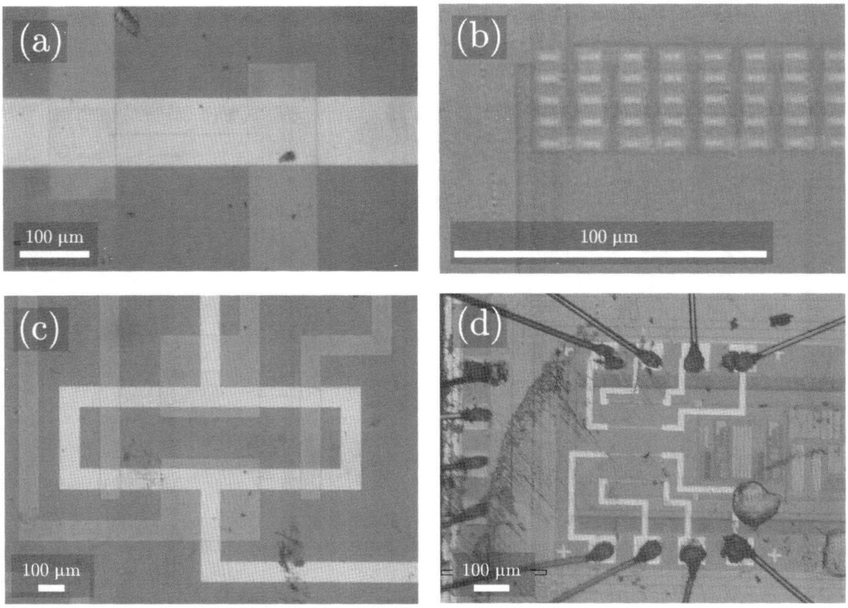

Fig. 6. Different devices

Geometric Arrangement

As described in Sec. 2, the magnetic field at any point of the space is given by the integral of the current, depending on the distance. When possible, the current path should be placed very close to the sensing elements. In addition, the current strips width must be as narrow as the Joule heating allows, in order to increase the magnetic field density. In [17] a systematic study of low current sensors performance as a function of the current strips width and their geometrical arrangement can be found.

When dealing with standard CMOS devices, the currents to be measured are recommendable to flow through higher metals current strips, in order to enhance the magnetic field density. Finite element method (FEM) simulations can be made in order to analyze different configurations [34].

The use of magnetic field flux guide concentrators has also been proposed in the literature for GMR sensitivity enhancement [13]. For IC current applications this approach is not recommendable.

Noise

In addition to the inherent thermal noise, GMR structures also display $1/f$ noise and shot noise (only MTJ). Although the TMR in MTJs is much higher than the MR in SVs, the intrinsic noise of a MTJ is also higher than the one of an SV. This way, the signal to noise ratio (SNR) is the key parameter that has to be considered.

For typical configurations of SVs and MTJs, it can be demonstrated ([13]) that:

$$\frac{\mathrm{SNR_{MTJ}}}{\mathrm{SNR_{SV}}} \cong 0.29 \frac{\mathrm{TMR}}{\mathrm{MR}} \tag{8}$$

this way, a MTJ will display better performance that a SV one only if his magnetoresistance level is, at least, more than three times higher.

3.2 Bandwidth

There are not intrinsic frequency limitations for GMR structures, so achieving theoretical bandwidths in the order of 1 GHz [25]. For typical linear applications, such a response is not restrictive. In real systems, frequency limitations arise from parasitic effects associated to the involved circuitry.

Regarding interconnections, due to the inductive character of the coupling, a 'zero' behavior in the transfer function has been reported for medium current sensors [6]. In integrated circuits, due to the closeness between the current straps and the contacting strips, a capacitive coupling (a 'pole' in the transfer function) is often found. For example, in Fig. 7, the frequency behavior of bridge based IC current sensors is shown. Nomenclature states 'N' for narrow current straps (10 μm) and 'W' for wide current straps. 'S' is for series and 'P' for parallel configurations, as defined in Fig. 5. As observed wider overlapping due to wider current straps produce major frequency limitations. These effects should can be modeled by means of FEM (or similar) based codes and can be included in a more complete SPICE model, by means of parasitic inductors or capacitors in series or in parallel with the intrinsic sensor resistances.

On the other hand, and regarding an electrical current sensor, a current strap drives the current flow close to the sensor. When high frequencies are involved,

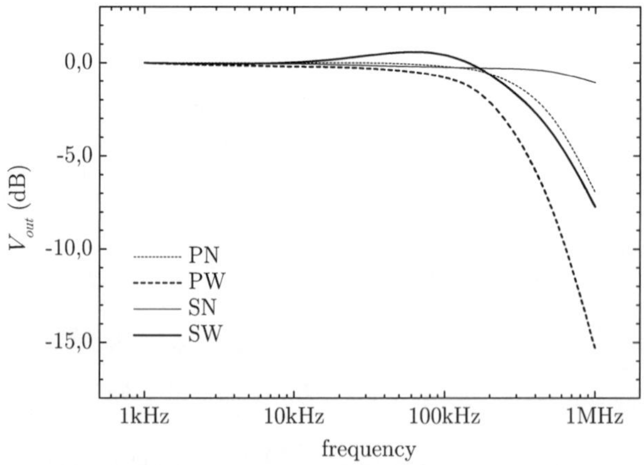

Fig. 7. Frequency limitations of different sensing configurations

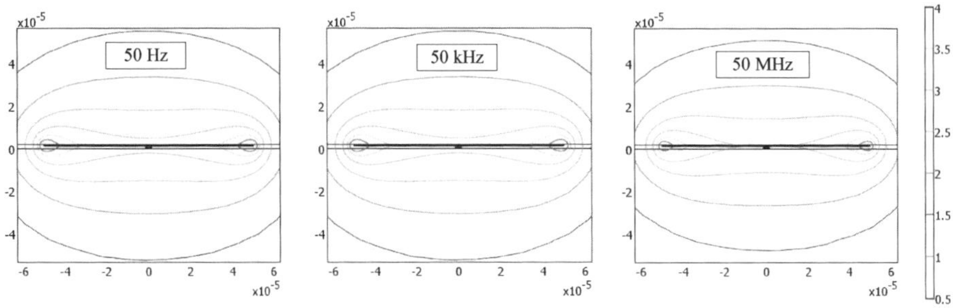

Fig. 8. Electrical current frequency dependence modeling: 'skin effect'

the current flow redistributes toward the more external edges of the strip, so modifying the magnetic field distribution and, in consequence, the magnetic field at the point where the magnetic sensor is placed. A frequency dependence response appears, and further studies must to be considered. In Fig. 8 a FEM modeling of the magnetic field in the cross section of a typical strap driven current is shown. At about 1 MHz, the current flow concentrates on the edges and the magnetic field at the hypothetical sensor position decreases [35].

3.3 Joule Heating

When integrated circuitry is involved, even low currents can produce Joule heating due to the small cross section of the current strips. GMR structures, lying near these current strips are also affected by so produced thermal drifts.

GMR sensors resistance, being common resistances, is a function of the temperature. For GMR based devices, and in the usual range of utilization, this dependence can be considered as linear, and can be defined by a temperature coefficient (TEMPCO) as following:

$$TCR(\%) = 100 \times \frac{1}{R_{T_0}} \frac{\Delta R}{\Delta T} \tag{9}$$

An analogue relationship can be defined for the thermal dependence of sensitivity, as:

$$TCS(\%) = 100 \times \frac{1}{S_{T_0}} \frac{\Delta S}{\Delta T} \tag{10}$$

When a full bridge configuration is considered, this thermal dependence is partially compensated and is expected to be low. Due to the inherent voltage offset of sensors configured as bridges, the temperature drift of the offset voltage must be specified:

$$TCV_{off}(\%) = 100 \times \frac{\frac{\Delta V_{off}}{\Delta T}}{V_{off,T_0}} \tag{11}$$

Moreover, the output voltage has also a thermal dependence, defined as:

$$\begin{aligned} TCV_o(\%) &= 100 \times \frac{1}{\Delta T}\frac{V_{o,T_i} - V_{o,T_0}}{V_{o,T_0}} \\ V_{o,T_i} &= V_{out,T_i} - V_{off,T_i} \end{aligned} \tag{12}$$

The TEMPCOs have been reported for SV based full Wheatstone bridge sensors [17], and the values are summarized in Tab. 2.

Table 2. Temperature coefficients for SV current sensors [17]

$TCR(\%/°C)$	$TCS(\%/°C)$	$TCV_{\text{off}}(\mu V/°C)$
$+0.11 \pm 0.01$	-0.14 ± 0.02	$+5 \pm 2$

As stated, even though several heat sources can be found in a electronic system, self heating is the most common one in ICs. In this case, their basis is to detect the magnetic field produced by a close current strap. The closer the strap is, the higher is the sensitivity ... and the heating produced by the Joule effect on the strap. An example of it is shown in Fig. 9 ([17]). By considering a Wheatstone bridge SV based current sensor with current straps of 10 μm and 100 μm width, a 50 mA step DC-current was set at a given time and switched off some time later, so avoiding an excessive overheating. Two kinds of results are shown: the input resistance and the output voltage.

It should be mentioned that the self-heating on GMR current sensing devices has been recently analyzed and the related effects have been included in a behavioral compact electrical model [36].

When necessary, various methods of temperature compensation have been reported in the literature addressed to reduce the thermal drift output of Wheatstone bridge type sensors. These methods can be differentiated as noninvasive [37] and invasive [22]. As noninvasive we mean technique consisting of the addition of different circuit elements in series or parallel to the bridge in order to change the bridge supply voltage due to the temperature variation, which produces a valid compensation. On the other hand, a Wheatstone bridge can also be temperature compensate by means of the modification of its original configuration. In this case, we should ensure that the terminals of the bridge are externally accessible, not always possible in IC schemes.

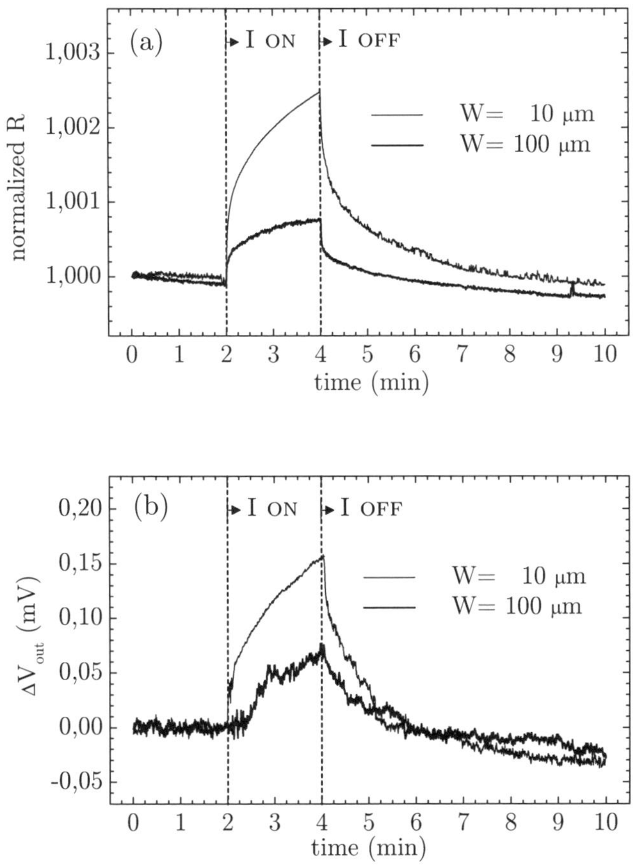

Fig. 9. Thermal drift due to moderate currents [17]

4 Current Measurement at the IC Level

The electrical current has been traditionally measured by means of shunt resistances, coils and solid state sensors. We will focus on the third option, where the magnetic field generated by a current flow is detected by a solid state GMR magnetic sensor. This general scheme can be applied to the measurement of a current driven by a wire or by a conductive strap in a printed circuit board or an integrated circuit. Both AC/DC currents can be measured in this way with small, cheap and contact-less systems. When dealing with GMR sensors, excellent sensitivities are achieved.

Regarding microelectronics, traditionally, currents flowing in integrated circuits (for example, I_{DDQ}) have been measured by the use of built-in current sensors (BICS) [7], with different sensing components: a resistor, a current mirror, a switched capacitor, ... Due to the inherent in-line characteristics of these approaches, some drawbacks appear such as the drop of the voltage or the increase of the silicon area. In this case, the GMR technology is expected to play a very important role.

4.1 Background on Medium Current Applications

GMR sensors have successfully applied to the measurement of electric current in different applications. For example, a specific spin valve sensor for industrial applications is described in [38]. The sensor is soldered onto a PCB current track and encapsulated chip-on-board. A full bridge (active in pairs, two-steps deposition) with crossed axis configuration was utilized. The sensor displayed a linear range up to 10 A. Then, a novel design principle is presented in [6]. The principle of operation is similar as this described in Fig. 5 in this chapter.

In addition, GMR sensors have also used for the measurement of differential currents, as presented in [39]. In this case, a commercial GMR sensor (AC004-01, from NVE) is placed into two Helmholtz coils, carrying the currents to be compared. When both currents are identical, the magnetic field in the middle point of the coils is zero, and so the output voltage of the sensor. The system was tested in a house-hold application, demonstrating to be useful for detecting differential currents below 30 mA.

Other applications such as the use of GMR based electrical sensors in switching regulators [16] or for monitoring the charging/discharging process in batteries [40] can also found in the literature.

The use of GMR based electric current sensors for the measurement of the electric power is particularly interesting. By using of an appropriate analog electronic multiplier, two signals proportional to the voltage and current, respectively, can be multiplied in real-time, as suggested in Fig. 10(a). Thus, the output of such a transducer is the instantaneous power of the signal defined as:

$$P(t) = i(t) \cdot v(t) \tag{13}$$

In this sense, GMR based electric sensors can be used as basic multipliers. The basic idea of using a MR element as an analog multiplier is very simple: the Wheatstone bridge of the MR sensor is supplied by a signal, which is proportional to the voltage of the measured signal. At the same time, current proportional to the current of the measured signal is led through a coil, generating magnetic field which is the Wheatstone bridge exposed to. The output (diagonal) voltage of the bridge is (linearly) dependent on the acting magnetic field, and at the same time, it is linearly dependent on the supplying voltage. As a direct consequence of these two facts, the output is dependent on the multiple of the two signals. A graphical scheme is presented in Fig. 10(b). The idea has been succesfully applied by using a KMZ51 AMR based commercial sensor [41].

4.2 Electric Current Monitoring at the IC Level

From the experience previously described, GMR based sensors have been applied to the measurement of electrical currents at the IC level.

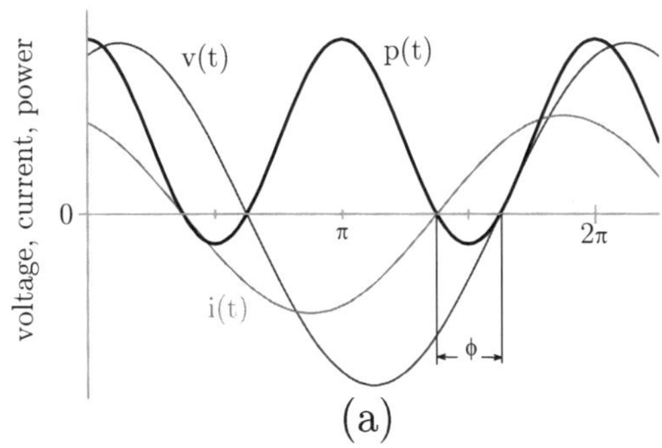

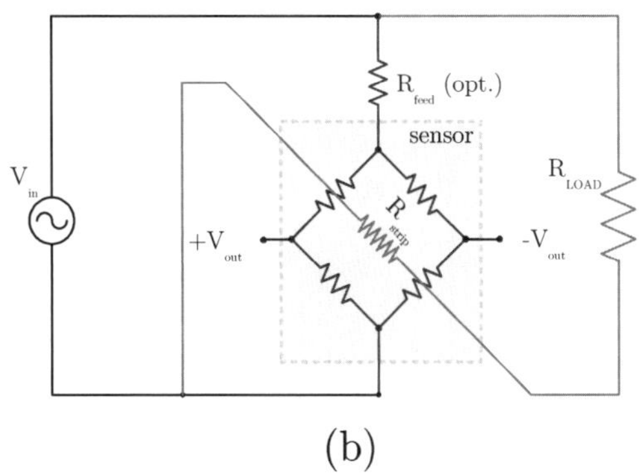

Fig. 10. Instantaneous power measurement by means of the GMR multiplying property: (a) concept and definitions, (b) practical realization for IC power monitoring purposes

The applicability of spin-valve structures to the measurement of low electric currents was firstly demonstrated in [29]. In this works, the concept was introduced and some fabrication parameters were established. The work of Pannetier-Lecouer and co-workers, also focused on this idea [30]. In parallel, Le Phan and colleagues [31,32] and then C. Reig and co-workers [33], demonstrated de possibility of using MTJ sensors for this purpose

In [17], the potentiality of SV based full Wheatstone bridges for low current monitoring at the IC level was demonstrated. A number of prototypes were specifically designed, fabricated and tested. The current lines were incorporated in the chip during the microfabrication process which reduces the separation to the sensing elements, leading to improved sensitivity. The characteristics and geometry of these current paths were considered as basic design parameters. Current ranges from $10\,\mu$A to $100\,$mA were covered with these sensors with excellent linearity and sensitivities above $1\,$mV/(VmA). Second order effects were also analyzed.

4.3 Power Consumption in ICs

By following the technological scheme exposed in a previous paragraph, we can translate the multiplier characteristic of GMR devices introduced previously in order to implement micro wattmeters suitable to be integrated jointly with the CMOS circuitry. As the power source (AC or DC), any active portion of the IC can be considered. As the load, any fed portion of the IC can be theoretically considered (R-L-C).

With the previous equations, the mathematical analysis of the circuit can be made. As a drawback, a not galvanically isolated approach needs to be considered. As an advantage, there is no need of an additional transformer. Then, the functionality of the proposal can be extended from DC to the bandwidth of the device. Nevertheless, a series resistance need to be added for matching impedances and limiting the sensor power consumption.

In Fig. 10(b), V_S is considered a power supply, but can be substituted by any active branch of the IC being V_S its equivalent Thévenin voltage without losing generality. R_L is an arbitrary resistive load that also can be substituted by any passive branch of the IC, being R_L its equivalent impedance. R_f is used for biasing the Wheatstone bridge and controlling the bias current.

If we consider an offset free full Wheatstone bridge we can write:

$$V_o = \Delta R \cdot i_b = MR \cdot I_L \cdot i_b \tag{14}$$

where I_L is the load or driven current, equal to the current undergoing through the sensor.

If we considere R_b as the sensor input impedance (from nodes a and b), we can assume:

$$i_b = \frac{V_S}{R_f + R_b}$$

and then

$$V_L = \frac{R_L}{R_L + r_s} V_S$$

We can, then, rewrite Eq. 14 as:

$$V_o = \frac{MR}{R_L} \frac{R_L + r_s}{R_f + R_b} (V_L \cdot I_L) \tag{15}$$

As assumed, $P_L = V_L \cdot I_L$.

By accepting $r_s \ll R_L$, we get a linear relationship between the power in the load and the output voltage of the sensor:

$$\frac{\mathrm{d}V_o}{\mathrm{d}P_L} = \frac{MR}{R_f + R_b} \tag{16}$$

so finally demonstrating the multiplying capability of the scheme and its property of giving an output voltage that is proportional to the power delivered to the load. A more detailed analysis, also considering the power delivered at the sensor and other second order effects can be found in [42].

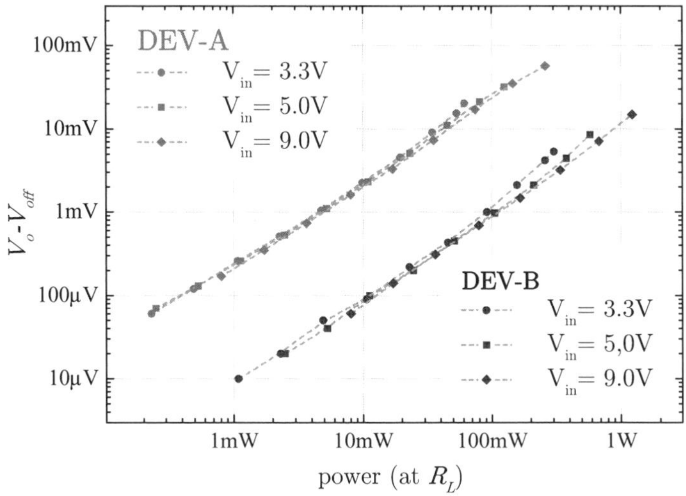

Fig. 11. Power measurement experimental results. DEV-A is a narrow-current-path (10 μm) full-Wheatstone-bridge spin-valve based sensor. DEV-B is a wide-current-path (100 μm) full-Wheatstone-bridge spin-valve based sensor.

Finally, in order to demonstrate the potentiality of the circuit for DC power measurement at the IC level, the voltage output (offset corrected) versus the power at the load (R_L) is plotted in Fig. 11.

In Fig. 11, DEV-A is a narrow-current-path full-Wheatstone-bridge, as previously described, and DEV-B is wide-current-path one. Due to its higher sensitivity, a higher output signal come from DEV-A. A power range from 100 μW to 1 W can be covered by combining both types of devices. Within this range, the device response is close to linearity. Deviations from linearity at higher powers are due to the effect of r_s. In any case, the driven currents are maintained within the ranges where self-heating effects are unappreciated. Slow deviations at low powers are due to the effect of the small voltage offset.

4.4 Analogue Isolators

Signal isolator devices are widely used in many electronics systems. The more commonly used isolators are optical isolators (optocouplers) and capacitive or inductive couplers (purely transformers). Some common disadvantages of these devices are that they are often limited in linearity and frequency performance; they need a notable power consumption and they display a considerable size and usually require hybrid packaging, a real handicap for integrated circuit fabrication.

In order to overcome these difficulties, the possibility of using GMR based full bridges in order to design analog magnetically coupled isolators was analyzed in [33]. To demonstrate the capability of GMR devices for acting as analogue

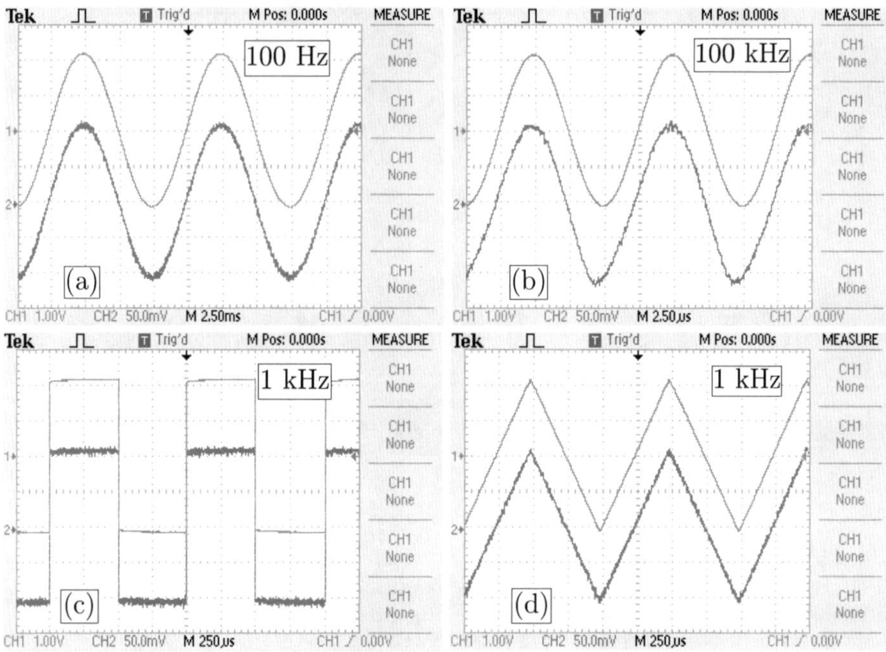

Fig. 12. Analogue isolation capability of GMR devices with different input current signals (current in orange, output voltage in green): (a) low frequency, (b) high frequency, (c) square wave, (d) triangular wave

isolators, different electrical signals were applied through the input terminals of specifically full bridge designed compact prototypes, with the help of a signal generator. Selected results are shown in Fig. 12.

4.5 Current-to-Time (I-to-t) Converters

Analogue voltage is the common magnitude for obtaining the output signal in a system with sensors in general and GMR devices in particular, as it is extracted from the bibliography. Nevertheless, when a digital interfacing is required, the conversion of the output voltage to a time dependent signal is, in some applications, highly recommended. Some people call them quasi-digital sensors. The resistance can be converted to time dependent signals by following different approaches. They are detailed in a specific chapter within this book.

Focusing on GMR based electric current sensors, they can be understood as current-dependent-resistances. This way, reported R-to-t converters can be easily transformed in I-to-t converters. In the next paragraphs, successfully designed I-to-t converter are described and analyzed.

I-to-*t* Converters for Single Resistances

Such converters are based on traditional oscillators, where the resistance of the *RC* constant is substituted by a current dependent sensing variable resistor. A typical implementation, based on the circuit presented in [43], is detailed in Fig. 13.

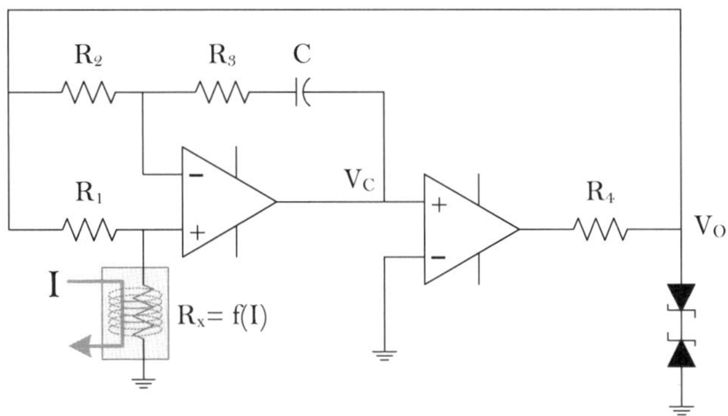

Fig. 13. *I*-to-*f* converter based on the *R*-to-*f* converter as described in [43]

The oscillator comprises a bridge amplifier and a comparator. R_1, R_2 and R_x, and the pair R_3-C are the arms of the bridge. If we consider ideal operational amplifiers, it can be theoretically demonstrated that oscillation period of the circuit is given by:

$$T = \frac{(1+\mu)^2 (R_2 \cdot R_x - R_1 \cdot R_3) C}{\mu R_1} \tag{17}$$

where $\mu = V_2/V_1$, being V_2 and V_1 the maximum and minimum output voltages, given by the breakdown voltage of the Zener diodes. For proper functionality, R_x must be grater than $(R_1 R_3)/R_2$. It can also being demonstrated that:

$$\delta T = \frac{(1+\mu)^2 R_2 C \delta R_x}{\mu R_1} \tag{18}$$

Last equation gives the sensitivity as a function of the resistance. In order to have it as a function of the current, the specific dependence of the considered R_x against the driven current should also being taken into account. By following eq. 18, the sensitivity can be adjusted with R_1 and R_2, and the period with R_3.

This scheme has been tested with real single spin-valve current sensors, and the results are displayed in Fig. 14, for three different values of the driven current.

In this Fig. 14, the channel #1 in the oscilloscope is displaying the signal V_c and the channel #2 is displaying the signal V_o, as defined in Fig. 13. Specific values for the considered components were: $R_1 = 1\,\text{k}\Omega$, $R_2 = 22\,\text{k}\Omega$, $R_3 = 12\,\text{k}\Omega$, $R_x = 1.1\,\text{k}\Omega$ (for $I = 0\,\text{mA}$), $R_4 = 680\,\Omega$ and $C = 0.33\,\mu\text{F}$. For this value, it can being obtained that the sensitivity is $40\,\mu\text{s/mA}$ or $0.25\%/\text{mA}$ in period.

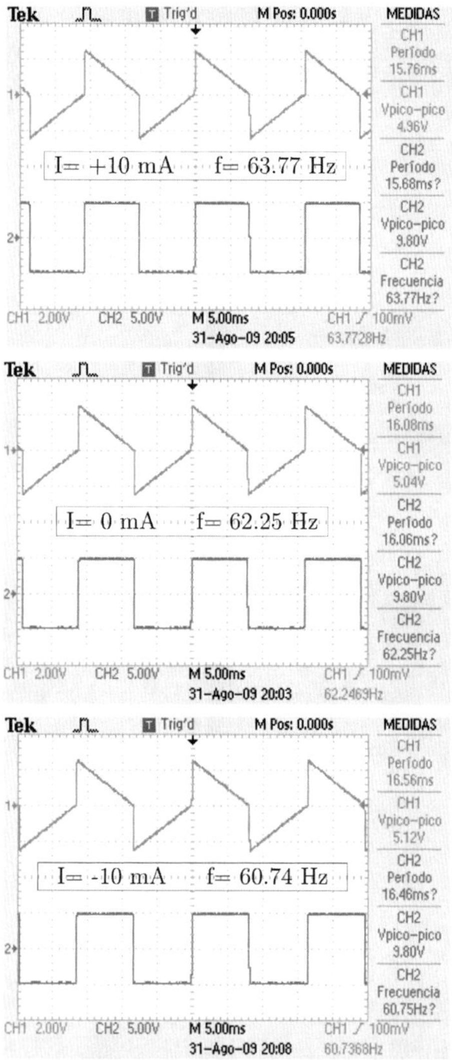

Fig. 14. Voltage waveforms for the I-to-t converter described in Fig. 13, for different driven current values

I-to-t Converters for Resistive Bridges

Similar schemes can be also implemented for bridge configurations, often found in this kind of applications. A typical circuit is shown in Fig. 15, introduced in [44].

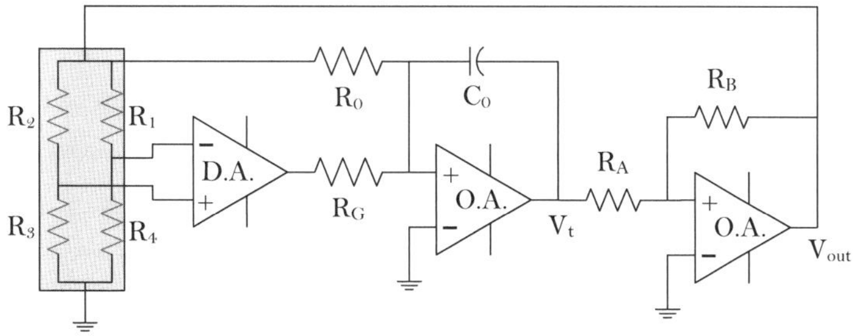

Fig. 15. V-to-f converter based in the circuit proposed in [44]

In this case, a differential amplifier (D.A.) is required for taking the output voltage from the bridge. With this configuration, it can be demonstrated that [44]:

$$f = \frac{1}{4R_0C_0}\frac{R_B}{R_A}\left[1 + \left(\frac{R_4}{R_4+R_1} - \frac{R_3}{R_3+R_2}\right)G\frac{R_0}{R_G}\right] \tag{19}$$

being G the gain of the D.A. The development of quasi-digital interfaces for IC current sensing based in this scheme is currently under evaluation.

5 Monolithic Integration: Trends and Challenges

The monolithic integration compatibility with standard CMOS processes is a capital issue in order to definitively consider GMR devices as useful design elements in real applications. In this sense, NonVolatile Electronics (NVE, [18]) has already demonstrated this fact on chips based on a 1.5 microns BiCMOS, in a semi dedicated process in which chips taken at an intermediate step of the process [45]. More recently, it has been proposed the more convenient option of using chips from generic CMOS processes. Thus, S. Han *et al.* [46] used chips made by 0.25 microns NSC BiCMOS technology, on applying a post process that uses reactive ion etching for the attack of the passivation. In Tab. 3, we show a selection of available technologies from Europractice [47] and others, that can be be considered for these purposes.

Table 3. Comparison of characteristics of different CMOS technologies available

	CNM25	C07M-A	C035U-A	C35B4C3
Foundry	CNM	On Semi	On Semi	AMS
Thecnology	2.5 um	0.7 um	0.35 um	0.35 um
Core-I/O voltages	5 V	5 V	3.3 V	5 V
Metalics layers	2	2	4	5
Cost: /mm^{2} $^{(2012)}$	TBD	360	670	580

As suggested, the compatibility of post-process actions on un-packaged diced electronics is the key point for the success of the device. Major problems in this process can be grouped in this way:

Temperature of post-process steps:
Microelectronics processes (oxidation, dopant diffusion, implantation, metal deposition, ...) are carried out at given temperatures, usually quite high (up to more than 1000°C, [48]). Post processes have to be selected in order to keep the temperature below some limits in order to not damage previously done processes. Fortunately, post-processes are typically depositions, etchings and lithography, operated at low temperatures. When annealing processes for enhancing GMR properties are required [14], the temperature must to be taken into account.

Samples handling
Usually, post-processes are not carried out at the sam facilities where the dices have been fabricated. When we are considering a dedicated run, so obtained wafers can be, with minor problems, directly used. If we deal with individual dies, these must be mounted in specifically fabricated holders, in cavities grooved with a precision milling machine. This allows the chip manipulation and loading in most of the processing machines.

Mask alignment
Because of the different machinery for the pre- and post-processes, it's possible to have some problems to solve regarding the masks alignment. Software can be different, the resolution of the masks aligners, ... So, additional alignment marks and references are commonly used.

In the next paragraphs we report on cases of success of GMR based devices fabricated onto previously processed dice, by highlighting on their inherent drawbacks, and demonstrating their potentiality.

5.1 CMOS 0.35 μm

In this case we worked on a processed unpackaged standard 0.35 μm dice from Europractice (AMS 0.35 μm 2P3M, Austria MicroSystems). The consortium Europractice [47] offers different CMOS technologies to develop fundamental and applied research, at European level, and associated countries. The die was mounted on a specifically hole arranged on a standard wafer in order to facilitate the postprocess.

Then, spin-valve structures were directly deposited and patterned (see Fig. 16(a)). In this figure, the underlaying electronics and the additional alignment marks (crosses) can be observed. A spin-valve with the associated metallic contacts are highlighted for clarity.

For checking the functionality of so deposited devices, the resistance against the external magnetic field was measured with the help of an automated system. The results are shown in Fig. 17. As observed, a linear response is displayed.

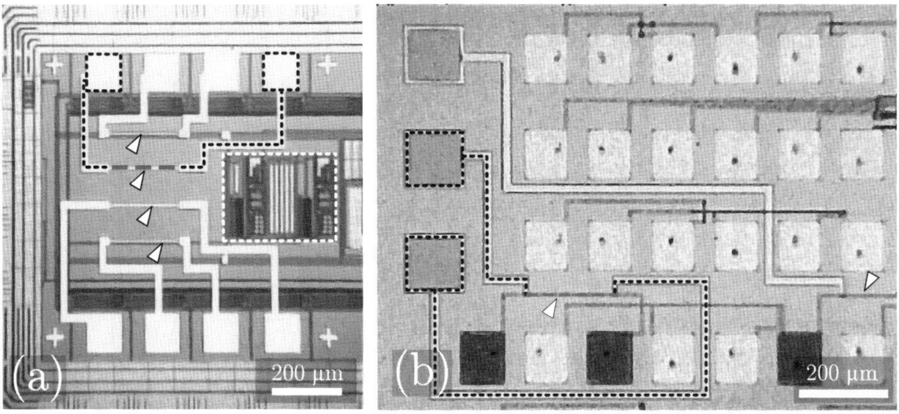

Fig. 16. Fabricated GMR devices on (a) 0.35 μm CMOS and (b) 2.5 μm CMOS. Contacts and spin-valve structures are highlighted.

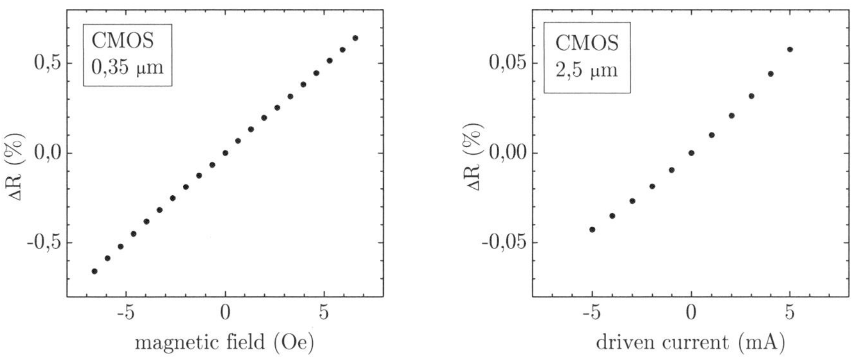

Fig. 17. Experimental results demonstrating the performance: (a) *versus* magnetic field, (b) *versus* driven current

5.2 CMOS 2.5 μm

CNM25 is a 2.5 μm technology developed at the *Institut de Microelectrònica de Barcelona* within the *Centre Nacional de Microelectrònica (CNM-IMB)*, with 2P and 2M layers onto 100 mm wafers of (110) epi-P-silicon. This is not a commercial process, so some variations can be accepted [49].

We have used a processed wafer in which some test structures (including current lines) were included. SV devices were then microfabricated (see Fig. 16(b)), by using the process previously described, on top of this current lines, for a local measurement of the current.

Then, the resistance against the driving current was measured. The results are shown in Fig. 17. As observed, a good sensing behavior is displayed.

6 Conclusions

GMR based sensors have been proven to be a real alternative for monitoring electric current at the IC level. This capability has been demonstrated through the realization of useful devices including single resistors and Wheatstone bridges, based on SV and MTJ technologies. The level of sensitivity, linearity, low hysteresis and bandwidth allow the consideration of multiple applications, behind the direct current measurement.

From a practical point of view, a straightforward application is their use as analogue isolators. As an interesting proposal, the instantaneous consumed power of integrated devices can also be measured by using the product feature of GMR devices. Delivered power below 1 mW can be measured in this way. In addition, quasi-digital interfaces can be designed based on standard resistance-to-time converters.

Finally, their compatibility with standard CMOS technology, both with dedicated processes and with general purposes technologies allocate GMR devices in an excellent position as basic tools for near future micro and nanoelectronics.

Acknowledgments. At the personal level, we should give thanks to E. Figueras, J. Madrenas and A. Yúfera for their kindness regarding standard IC's. Also thanks to A. Roldán and J. B. Roldán for their help in developing electrical models. The authors are permanently grateful for the very fruitful collaborations with the INESC-MN. Part of the work has been carried out under projects: HP2003/0123 (*Ministry of Science and Technology, Spain*), GV05/150 (*Valencian Regional Government*), ENE2008-06588-C04-04 (*Ministry of Science and Innovation, Spain and European Regional Development Fund*), UV-INV-AE11-40892 (*Universitat de València*) and NGG-229 (2010).

References

[1] Ziegler, S., Woodward, R.C., Iu, H.H.-C., Borle, L.J.: Current Sensing Techniques: A Review. IEEE Sensors Journal 9(4), 354–376 (2009)
[2] Ripka, P.: Electric current sensors: a review. Measurement Science & Technology 21(11), 1–23 (2010)
[3] Reig, C., Cubells-Beltrán, M.-D., Ramírez Muñoz, D.: Magnetic Field Sensors Based on Giant Magnetoresistance (GMR) Technology: Applications in Electrical Current Sensing. Sensors 9(10), 7919–7942 (2009)
[4] Lenz, J.E.: A review of magnetic sensors. Proceedings of the IEEE 78(6), 973–989 (1990)
[5] Chomsuwan, K., Yamada, S., Iwahara, M.: Bare PCB inspection system with SV-GMR sensor eddy-current testing probe. IEEE Sensors Journal 7(5-6), 890–896 (2007)
[6] Reig, C., Ramírez, D., Silva, F., Bernardo, J., Freitas, P.: Design, fabrication, and analysis of a spin-valve based current sensor. Sensors and Actuators A-Physical 115(2-3), 259–266 (2004)

[7] Ferré, A., Isern, E., Rius, J., Rodriguez-Montañés, R., Figueras, J.: I_{DDQ} testing: state of the art and future trends. Integration, the VLSI Journal 26, 167–196 (1998)
[8] Vazquez, J.R., Pineda de Gyvez, J.: Built-In Current Sensor for ΔI_{DDQ} Testing. Solid-State Circuits, IEEE Journal of 39(3), 511–518 (2004)
[9] Baibich, M.N., Broto, J.M., Fert, A., Vandau, F.N., Petroff, F., Eitenne, P., Creuzet, G., Friederich, A., Chazelas, J.: Giant magnetoresistance of (001)Fe/(001)Cr magnetic superlattices. Physical Review Letters 61(21), 2472–2475 (1988)
[10] Binasch, G., Grunberg, P., Saurenbach, F., Zinn, W.: Enhanced magnetoresistance in layered magnetic-structures with antiferromagnetic interlayer exchange. Physical Review B 39(7), 4828–4830 (1989)
[11] Berkowitz, A.E., Mitchell, J.R., Carey, M.J., Young, A.P., Zhang, S., Spada, F.E., Parker, F.T., Hutten, A., Thomas, G.: Giant magnetoresistance in heterogeneous Cu-Co alloys. Physical Review Letters 68(25), 3745–3748 (1992)
[12] Arana, S., Arana, N., Gracia, R., Castaño, E.: High sensitivity linear position sensor developed using granular Ag-Co giant magnetoresistances. Sensors and Actuators A – Physical, 116–121 (2005)
[13] Freitas, P.P., Ferreira, R., Cardoso, S., Cardoso, F.: Magnetoresistive sensors. Journal of Physics-Condensed Matter 19(16), 21 (2007)
[14] Vadala, M., Zhernenkov, K., Wolff, M., Toperverg, B.P., Westerholt, K., Zabel, H., Wisniowski, P., Cardoso, S., Freitas, P.P.: Structural characterization and magnetic profile of annealed CoFeB/MgO multilayers. Journal of Applied Physics 105(11) (2009)
[15] Roldán, A., Reig, C., Roldán, J.B., Cano-Abellán, A., Ramírez, D.: Quasi-static electrical model for magnetoresistive current sensors. In: International Conference on Renewable Energies and Power Quality (ICREPQ 2010), Granada (2010)
[16] Pelegrí, J., Ramírez, D., Freitas, P.P.: Spin-valve current sensor for industrial applications. Sensors and Actuators A-Physical 105(2), 132–136 (2003)
[17] Cubells-Beltrán, M.D., Reig, C., Ramírez, D., Cardoso, S., Freitas, P.P.: Full Wheatstone bridge spin-valve based sensors for IC currents monitoring. IEEE Sensors Journal 9(12), 1756–1762 (2009)
[18] NVE CORPORATION. AA and AB-Series Analog Sensors (Datasheet) (2003)
[19] Vopalensky, M., Ripka, P., Kubik, J., Tondra, M.: Alternating biasing of SDT sensors. Sensors and Actuators A-Physical 110(1-3), 182–186 (2004)
[20] Cardoso, F.A., Ferreira, R., Cardoso, S., Conde, J.P., Chu, V., Freitas, P.P., Germano, J., Almeida, T., Sousa, L., Piedade, M.S.: Noise characteristics and particle detection limits in diode plus MTJ matrix elements for biochip applications. IEEE Transactions of Magnetics 43(6), 2403–2405 (2007)
[21] Guerrero, R., Pannetier-Lecoeur, M., Fermon, C., Cardoso, S., Ferreira, R., Freitas, P.P.: Low frequency noise in arrays of magnetic tunnel junctions connected in series and parallel. Journal of Applied Physics 105(11) (2009)
[22] Muñoz, D.R., Moreno, J.S., Berga, S.C., Montero, E.C., Escrivà, C.R., Anton, A.E.N.: Temperature compensation of Wheatstone bridge magnetoresistive sensors based on generalized impedance converter with input reference current. Review of Scientific Instruments 77(10), 6 (2006)
[23] Kim, H., Chung, W.-S., Kim, H.-J., Son, S.-H.: A Resistance Deviation-to-Pulsewidth Converter for Resistive Sensors. IEEE Transactions on Instrumentation and Measurement 58(2), 397–400 (2009)
[24] Chung, W.-S., An, M.-Y., Son, S.-H.: A simple resistance-to-time converter for resistive bridge sensors. IEICE Electronics Express 5(9), 310–315 (2008)

[25] Ziese, M., Thornton, M.J. (eds.): Spin Electronics. Lecture Notes in Physics. Springer, Berlin (2001)
[26] Hartmann, U. (ed.): Magnetic Multilayers and Giant Magnetoresistance: Fundamentals and Industrial Applications. Surface Sciences. Springer, Berlin (1999)
[27] Hirota, E., Sakakima, H., Inomata, K.: Giant Magnetoresistance Devices. Surface Sciences. Springer, Berlin (2002)
[28] Freitas, P.P., Cardoso, S., Ferreira, R., Martins, V.C., Guedes, A., Cardoso, F.A., Loureiro, J., Macedo, R., Chaves, R.C., Amaral, J.: Optimization and integration of magnetoresistive sensors. SPIN 1(1), 71–91 (2011)
[29] Reig, C., Ramírez, D., Li, H.H., Freitas, P.P.: Low-current sensing with specular spin valve structures. IEE Proceedings-Circuits Devices and Systems 152(4), 307–311 (2005)
[30] Pannetier-Lecoeur, M., Fermon, C., de Vismes, A., Kerr, E., Vieux-Rochaz, L.: Low noise magnetoresistive sensors for current measurement and compasses. Journal of Magnetism and Magnetic Materials 316(2), E246–E248 (2007)
[31] Le Phan, K., Boeve, H., Vanhelmont, F., Ikkink, T., Talen, W.: Geometry optimization of TMR current sensors for on-chip IC testing. IEEE Transactions on Magnetics 41(10), 3685–3687 (2005)
[32] Le Phan, K., Boeve, H., Vanhelmont, F., Ikkink, T., de Jong, F., de Wilde, H.: Tunnel magnetoresistive current sensors for IC testing. Sensors and Actuators A-Physical 129(1-2), 69–74 (2006)
[33] Reig, C., Cubells-Beltrán, M.-D., Ramírez, D., Cardoso, S., Freitas, P.P.: Electrical Isolators Based on Tunneling Magnetoresistance Technology. IEEE Transactions on Magnetics 44(11), 4011–4014 (2008)
[34] Cubells-Beltrán, M.D., Reig, C., Ramírez, D., Cardoso, S., Freitas, P.P.: 2D-FEM model for integrated GMR current sensors. In: 4th International Conference on Sensing Technology (ICST 2010), Lecce (2010)
[35] Beltrán, H., Reig, C., Fuster, V., Ramírez, D., Cubells-Beltrán, M.D.: Modeling of magnetoresistive-based electrical current sensors: A technological approach. IEEE Sensors Journal 7(11-12), 1532–1537 (2007)
[36] Roldán, A., Roldán, J.B., Reig, C., Cubells-Beltrán, M.-D., Ramírez, D., Cardoso, S., Freitas, P.P.: A DC behavioral electrical model for quasi-linear spin-valve devices including thermal effects for circuit simulation. Microelectronics Journal 42, 365–370 (2011)
[37] Gakkestad, J., Ohlckers, P., Halbo, L.: Compensation of sensitivity shift in piezoresistive pressure sensors using linear voltage excitation. Sensors and Actuators A – Physical 49(1-2), 11–15 (1995)
[38] Sebastiá, J.P., Munoz, D.R., de Freitas, P.J.P., Ku, W.J.: A novel spin-valve bridge sensor for current sensing. IEEE Transactions on Instrumentation and Measurement 53(3), 877–880 (2004)
[39] Pelegrí-Sebastiá, J., Ramírez-Muñoz, D.: Safety device uses GMR sensor. EDN 48(15), 84–86 (2003)
[40] Ramírez, D., Pelegrí, J.: GMR sensors manage batteries. Edn 44(18), 138 (1999)
[41] Vopalensky, M., Platil, A., Kaspar, P.: Wattmeter with AMR sensor. Sensors and Actuators A-Physical 123-24, 303–307 (2005)
[42] Roldán, A., Reig, C., Cubells-Beltrán, M.-D., Roldán, J.B., Ramírez, D., Cardoso, S., Freitas, P.P.: Analytical compact modeling of GMR based current sensors: Application to power measurement at the IC level. Solid-State Electronics 54, 1606–1612 (2010)
[43] Kaliyugavaradan, S.: A simple resistance-to-time converter for signal conditioning of resistive transducers. Meas. Sci. Technol. 11, N73–N75 (2000)

[44] Ferrari, V., Ghidini, C., Marioli, D., Taroni, A.: A conditioning circuit for resistive sensors combining frequency and duty-cycle modulation of the same output signal. Measurement Science and Technology 8(7), 827 (1997)
[45] Brown, J.L.: A universal low-field magnetic field sensor using GMR resistors on a semicustom BiCMOS array. In: Cameron, G., Hassoun, M., Jerdee, A., Melvin, C. (eds.) Proceedings of the 39th Midwest Symposium on Circuits and Systems, pp. 123–126 (1996)
[46] Han, S.-J., Xu, L., Yu, H., Wilson, R.J., White, R.L., Pourmand, N., Wang, S.X.: CMOS integrated DNA Microarray based on GMR sensors. In: 2006 International Electron Devices Meeting, International Electron Devices Meeting, pp. 451–454 (2006)
[47] EUROPRACTICE IC Service, `http://www.europractice-ic.com/` (accessed on September 21st, 2012)
[48] Jaeger, R.C.: Introduction to Microelectronic Fabrication. Modular series on solid state devices. Addison-Wesley (1988)
[49] Villarroya, M., Figueras, E., Montserrat, J., Verd, J., Teva, J., Abadal, G., Perez-Murano, F., Esteve, J., Barniol, N.: A platform for monolithic CMOS-MEMS integration on SOI wafers. J. Micromech. Microeng. 16, 2203–2210 (2006)

GMR Sensors in Automotive Applications

Konrad Kapser[1], Markus Weinberger[2], Wolfgang Granig[2], and Peter Slama[2]

[1] Infineon Technologies, Am Campeon 1-12, 81726 Munich, Germany
Konrad.Kapser@infineon.com
[2] Infineon Technologies, Siemensstr. 2, 9500 Villach, Austria

Abstract. In this chapter, the various automotive applications for GMR based sensors are presented. The different applications have different requirements which are reflected in the individual sensor concept and features. The advantage of GMR based magnetic sensors over conventional silicon based Hall sensors is shown.

1 Introduction

Although GMR based sensors are relatively new, they have already entered various automotive applications. Due to the magnetic measurement principle, the high sensitivity and low noise, they are very well suited for a wide range of different automotive applications in the field of speed sensing as well as angle sensing. GMR sensing elements can be integrated with a standard CMOS silicon circuit and therefore a monolithic sensor solution is possible which benefits from all the advantages known from other silicon based sensors (e.g. Hall sensors). GMR based sensors offer the highest sensitivity compared to Hall and AMR devices and the lowest noise and jitter. Mass production combined with high reliability even in harsh environment meets all the automotive quality requirements and provides a cost effective solution with high performance.

Among the target applications in an automotive environment are:

- Steering angle measurement.
- Rotor position measurement for motor commutation in a brushless-DC motor (e.g. EPS motor).
- Speed sensing for wheel speed measurement (ABS Sensor).
- Crank shaft speed and position sensing with direction information.

To address above applications, two different configurations of GMR sensors are needed: speed sensors and angle sensors. They use the same measurement principle but whereas a speed sensor works in the linear range of the GMR characteristics, an angle sensor is in the saturated region (Fig. 1).

For the automotive GMR sensors a spin-valve (SV) stack type is used basically consisting of two ferromagnetic layers, where one's layer magnetization is fixed by special measures into a defined direction and the other's magnetization is free to be rotated within an external in-plane magnetic field. The real SV stack consists of about

C. Reig et al.: *Giant Magnetoresistance (GMR) Sensors*, SSMI 6, pp. 133–156.
DOI: 10.1007/978-3-642-37172-1_6

nine different layers with materials like CoFe, NiFe as ferromagnetic composition, Cu, Ru and Ta as non-magnetic material as well as PtMn as a natural anti-ferromagnetic material. The thickness of each layer is in the range between <1 nm and a few 10 nm. In order to deposit those thin films with a high homogeneity over the wafer a magnetron sputter technique is applied to grow the films. The Cu and the Mn in the PtMn are known to be mobile at elevated temperatures. As a consequence, all temperatures following the deposition of the GMR thin film stack have to be lower than ~350°C in order to avoid stack degradation. Furthermore, due to the small thickness of most of the single layers also the surface roughness plays an important role for the GMR properties and must be well controlled for a successful and reliably integration concept [1, 2].

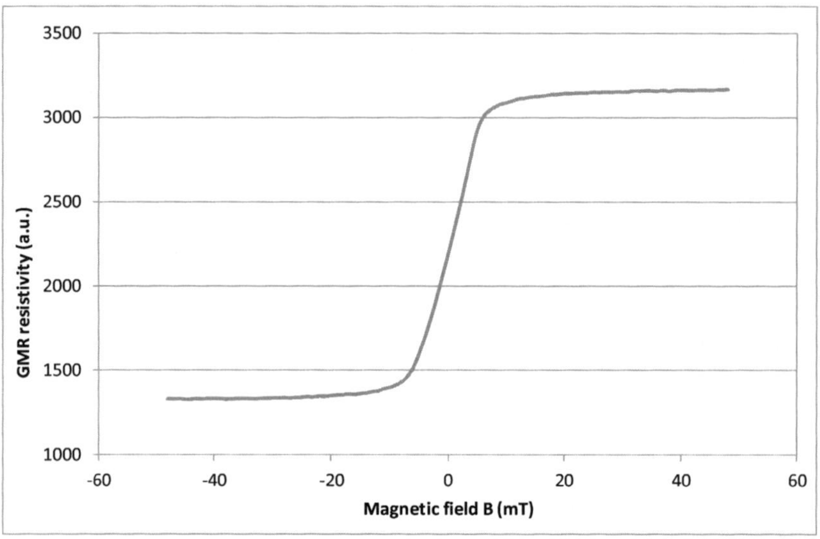

Fig. 1. Characteristics of a GMR resistor with linear and saturated region

The following chapters show more details of the different applications and sensing concepts.

2 GMR Angle Sensing Applications

A GMR angle sensor consists of GMR resistors connected to a bridge circuit with one bridge for each orthogonal magnetic direction and each bridge consisting of four GMR resistors. These resistors contain an anti-ferro magnet made up of platinum and form the pinned magnetic direction, which is 180° opposite to the GMR reference direction. This magnet must be generated in the production process where the magnetic orientation can be programmed in any desired direction. The resistors layout has meander shapes to increase the total resistance in the range of 2kΩ for power-consumption reasons (Fig. 2).

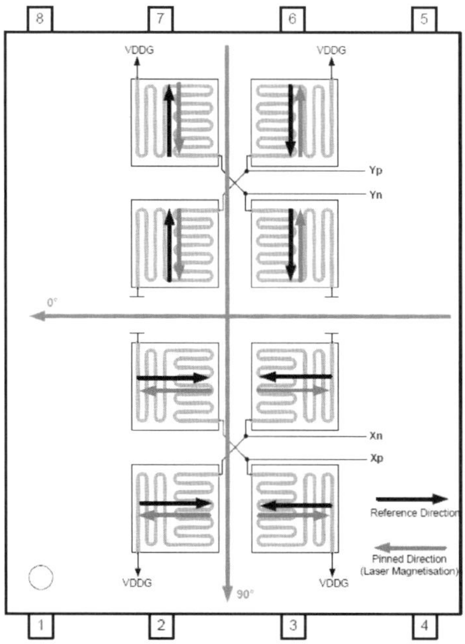

Fig. 2. Bridge layout of a GMR angle sensor consisting of two bridges with four resistors each

The angle sensor needs a certain magnetic field strength to drive the GMR characteristics into the saturated region. In this case, the resistivity of the GMR element and thus the bridge output voltage depends only on the direction of the external magnetic field in respect to the internal reference magnetization. This provides a sine and a cosine signal which can be used for calculation of the absolute angle of the magnetic field vector. Due to this measurement principle, only the field direction, not the field magnitude is relevant. The use of the GMR principle allows a measurement of angles in the full range of 360° in contrast to AMR based sensors which cover only 180°.

The sensor concept of an integrated angle sensor converts the raw signals of the GMR bridges into the digital domain as early as possible to perform there all signal processing and error compensation. Additional features can be implemented as for example an auto-calibration routine to increase the angle accuracy over temperature and lifetime (see chapter 2.3). Flexible interfaces (eg. SPI, PWM, Incremental) can be configured and several settings (update rate, internal filters, ...) are also accessible by the user to adjust the device according to the application requirements. Some of the automotive angle sensing applications are safety relevant (for example steering angle measurement or motor commutation of the BLDC motor for the electric power steering (EPS) system). To support such applications, additional features can be implemented to account for the requirements of the ISO26262 covering functional safety aspects [3].

2.1 BLDC Rotor Position Measurement

Due to the hybridization and change of hydraulic systems into electrical systems, the number of electric commutated motors in vehicles is increasing. This trend towards brushless motors reflects their advantage of having no wearing parts (brushes), low acoustic noise and very low torque ripple. In order to ensure exact commutation of electric motors the position of the rotor has to be measured very accurate over a wide speed range. The position is transferred to the control unit, which generates the necessary commutation sequence for the motor (Fig. 3). For optimized efficiency and lowest torque ripple, a field oriented control algorithm is used which needs a very accurate information of the rotor position.

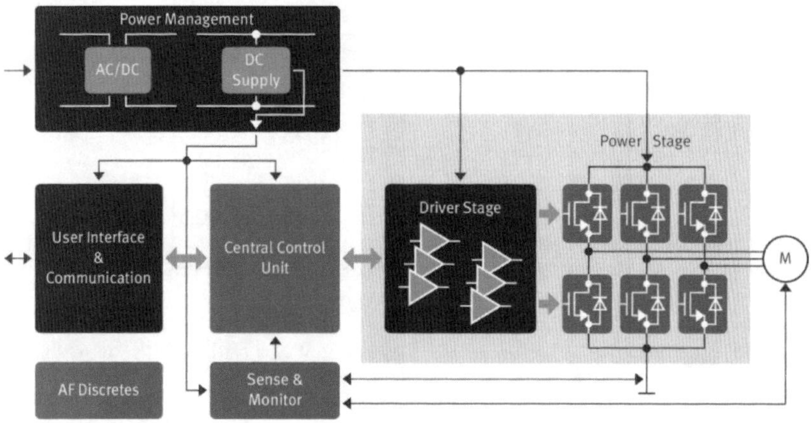

Fig. 3. Principle of a motor control system with a rotor position sensor

GMR angle sensors can be used for various motor topologies with different numbers of pole pairs. Figure 4 depicts the setup with the GMR angle sensor and a diametrically magnetized magnet.

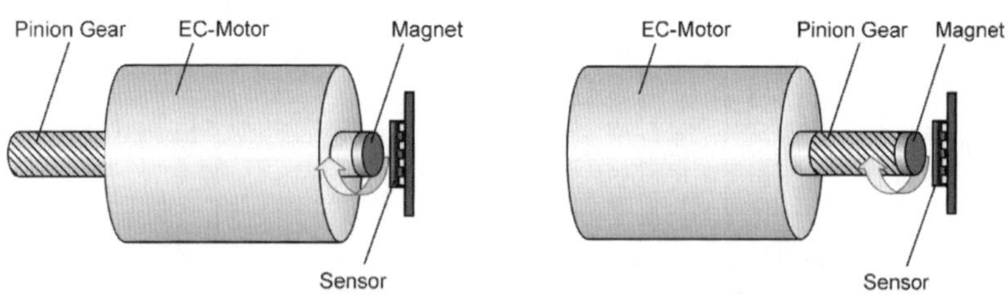

Fig. 4. End of shaft configuration of magnet and GMR angle sensor

As GMR angle sensors have a measurement range of 360°, even and odd numbers of pole-pair motors can be used. AMR based sensors in contrast, due to their limited measurement range of 180°, can only be used for even pole-pair motors. Compared to classical resolvers the GMR based systems have the advantage of a small volume and weight and can offer a very cost-effective solution. Monolithic integration of the GMR sensor element with standard CMOS technology opens up further possibilities of advanced signal processing, error compensation algorithm and flexible, customer programmable interfaces on the chip.

There are several possibilities of the sensor concept especially whether the raw data or the calculated angle is transmitted to the microcontroller. Which option is preferred depends on the complete system architecture. In case the raw data, i.e. sine and cosine values, are transmitted, the angle calculation has to be done in the microcontroller. This can be performed by the use of the CORDIC algorithm (COordinate Rotation DIgital Computer), which calculates the arctan of the ratio of the sine and cosine value and thus obtains the absolute angle. Another possibility avoiding the CORDIC algorithm is a tracking loop which minimizes the angle error calculated out of the measurement data (sine and cosine) and an estimated angle value. Details of this approach can be found in the literature [4].

Even if only the raw values are used, it is of advantage to have some signal conditioning on the GMR sensor. A voltage regulator to maintain a constant bridge supply voltage, an amplifier to adjust the bridge output signal to the input range of a microcontroller ADC and a temperature compensation to minimize temperature effects and increase the angle accuracy.

In Fig. 5 the block diagram of the Infineon GMR angle sensor TLE5009 is shown. It contains two Wheatstone bridges (X-GMR, Y-GMR) consisting of four GMR resistors each, an amplifier for each bridge, a power management unit (PMU) and a temperature compensation to eliminate the GMR temperature effects.

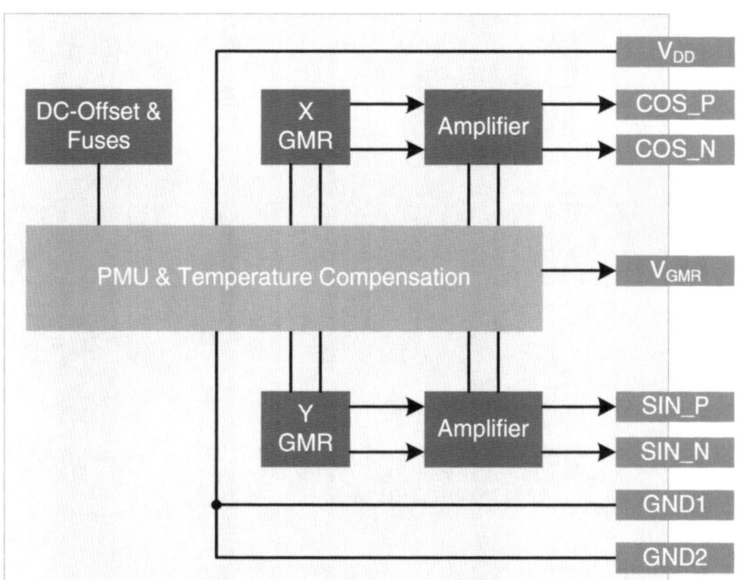

Fig. 5. Block diagram of Infineon TLE5009 GMR angle sensor

The sensor provides sine and cosine output signals which can be used either single ended or in a differential configuration. An additional pin V_{GMR} is available which provides the chip temperature in case additional temperature compensation needs to be done in the microcontroller. Furthermore, this pin has also a diagnostic function, in case of internal failure e.g. of the supply voltage, the signal is pulled down below 0.39V.

Such a device can be used for rotational speed up to 30.000rpm with an angle accuracy of < 0.6° (typ. value) and a short signal delay of only 9µs. To enhance angle accuracy, a calibration algorithm can be implemented in the microcontroller to eliminate temperature and lifetime drifts and increase overall angle accuracy. Details for this procedure are described in chapter 2.3.

As the GMR technology allows integration with standard CMOS circuitry, further features and functionalities can be monolithically implemented in the sensor. The calculation of the absolute angle value can be done directly on the chip using the CORDIC algorithm, no need to do this in a microcontroller. Depending on the application, different interfaces for transmission of the angle value might be required. Analog or various digital protocols as e.g. PWM or incremental interface are in use. A communication interface with the sensor to select the different options and to configure the device according to the application requirements is also helpful.

Such a sophisticated, flexible device is Infineons TLE5012B GMR angle sensor which can provide the sine and cosine raw data as well as the calculated absolute angle value with different interface options and a SPI chip communication. Due to internal temperature compensation, an accuracy below 1.6° in the full temperature range can be obtained. The block diagram of this sensor is shown in figure 6.

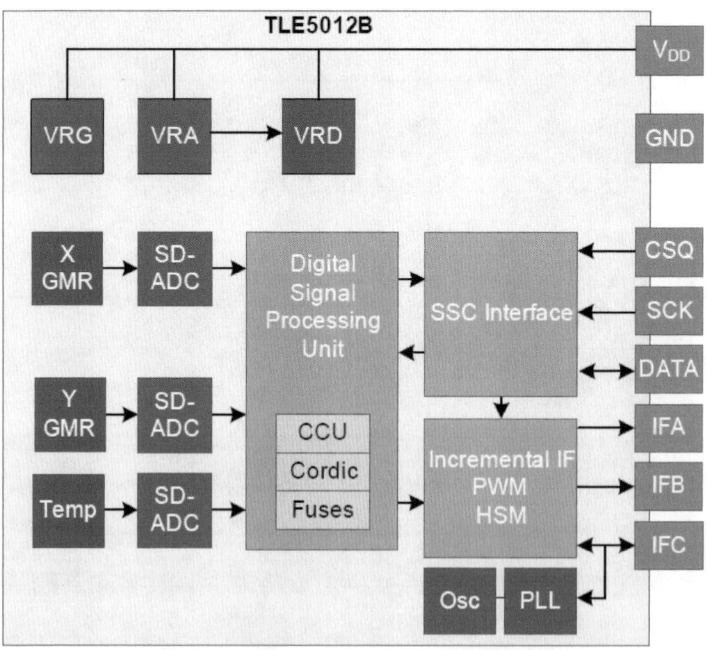

Fig. 6. Block diagram of Infineon TLE5012B integrated GMR angle sensor

The raw signals from the bridges are digitalized with a Sigma-Delta ADC and all further signal conditioning is done in the digital world. As also the chip temperature is measured, compensation of the signals is performed to increase accuracy. Two parallel interfaces are available: SSC (synchronous serial interface, compatible to standard SPI interface) for communication and configuration of the chip as well as for digital read-out of the calculated angle value with a 15bit word. Parallel to the SSC there are also other interface options available as PWM or incremental interface. They provide either the calculated angle (PWM coded) or increments similar to an encoder interface with phase shifted A and B signals and an index signal Z for absolute position (once per revolution) (Fig. 7). The resolution (number of increments per revolution) can be configured via SSC from 9bit to 12bit.

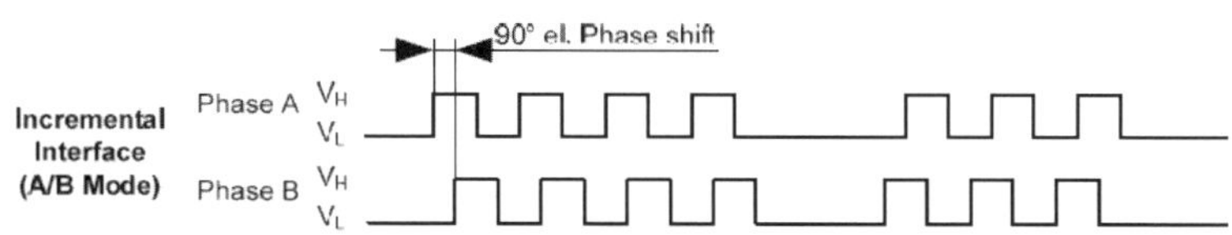

Fig. 7. Output signal in the incremental mode: Two phase shifted channels A / B

A special feature is the HSM mode (Hall switch mode). This mode can be used for block commutated motors (6-step commutation) and substitutes three Hall switches. The three output signals (Fig. 8) can be directly used as an input for the microcontroller. To account for different motors, the number of pole pairs of the motor can be configured in the angle sensor.

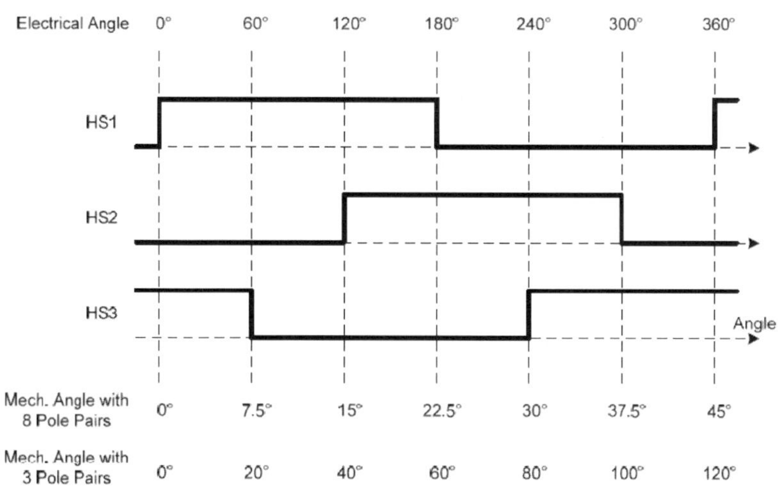

Fig. 8. Output signals in the Hall switch mode for 6-step commutation (block commutation)

Further improvement yields the implemented auto-calibration routine. This algorithm is based on the calibration algorithm described in chapter 2.3 and reduces the overall angle error in the temperature range up to 150°C and within the complete lifetime to a value below 0.6° (typ. value). This feature can be enabled via an SPI command and needs a continuously rotating application over the full angle range of 360°. The rotor position measurement for a BLDC motor is an ideal application for this feature.

This auto-calibration continuously determines the amplitude and offset of the raw signals during operation and compensates drifts and temperature variation when calculating the new absolute angle value. Figure 9 shows the angle error over temperature for different external magnetic fields. Activating the auto-calibration feature significantly reduces the angle error to approx 0.2° independent of temperature.

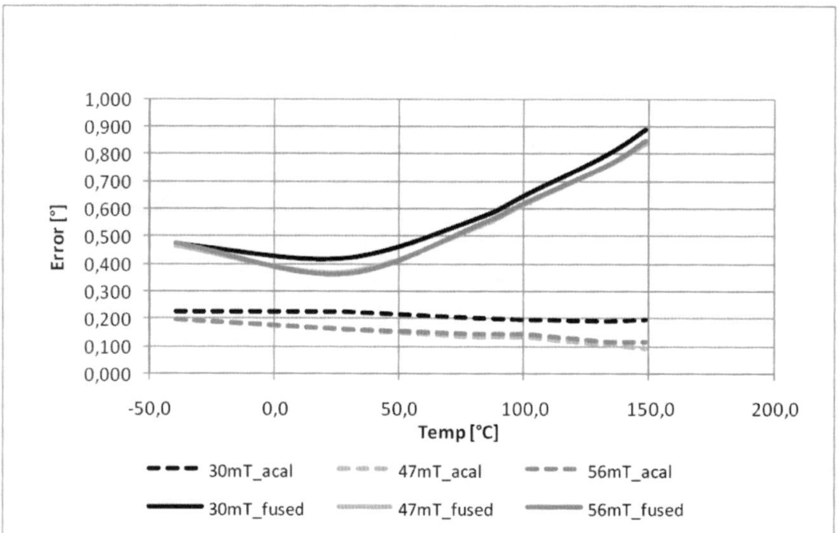

Fig. 9. Total angle error over temperature with and without enabled auto-calibration for different magnetic field strength

For angle sensors used in motor commutation applications, one important parameter is the signal delay, the time between the measurement of the angle value and the availability of this value at the interface. This delay time is equivalent to a dynamic angle error and especially for motors with fast rotation speed this cannot be neglected. Compensation in the microcontroller is possible in principle but requires the knowledge of the delay time as well as the motor speed. To avoid this external compensation, the TLE5012B is equipped with a special "prediction" mode where the angle value is corrected for the internal delay taking into account the actual motor speed. This linear prediction extrapolates the expected angle value at the time (t+1) from the angle values at t and (t-1) (Eq. 1).

$$\alpha(t+1) = 2\alpha(t) - \alpha(t-1) \quad (1)$$

In figure 10 the mechanical angle and calculated angle for a direction change is shown. Without prediction the sensor will follow the mechanical rotation slowly. The advantage of the prediction can be seen. With activated prediction the sensor will follow the mechanical angle much better than without prediction.

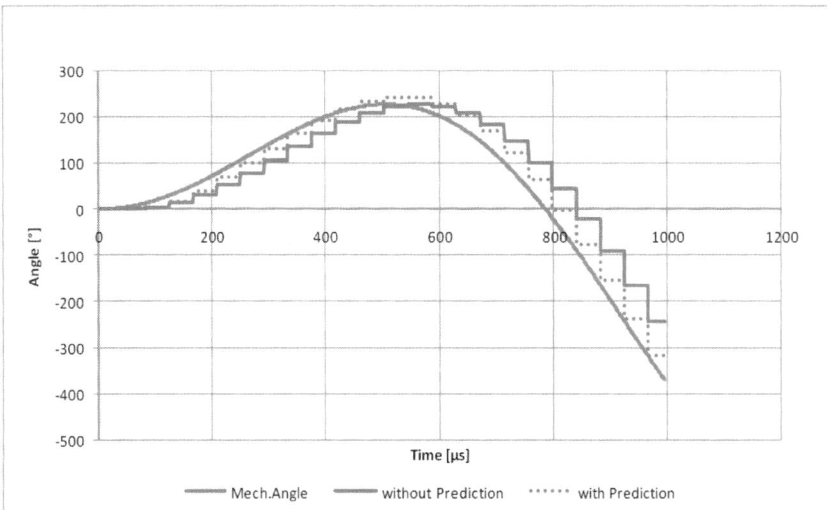

Fig. 10. Calculated angle vs. real angle (mech. angle) with and without prediction enabled (update rate 42μs)

2.2 Steering Angle Application

The Steering Angle Measurement is one of the most important criteria for controlling vehicle dynamics. It is the only driver information used by the Electronic Stability Control (ESC) to detect the desired vehicle driving direction. The steering angle is, therefore, a safety relevant information and must be generated in a safe and redundant way. This requires adding backup logic and circuits to this system and measuring the same value under different conditions to ensure feasible results. This information is also used for advanced front lighting, lane departure warning and active suspension. Typically a steering wheel can be turned for ±2 revolutions, which increases the effort required to measure the absolute steering angle. One method for measuring steering angle is to use gears to transfer the ±2 revolutions to <1 revolution. Systems using potentiometric angle measurement are often not reliable enough for today's functional safety requirements.

Basically, two different places in a car are used to detect the steering angle:

- Passenger Compartment: Sensor is located behind the steering wheel; reduced requirements on temperatures ($T_A < 85°C$), vibration and dust
- Engine Compartment: Sensor is mounted on the steering shaft near the engine; high robustness against temperature ($T_A > 125°C$), vibration and dust required

To meet the safety requirements and multi-turn capability of the system, different implementations can be used. It is quite common to transmit the raw signals of the GMR sensor bridges X, Y and not the calculated angle. This can be done either with the TLE5012B using the SPI interface or a dedicated sensor which consists only of two bridges with some compensation electronics but no digital signal conditioning (TLE5009 of Infineon, Fig. 5). In this way, a correlation of the raw signals X and Y can be done in the microcontroller. For example, the vector length S has to be constant and independent of the angle.

$$S = \sqrt{X^2 + Y^2} \tag{2}$$

This gives a control value which enables the detection of a malfunction of the sensor or data transmission.

Usually two sensors are used in a steering angle measurement due to the safety level of the application and the need to have a multi-turn capability of measuring absolute angles exceeding 360°. This can be done for example using two gears with different ratio each carrying the magnet for the corresponding sensor. One gear can have a ratio in that way that it transfers the steering angle 1:1 on the sensor thus giving absolute angle position. To get the multi-turn capability the second gear has a

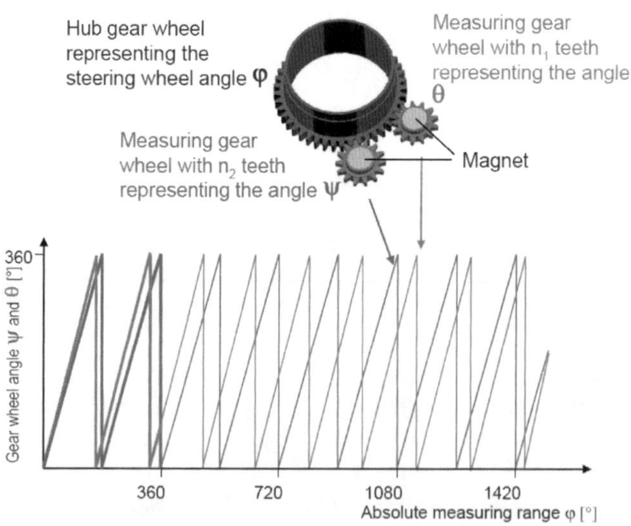

Fig. 11. Steering angle measurement with the Nonius principle

ration 1:4 so that ±2 turns (= ±720°) are mapped to a 0° … 360° magnetic input signal. In this way, the absolute angle information in the full range of ±720° can be calculated. Furthermore, the two sensor signals can be used for a plausibilization and redundancy of the measured data.

Another way of implementation is the so call Nonius principle. This uses two gears with slightly different ratio (= number of teeth, n_1 and n_2). By evaluating the two sensor signals and their phase difference, the absolute angle in a range >360° can be determined (Fig. 11)

2.3 Angle Error Calibration

To achieve the best performance of a GMR angle sensor, a calibration of the device is necessary. The following chapter shows how to calibrate a GMR angle sensor.

Due to process tolerances, temperature and packaging effects, the raw signals coming out of the Wheatstone bridges are not ideal. They exhibit an offset, the amplitudes of the X and Y signals might not be identical and also an orthogonality error can exist in the way that the X and Y signals are not perfectly shifted by 90°.

In general, the raw signals X and Y can be written as follows (Eq. 3):

$$\begin{aligned} X &= Ax \cdot cos(\alpha) + Ox \\ Y &= Ay \cdot sin(\alpha + \varphi) + Oy \end{aligned} \tag{3}$$

With:

Ax:	amplitude of X (cos) signal
Ay:	amplitude of Y (sin) signal
Ox:	offset of X (cos) signal
Oy:	offset of Y (sin) signal
φ:	orthogonality error

The three parameters amplitude, offset and phase will affect the angle calculation and yield an angle error when calculating the angle α out of Eq. 4:

$$\alpha = \arctan\left(\frac{Y}{X}\right) \tag{4}$$

By drawing the X and Y signal in the X-Y plane for a complete rotation of 360° yields in general an elliptical shape which is shifted and tilted and not a circle centered at 0/0. (Fig. 12). This shape reflects the error in offset, amplitude and orthogonality.

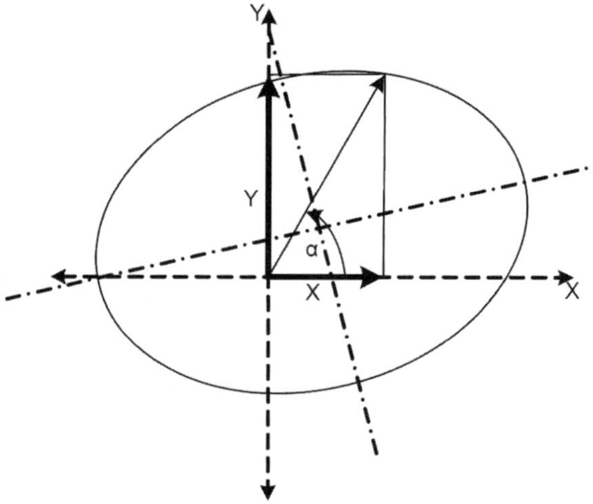

Fig. 12. X and Y signals with offset, amplitude and orthogonality error

The ideal shape would be a circle centered at 0/0. This corresponds to an X and Y signal with no offset, identical amplitude and 90° phase shift to each other. The task for the calibration now is to find correction values for the X and Y data to transform the elliptical shape to the circle shown in Fig. 13.

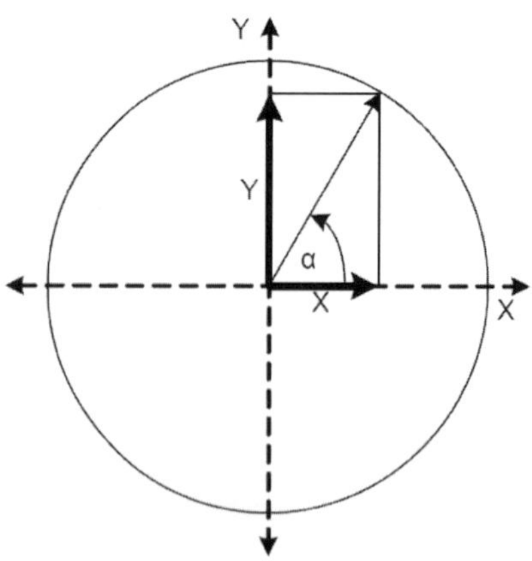

Fig. 13. X and Y signals without offset, amplitude and orthogonality error (after calibration)

The basic idea for the calibration algorithm is to measure the elliptical shape, determine the values for offset, amplitude and phase error and calculate corrected signals Xc and Yc.

To achieve this, a full rotation of 360° is required. The signal has to be sampled and the minima and maxima of X and Y respectively, have to be determined (Fig. 14).

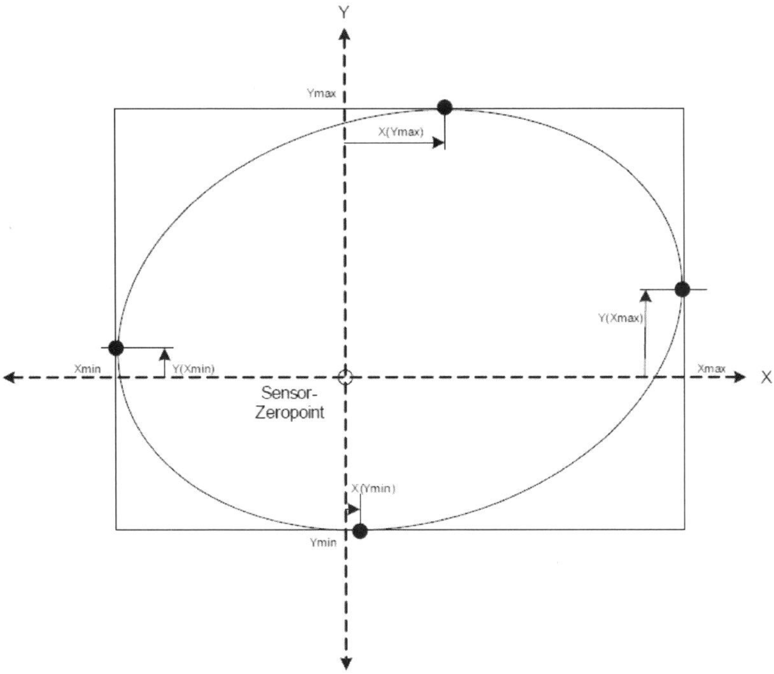

Fig. 14. Determination of the minimum and maximum values of the X and Y signal for one full rotation

The amplitude Ax, Ay and offset Ox, Oy of the X and Y signals can now be calculated according to Eq. 5

$$Ax = \frac{Xmax - Xmin}{2} \tag{5}$$

$$Ay = \frac{Ymax - Ymin}{2}$$

$$Ox = \frac{Xmax + Xmin}{2}$$

$$Oy = \frac{Ymax + Ymin}{2}$$

To determine the orthogonality error φ, two angle values which are 90° shifted must be used, e.g. 45° and 135°. The values for X and Y are measured at this position and the orthogonality error are calculated according to Eq. 6

$$M_{45} = \sqrt{(X_{45})^2 + (Y_{45})^2} \tag{6}$$

$$M_{135} = \sqrt{(X_{135})^2 + (Y_{135})^2}$$

$$\varphi = 2 \cdot arctan\left(\frac{M_{135} - M_{45}}{M_{135} + M_{45}}\right)$$

To account for possible errors due to hysteresis effect, the determination of offset, amplitude and orthogonality should be done twice, for clock-wise (cw) and counter-clock-wise (ccw) rotation. The effective value for amplitude and offset for the respective X and Y signal is the mean value of the cw and ccw values. These values must be used for further signal correction.

Now the offset and amplitude error can be compensated and corrected values Xc and Yc can be obtained (Eq.7)

$$Xc = \frac{X - Ox}{Ax} \qquad (7)$$

$$Yc = \frac{Y - Oy}{Ay}$$

The orthogonality error is compensated in the Yc signal as follows (Eq. 8)

$$Yc' = \frac{Yc - Xc \cdot sin(-\varphi)}{\cos(-\varphi)} \qquad (8)$$

The corrected angle value α_c can now be determined with the arctan function using Xc and Yc' (Eq. 9)

$$\alpha_c = arctan\left(\frac{Yc'}{Xc}\right) \qquad (9)$$

When using a GMR angle sensor, this calibration routine should be performed to optimize the performance of the device. The determined calibration parameters can be stored in the microcontroller and used for a corrected angle calculation.

The TLE5012B angle sensor of Infineon is delivered already with stored calibration values for offset, amplitude and orthogonality to achieve high angle accuracy. The temperature dependency of offset and amplitude can be measured and temperature depending compensation can be performed implementing a specific algorithm in the microcontroller.

For applications which have a continuously rotating magnetic input signal, as for example rotor position measurement in a BLDC motor, this calibration algorithm for offset, amplitude and orthogonality can be implemented in a way that the parameters are ongoing updated (e.g. once per revolution) during operation and all temperature and lifetime effects are almost perfectly compensated. In such a case the achievable angle accuracy is well below 1° over the full temperature range also including lifetime drift effects. This algorithm is implemented as "auto-calibration" feature in the TLE5012B.

3 GMR Speed Sensing Applications

The main advantage of GMR based sensors in a speed application is the high sensitivity and a high S/N ratio. This gives on the one hand a large air gap capability which cannot be realized with a silicon based Hall sensor. On the other hand, the GMR sensing principle has a low noise which gives a very good jitter performance in the application. This provides superior performance for a crankshaft sensor in engine management. In this application, not only the speed is measured but also a precise knowledge of the crankshaft position is required. With the low jitter of a GMR sensor, crankshaft position repeatability of 0.015° is possible. Also for wheel speed sensors used for the ABS and ESC system, a low jitter is of advantage [5]. With the wheel speed information available for different applications in the car (not only ABS/ESC), for example an indirect tire pressure monitoring system can use this data for tire pressure calculation. Due to the system requirements, a low noise wheel speed signal is necessary.

3.1 Crankshaft Speed and Position Measurement

Active magnetic sensors in advanced engine management systems provide a digital switching signal, which maps the mechanical teeth profiles or the magnetic domains of a passing pole wheel. Subsequent processing in the microprocessor determines the current speed or angle position of the target wheel from this switching signal. This data is further processed for accurate ignition control and misfire detection.

The sensor needs to provide a phase accurate output signal for magnetic input frequencies from 0 to 10 kHz and over an amplitude range of approximately 1 to 100. The required temperature range is from -40°C to 150°C with an air gap range of 0 to 2.5 mm.

Optimization of engine management systems shall improve CO_2 emissions. Target is minus 30% for gasoline and minus 10% for diesel engines. In order to meet this challenging goal the sensing system has to provide more accurate information and provide additional features for start-stop, hybrid and electric vehicles. The task of the sensor is therefore to provide the switching edge with good reproducibility and high angle accuracy relative to the target wheel. Nevertheless the mounting tolerance of the sensor should be relatively large to keep the system costs low.

The goal of future crankshaft sensors is delivering even more accurate digital signals with respect to phase jitter and phase accuracy. The challenging target is $< \pm$ 0.015 [°crank] phase jitter (repeatability) and $< \pm$ 0.3 [°crank] absolute phase accuracy.

Compared to other technologies the GMR effect has a higher sensitivity and a superior signal to noise ratio. Due to these characteristics the new active Infineon monolithic integrated GMR sensor can cope with these upcoming ambitious demands.

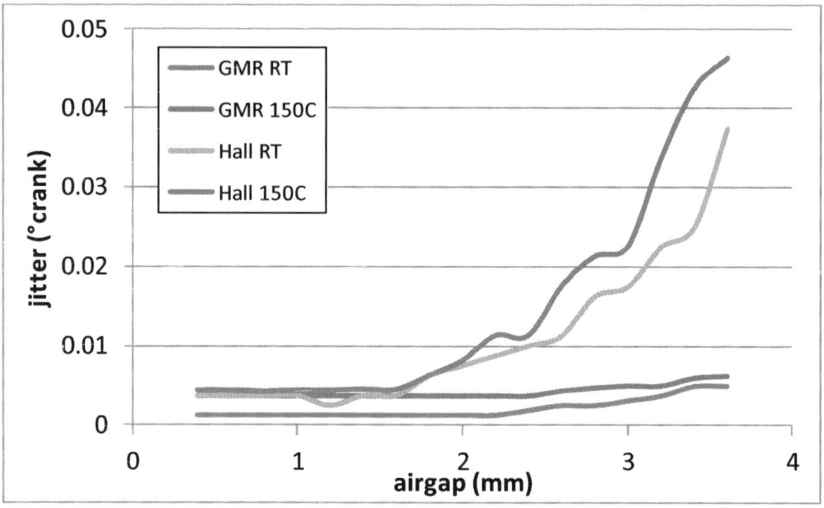

Fig. 15. Phase jitter (repeatability) comparison of Hall and GMR based sensors for different air gap and temperature

3.1.1 Chip Concept

The characteristic of a GMR resistor is shown in Fig. 1. It exhibits a linear region ranging from typically H_k = -5mT to +5mT and a saturation of the resistivity for magnetic values exceeding that field. The change of resistivity can be as much as 10% and is therefore considerably higher than that of an AMR based device. The spatial distance of the bridge resistors on the chip is 2.5mm. With this Wheatstone bridge configuration only differential magnetic fields are measured, magnetic offsets are

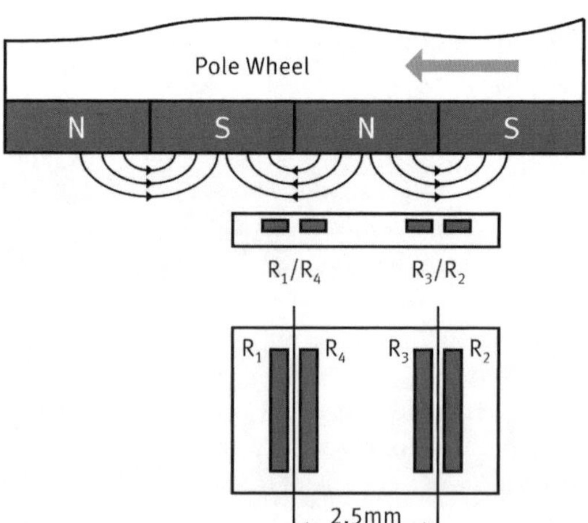

Fig. 16. Wheatstone bridge with four GMR resistors integrated monolithically on the signal conditioning ASIC

suppressed. The pitch of the pole wheel has therefore to be adjusted to the distance of the GMR elements with an optimum matching at 5mm pole wheel pitch. In this case the maximum bridge voltage is generated. Each side of the bridge experiences the maximum magnetic field but with opposite direction and is thus in the low resistivity or high resistivity region, respectively (Fig. 16).

The resulting bridge voltage is amplified and digitalized with AD converter. All further signal conditioning as offset correction and calculation of switching thresholds is done in the digital domain. The block diagram of the GMR speed sensor is shown in Figure 17.

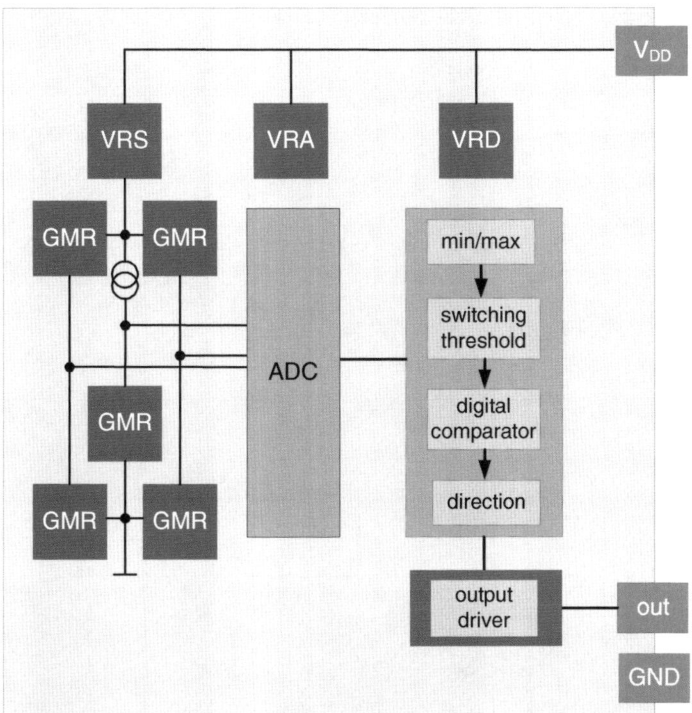

Fig. 17. GMR based crank sensor block diagram (Infineon TLE5028)

3.1.2 Start-Up Behavior and Fast Calibration

With respect to optimized engine control a faster initial calibration of the system at start-up is an important factor. The engine management system can earlier determine the exact position of the pistons. As a consequence the combustion cycle can precisely be controlled from earlier phase on. This will lower both (CO_2-) emissions and fuel consumption.

The main advantage of the GMR sensor is that the initial calibration phase is shorter compared to other sensing technologies. Prior to issuing the second output pulse the sensor calibration phase was finished. Therefore the second pulse delivers high accurate phase information to the engine management system. Compared to this a Hall based sensor needs (maximum) 6 edges for calibration.

The rotational direction information is transferred via a pulse width modulated (PWM) sensor output protocol. An advantage of the GMR sensor is that from the first output pulse onwards the correct rotational direction information will be transferred (at first start-up per default).

Because of the specific GMR characteristic the electrical input signal can perfectly be adapted to the analog-to-digital converter input range. Any kind of gain control or PGA (programmable gain amplifier) as it is needed for Hall sensors is not required. The consequence out of that is that the sensor calibrated mode can be entered at the third edge of the differential input signal $B_{x,diff}$. (Fig. 18)

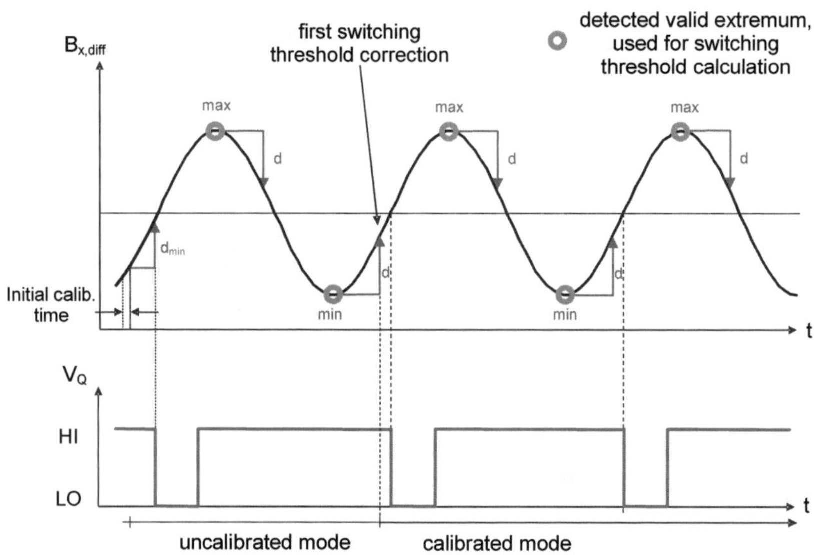

Fig. 18. Sensor start-up at magnetic rising edge

In order to calculate the calibrated mode switching threshold the input signal has to be tracked and the extreme values have to be determined. After an initial calibration time t_{on} the magnetic signal is tracked by an ADC and monitored within the digital circuit. For detection the signal needs to exceed an internal threshold (digital noise constant d_{min}). After power on, when the signal slope is identified as a rising edge and the signal change exceeds d_{min}, the first pulse is issued at the output. The first output pulse includes the correct direction information. As soon as the second extrema is detected, the switching threshold can be calculated. All following output switching are done at zero crossing of the signal on the magnetic rising edge. The sensor is in calibrated mode from that point. From 2nd output pulse on the advanced GMR sensor provides accurate position information of the crankshaft.

In case the start-up of the sensor occurs at the falling magnetic edge, the first output pulse will be issued on the first rising magnetic edge. A "virtual zero" switching point has to be calculated as only one extreme value is determined so far. Depending on start-up position this switching point is close to the calibrated switching point. (Fig. 19).

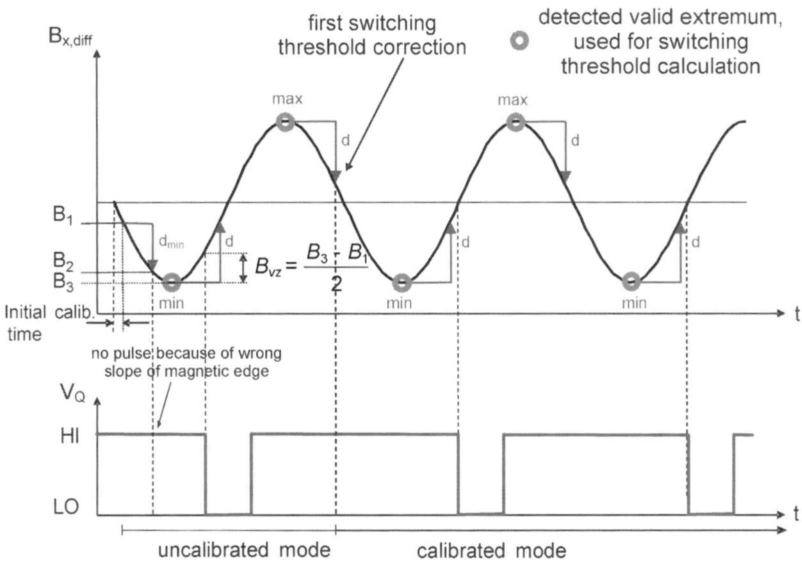

Fig. 19. Sensor start-up at magnetic falling edge

3.1.3 Hybrid and Start/Stop Engine Requirements

Hybrid vehicles and start/stop engines significantly reduce CO_2 emissions compared to normal combustion engines. The worldwide trend towards start/stop engines is ongoing. Start/stop engine implies that the combustion engine is stopped at vehicle stand still. The engine is re-started when the driver intends to continue driving. This concept is considered to be most efficient in city traffic ("stop and go mode") for emission reduction and fuel consumption saving. A start/stop engine can reduce CO_2 emissions up to -10 %. Depending on the version hybrid vehicles can lower CO_2 emissions even more up to -30 %.

These two concepts require besides higher accurate sensing systems a crankshaft sensor with rotational direction detection feature. During engine stop mode it happens, that the crankshaft rotates backwards because of mass moment of inertia. With this new feature the sensor detects a backward rotation of the crankshaft and the engine management system does not lose the exact position of the pistons. After standstill of the target wheel (no differential magnetic input signal), there is no timer present which resets the sensor. As soon as the target wheel moves again (a differential signal appears) the output pulses are triggered by the last calculated threshold (zero crossing calculated with last detected offset). At re-start of the engine the engine management system is still in calibrated operation mode, wrong injections can be avoided.

For rotational direction determination Infineon uses an additional sensing element. This sensor in the center of the IC is used to provide a typically 90 ° phase shifted signal (Fig. 17). Depending upon the rotation direction of the target wheel, the signal of the center probe anticipates or lags behind for 90 °. This phase relationship can be evaluated and converted into rotation direction information by sampling the signal of the center probe in the proximity of the zero crossing of the "speed" bridge signal. The evaluation of the rotation direction is interesting only at low rotation speed, since a direction of rotation reversal can take place only there. Hence at low rotation the already existing ADC can be used in time multiplexing to convert also the “direction” signal of the center probe without losses of phase accuracy in the output signal. After a change of the rotational direction a 90μs output pulse is being issued at the magnetic falling edge. In this way the reference to the mechanical edge of the gear tooth is still consistent.

3.2 Wheel Speed Measurement for ABS and ESC Systems

Speed sensors are widely used in modern cars especially for safety applications as ABS (antilocking system) and ESC (electronic stability control) to monitor the speed of each wheel. These systems control actively the wheel speed and avoid a blocking of the wheel or instable driving conditions by braking individual wheels to stabilize the car and reduce braking distance. The wheel speed information, however, is used for a large variety of further systems as for example parking assistant or hill holder. An upcoming application is the tire pressure monitoring, which is already mandatory in several countries. In contrast to the widely used direct measurement systems which are implemented in or close to the tire, the systems based on wheel speed sensors don’t need additional hardware. They use the available wheel speed information to calculate tire pressure by pure software algorithm. These methods are usually based on tire resonance frequency measurement and their dependency on tire pressure. For this purpose, a speed signal is required which has very low jitter to be able to extract the desired information. GMR based speed sensor can meet this requirements with their superior signal to noise ration whereas the use of Hall based sensors has much more limitations in this application.

The GMR based wheel speed sensor is very similar to the GMR based crank sensor described in chapter 3.1. Main difference is that speed sensors use a 2-wire interface as output signal. As wheel speed sensors are mounted directly at each wheel remote from the ECU, this interface can save cost for wiring. The speed signal is transmitted by a modulated supply current which is detected in the ECU. For sensors without the requirement of providing a direction signal, the current switches between 7mA and 14mA. The frequency of the output pulses corresponds directly to the north/south pole transition of the pole wheel and thus to the wheel speed.

Similar sensors exist which can also determine the rotation of direction of the wheel. In this case the communication to the ECU is done with a PWM coded signal using different duty cycles for the rotation direction. Beside this also some diagnosis information can be transmitted. A newly agreed standard between Europe OEMs uses a Manchester coded 2-wire protocol for transmitting the wheel speed as well as rotation direction and additional status information.

3.3 Magnetic Back Bias of GMR Sensors

As shown, magnetic speed sensors are widely used in various automotive applications. Two different basic configurations are possible (Fig. 20):

- To generate the measurement signal a magnetic pole wheel with alternating south and north pole magnetization can be used. The sensor is mounted in close proximity to this pole wheel and the rotational speed can directly be measured by monitoring the output frequency of the sensor.
- In a second approach, a ferromagnetic target wheel is used. In such a configuration an additional magnet is necessary for generating the magnetic field. The field is then modulated by the teeth and notches of the rotating target wheel and the field changes are measured with the sensor. As this external magnet is usually attached to the backside of the sensing element, this configuration is referred to as "back bias" configuration. The magnet is called a "back bias magnet". For Hall based sensors this approach is well known and widely in use. For a GMR based magnetic sensor this back bias configuration is more difficult to implement.

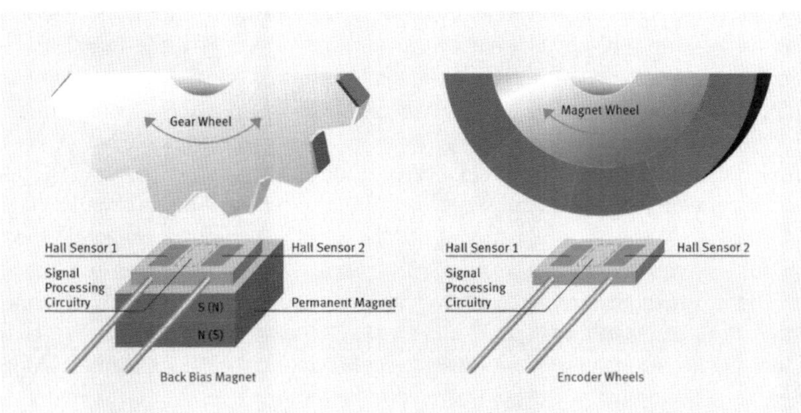

Fig. 20. Application of Hall and GMR based sensors with ferromagnetic target wheel and magnetized pole wheel

In general, sensors with a differential measurement principle are used. This means that there are two sensing elements on the chip with a certain distance to each other (e.g. 2.5mm). The measurement signal is the difference of the signal of these two individual elements. Due to the different physical principle of Hall based and GMR based sensors, the back bias application for GMR sensors has different requirements on dimension, geometry and assembly accuracy of the magnet.

The magnet produces a field which points mainly in the z-direction (Fig. 21)

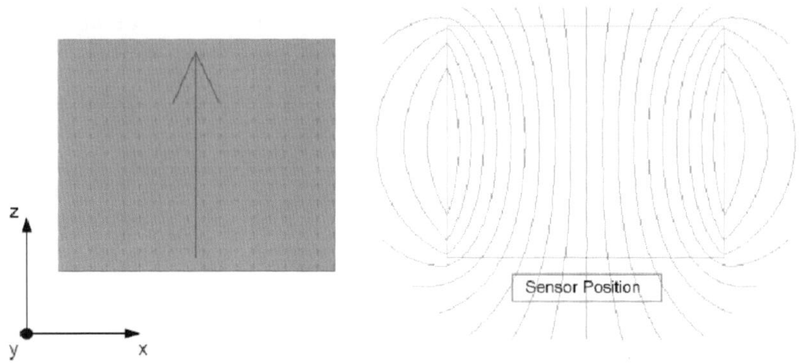

Fig. 21. Magnetic field of a homogeneous in z-direction magnetized magnet

A Hall sensor has a sensitivity axis perpendicular to the surface of the IC, i.e. z-axis. Due to the differential measurement principle and an implemented offset calibration algorithm, this offset field is canceled. X-field components, which may exist due to the position of the Hall elements being not in the center of the magnet, do not have any effect as they are not detected by the Hall effect. Variation in the z-component of the magnetic field from left to right sensing element caused by field modulation of the target wheel gives the measurement signal, which is then further internally processed.

In contrast, a GMR based sensor has a sensitivity axis in the x-direction, i.e. parallel to the chip surface. The back bias field, generated by the external magnet, is perpendicular to the sensitivity axis and hence "invisible" for the sensor. The approaching tooth of the target wheel gives rise to a field-bending and an x-component of the magnetic field is generated, which can then be detected.

Another main difference, beside the orientation of the sensitive axis, is the sensing element characteristics. The Hall effect is a linear effect and the internally generated Hall voltage is directly proportional to the applied external magnetic field (in z-direction). No saturation effect occurs. In contrast, the GMR sensor exhibits a saturation effect where the output signal (resistivity) remains constant independent of applied external magnetic field.

A further difference is the much larger sensitivity of the GMR element compared to the Hall sensor which is expressed in the larger slope of the linear region in the sensor characteristic.

As seen in Figure 21 a perfect back bias field only would have only a z-field component in the center of the magnet. However, as the sensing elements are spaced with 2.5mm distance to each other, they are located in a region of the magnet where already an in-plane field component occurs. These x-components of the magnetic field can easily be in the range of several mT and therefore will shift the working point of the sensor. In worst case the working point can be close to or even within the saturated region of the sensor characteristic and completely disturb the sensor behavior.

As a Hall sensor has a linear characteristic without any saturation effect, such a shift of the working point is not critical and completely compensated by the differential principle and offset calibration algorithm.

For the GMR based sensors this compensation is not possible due to the non-linear sensor characteristic.

A back bias configuration with a GMR sensor therefore has the essential requirement to reduce any in-plane field component (x-field component in the direction of the sensitive axis) as much as possible to minimize the shift of the working point and keep the symmetric behavior of the sensor characteristic. This is not possible with the standard approach using a conventional magnet attached to the backside of the sensor.

This problem is solved by a modified magnet design which shapes the magnetic flux line in a way that there are no in-plane field components which could shift the working point of the sensor (Figure 22).

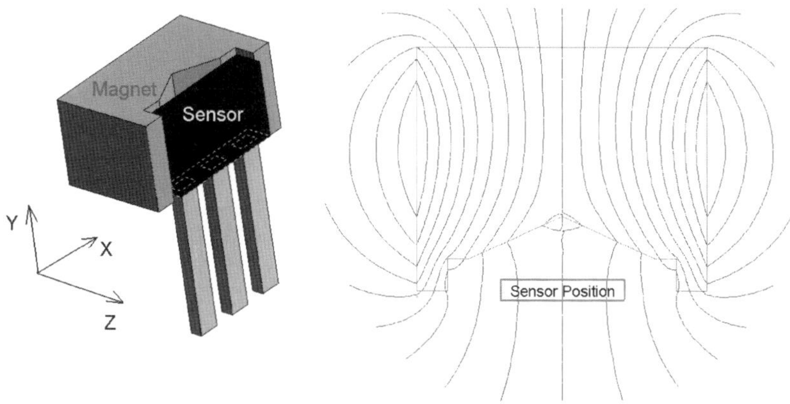

Fig. 22. Special magnet shape to avoid in-plane field components for a back bias application with a GMR speed sensor

Infineon offers GMR based sensors which have already this magnet integrated in the package. Thus, no alignment inaccuracy has to be taken into account and the sensor can be directly used with a ferromagnetic target wheel. In the manufacturing process a magnetic mould compound is used to mold the magnet together with the package thus providing very low assembly tolerances and high precision (Fig. 23).

Fig. 23. GMR based speed sensors with integrated back bias magnet and additional capacitors on the leads to enhance EMC performance

4 Summary

GMR based magnetic sensor are rather new but already widely used in various automotive applications. The advantage of monolithic integration with a signal conditioning ASIC together with the high sensitivity and low noise are an ideal combination to realize cost-effective and high performance sensing solutions. Angle and speed measurements in different applications can be performed with dedicated products targeting the individual requirements. Millions of GMR sensors in the field have already proven the high reliability of this technology which is mandatory for automotive applications. Continuous development efforts to optimize performance and reduce system costs are being done which will lead to a further increase of market share for this kind of GMR based sensors.

References

[1] Rössler, W., Zimmer, J., Bever, T., Prügl, K., Granig, W., Hammerschmidt, D., Katzmaier, E.: Integrated Giant Magneto Resistors – a new Sensor Technology for Automotive Applications. In: Advanced Microsystems Automotive Applications 2006, p. 323. Springer (2006)

[2] Slama, P., Weinberger, M., Zimmer, J., Bever, T., Raberg, W., Schmitt, S.: Speed and Angle Sensors for Automotive Powertrain Applications based on Integrated GMR Technology. In: 11th MR-Symposium, Wetzlar (2011)

[3] Granig, W., Weinberger, M., Reidl, C., Bresch, M., Strasser, M., Pircher, G.: Integrated GMR Angle Sensor for Electrical Commutated Motors including Features for Safety Critical Application. Eurosensors (3187) (2010)

[4] Burke, J., Moyhihan, J.F., Unterkofler, K.: Extraction of high resolution position information from sinusoidal encoders. In: PCIM 2000 Conference (Nürnberg) Intelligent Motion Proceedings, pp. 217–222 (2000)

[5] Kapser, K., Zaruba, S., Slama, P., Katzmaier, E.: Speed Sensors for Automotive Applications based on Integrated GMR Technology. In: Advanced Microsystems Automotive Applications 2008. Springer (2008)

Compass Applications Using Giant Magnetoresistance Sensors (GMR)

Michael J. Haji-Sheikh

Electrical Engineering Department
Northern Illinois University, DeKalb Illinois, USA
mhsheikh@niu.edu

Abstract. The use of giant magnetoresistors, or GMR, for compass magnetometers is a recent trend in Earth field sensing. Thin film compass devices are often used in applications in global positioning systems (GPS) to aid in navigation. A GPS system can only tell a user about the direction of travel as long as the user is in motion. A compass indicates static orientation of the user which, with the addition of gyroscopic information along with GPS data can give precise location and orientation to users. Modern digital devices often incorporate compass devices with location applications to better aid consumers in locating services.

1 Introduction

Thin-film magnetoresistive compass magnetometers have been in existence since the 1980's. Prior to the thin-film devices, magnetometers were large and relatively bulky. Early magnetometers were often used to detect the presence of submarines and surface ships for military applications. Large iron objects, such as a ship or a submarine, will distort the Earth's magnetic field thereby making magnetometry a valuable method of remote sensing. These systems, flux-gate and proton magnetometers, were sensitive enough to help change our perceptions on the natural world [1,2]. In the late 1940's, geologists used Naval ships to map the ocean floor using magnetometers. Starting in the 1950's, magnetometry made the jump into orbit - the United States and the Soviet Union both launched magnetometers into space. These space based devices taught us about how the Earth's magnetosphere is configured. The earliest forms of these solid state devices were used in military applications during the cold war. These devices were made from an anisotropic magnetoresistor (AMR) material called permalloy. Permalloy, generally made from eighty percent nickel and twenty percent iron, was originally used as a transformer core material in the 1930's due to the permalloy's high permeability. These thin film devices have been combined with global positioning to make systems that accurately locate hikers in the back areas. Recently, due to the demands of the smart-phone industry, novel magnetometers using giant magnetoresistive (GMR) materials have been introduced into the market.

C. Reig et al.: *Giant Magnetoresistance (GMR) Sensors*, SSMI 6, pp. 157–180.
DOI: 10.1007/978-3-642-37172-1_7

The work by S. Tumanski [3] published in 2001 discusses in detail using anisotropic magnetoresistive barber-pole magnetometers to sense the Earth's magnetic field. In it, Tumanski details both how an anistropic magnetoresistor operates and how to sample and amplify the signal produced. A critical step in designing a magnetometer is to realize that the magnetometer is a vector sensor. A vector sensor must give an output that can look at the magnetic environment in three dimensions. By knowing all three dimensions we eliminate the need for a consistent level surface. This is important since hand held devices are rarely held level. The electronic compass must be able to translate this information into a usable form. This three axis compass needs to be connected to the relative direction of the observer's device, whether the device is a cell phone, an automobile or a global positioning system. Another important piece of information needed to understand the difficulty of the application is that the Earth's magnetic field is non-uniform globally. To take advantage of the sensors, users must include magnetic zone information in their software. Users would also like to be able to tilt their phones and still get directional information. Most suppliers of magnetometers for compass applications also supply algorithms for their customers to use in their devices. The algorithms use "flattening equations" which relate yaw, pitch and roll tensors (rotations around the three different axis) into a two dimensional direction vector which are connected to the compass rose.

The layout of the sensors has to take advantage of the odd-function behavior of certain configurations of thin-film sensors to be able to discern the orientation of the user to any degree of accuracy. The description of the odd-function is that the sensor has a different behavior in the right-half plane than in the left-half plane. Figure 1 [4] shows the basic types of GMR structures. Standard multilayer GMR sensors, such as a copper-cobalt multi-layer, are even function devices with transfer functions which have left-half plane and right-half plane behaviors that are identical (except for a small hysteresis). Devices made with AMR films can be configured as an odd-function device by geometry, where GMR type films use the interaction of layers of magnetic and non-magnetic thin films. The types of GMR films that have an odd-function response are the spin-valve (SV) and the magnetic tunnel junctions (MTJ). These sensing films rely on their directionality i.e. the odd function behavior on the pinned layer. The concept of the magnetic tunnel junction is quite interesting. The MTJ is constructed with two magnetic layers separated by a very thin non-conductor. When the two magnetization vectors of the magnetic layers are aligned in the anti-ferromagnetic orientation, the current through the alumina is at a minima and when they are aligned in a ferromagnetic orientation the current is at a maxima. A common non-conductor used in this type of design is thin film deposited alumina. To create the odd-function response, one of the films must have its orientation pinned to a reference orientation. If this orientation ever becomes unpinned, the sensor will lose its sense of direction.

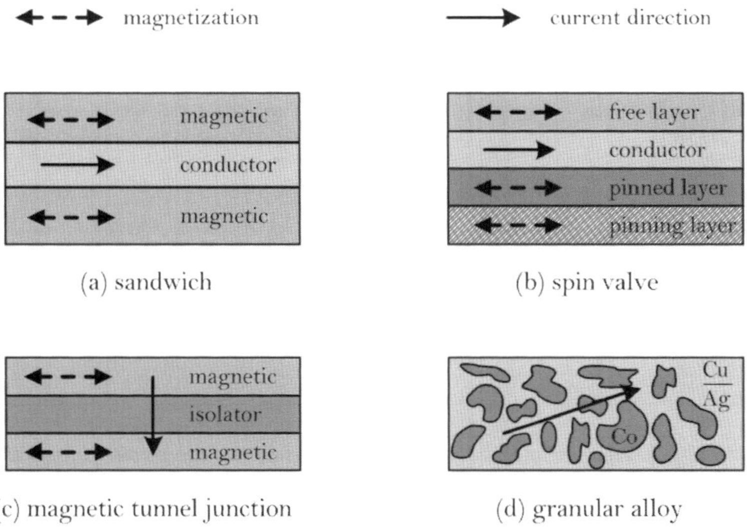

Fig. 1. Basic GMR structures from C. Reig , M.-D. Cubells-Beltrán and D. R. Mũnoz [4]

This loss of sense of direction is not exclusive to GMR sensors since AMR sensors have a similar issue. A survey of high sensitivity magnetometers by Robbes [5] compares the superconducting quantum interference devices or SQUID with GMR and anisotropic magnetoresistance devices or AMR. The author's conclusion is that the GMR devices are promising as a highly sensitive magnetometer. Marchesi [6] summarizes the different technologies used in compass applications, though the goal of his work is to develop a planer fluxgate magnetometer.

Commercial entities have introduced new and innovative thin film magnetometers for compass applications. The market for compass devices now includes magnetic tunnel junctions (MTJs) and spin-valve devices along with pole piece devices. Companies that are supplying magnetometers for Hall compass applications are Aichi Steel, Honeywell, Philips, AK, MEMSIC, Yamaha, Freescale and others. A quick survey of commercially available compass devices is shown in Table 1. These compass devices are specifically designed to minimize the space required to fit inside of cellular devices. Most of the devices are surface mounts. The performance standard for these devices is the Honeywell HMC series which is based on the barber-pole permalloy devices. All the other devices are measured against this sensor. The most comprehensive of these devices are packaged with analog to digital conversion circuits. The chip manufacturers also supply algorithmic information to aid in the use of these devices since this information, as stated earlier, relates a three dimensional vector onto a two dimensional plain.

Table 1. Commercially available thin film AMR and GMR compass products

GMI	**GMI**	**AMR**		**GMR**	
Aichi Steel	***Aichi Steel***	***MEMSIC***	***Honeywell***	***Yamaha***	
AMI306	AM603	MMC314XMS	HMC5883L	YAS529	

TMR	**TMR**	**TMR**			**AMR**
Yamaha	***Yamaha***	***Freescale***	***Multidimension#***	***Baolabs***	***SENSITEC***
YAS530	YAS532	Mag3110	MMC3031 (target)	BLBC3-D / BLB3-B	AFF756

photo not available.

2 The Earth's Magnetic Field

Any discussion of compass applications requires an understanding of the Earth's magnetic field. The Earth's magnetic structure has a definite effect on how much sensitivity required in a compass device and how to interpret the information. The Earth's composition simply is a large ball of liquid rock and metal with a very hot solid iron core and a thin cool skin. Significant research on the structure and shape of the Earth's magnetic environment performed by Glatzmaier and Roberts [7,8,9] has been performed in the last twenty years. Song [10] describes the structure of the Earth and the differential rotation of the core and mantle with respect to each other. Figure 2 [11] shows a cross-section of the Earth describing the differential rotation. This rotation was analyzed by seismic studies and determined to be as much as 1.1° per year of eastward rotation about the rotational axis. The Earth has gone through periods of time in which the magnetic poles reverse their direction. It is notable also that the core spins at a slightly faster rate than the rest of the planet which causes shear forces in the surrounding molten iron layer. The research shown in Figure 3 [11] was driven by the measurable variances in the location and direction of the local north and south poles and the magnetic reversals measured in the rocks of the Atlantic seabed. According to Glatzmaier and Roberts, this interaction between the central iron core and the molten iron core drives these field reversals. This interaction also drives the local variations in magnetic field. Their model treats the Earth's core like a dynamo i.e. an electric motor. In their model, the outer molten core acts like a stator and the core acts like the rotor. In a geologic time frame, the instability of the system results in a pole reversal every 100,000 years but only is a minor problem for compass applications in

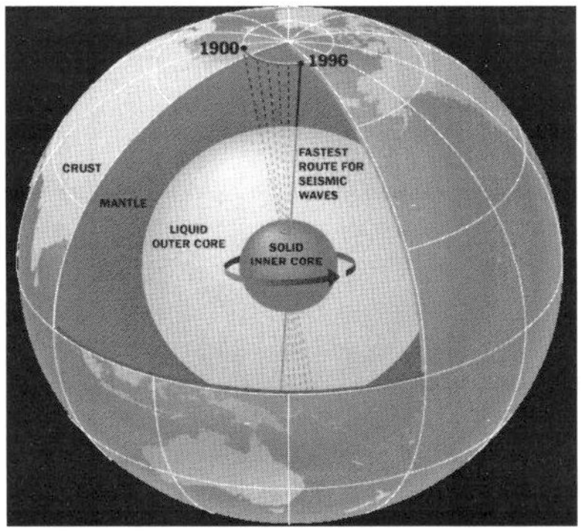

Fig. 2. Cross-section of the Earth [11] showing the direction of the solid cores rotation and the precession of the rotational axis of the core from 1900 to 1996

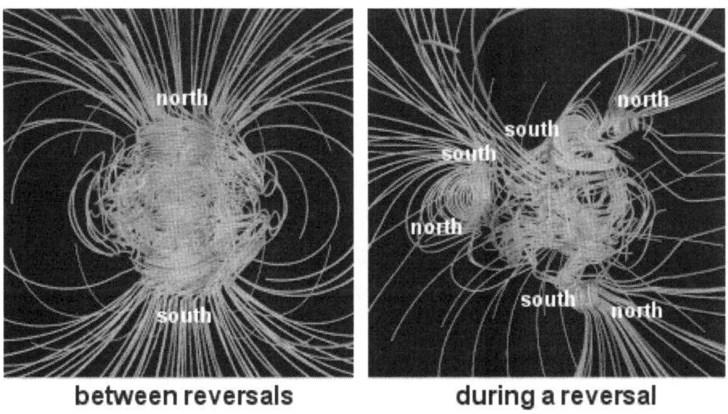

Fig. 3. Magnetic field models [11] representing a period like we are presently in and how it would look like during a reversal. This image represents 9000 years of simulation; a) represents the initial magnetic state, b) represents the midpoint. The images are from results of modeling from Glatzmaier and Roberts [9].

the short term. This has one notable exception; there are regions in the planet where compasses do not behave as expected.

The map of the Earth's present field conditions shows the difficulty in developing a compass device. The Earth's magnetic field does not always allow compasses to point north worldwide. This makes developing an application, which can take advantage of the high sensitivity of the magnetometer device, difficult at best. To improve the understanding of compass magnetometry, the United States Geological Survey (USGS) has an education page and field declination map available [12] that is only useful for the continental United States. Figure 4 [13] shows the lines of declination from the USGS. There has been, for centuries, an acknowledgment that a compass has to have augmented information for accuracy. Prior to the electronic age, the compass information was augmented by using the stars and clocks were used by mariners to navigate. After the advent of global positioning, navigation can be performed by the use of hand held GPS units. Many digital compasses use built in zone maps to correct for these magnetic variations.

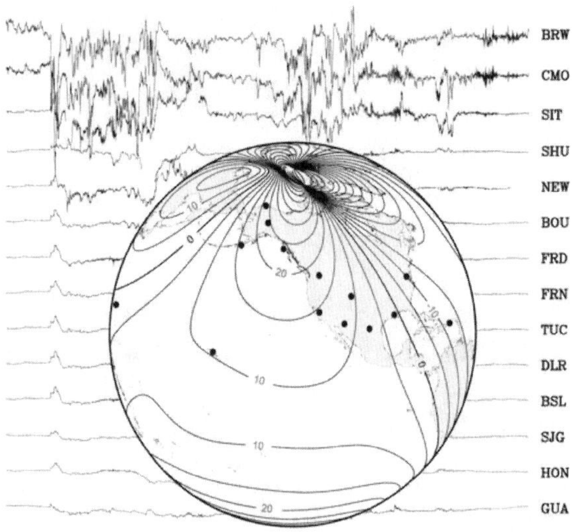

Fig. 4. Magnetic field map of the northern hemisphere as viewed from the Pacific ocean from the USGS [12]. It is obvious from this image that the Earth's magnetic field is not a simple north south reference.

3 Compass Concepts

Compass devices have been produced in two different modes. The first is the 2-D compass which relies on a consistent z-axis orientation and the other type is the 3-D magnetic sensor, which does not depend on a consistent z-axis. For the 3-D device, it is important to project the three dimensional magnetic field vector $\vec{B}$ onto a two dimensional plane to be able to ascertain user orientation. The components of the field can be broken into the three Cartesian directions as shown in Figure 5. By convention, the magnetic field normal to the Earth's surface is designated as the z - direction. A compass, (spinning magnet type) is generally set flat on a surface and the with the z direction normal to the plane of orientation. On a boat or an aircraft the compass is usually gimbal mounted to keep the compass level. This makes the measurement a pure two dimensional problem and can be handled by a two dimensional axis rotation matrix. The problem can then be solved by finding the angle in which the y direction is maximized and the x direction is zero. At that point the y direction is south and a reference direction is set. It then becomes relatively simple to orient the user to a compass rose. The angle is then found using trigonometry.

A common way to orient and calibrate an electronic compass requires that the compass is placed in a mode in which the user either drives his automobile in a circle or waves the compass in a figure eight by hand. It is possible to make a calculation without the rotation but there will be an error due to the small variations in the voltage offset and sensitivities involved in the manufacture of these sensors. To solve this 2-D angle, we designate that there are two vectors, X and Y, which designate the bridge voltage signal value. The magnitude of this signal is

$$V_{mag} = \sqrt{X^2 + Y^2} \ . \tag{1}$$

The result is then used to determine magnetic north

$$Y_{north} = -V_{mag} \ , \tag{2}$$

and the value of $X=0$, since most compass devices are vector sensors and have an odd-function transfer curve which gives a positive output in the right-half plane and a negative output in the left-half plane. The Y sensor will have a positive output on the top-half and a negative output on the bottom half of the plane. This gives a very simple vector rotation problem where the angle is

$$\cos(\theta) = \frac{Y}{V_{mag}} \ . \tag{3}$$

This analysis completely leaves out errors due to local field variations and actual sensor voltage-offsets. Unfortunately this only works for limited cases and most real sensors have variations in sensitivity and offsets. Most of the newer applications need three axis (*x*, *y*, and *z*) of information since they are mostly in hand held devices. It is easy to see the difficulty in figuring out the directions by running a simple thought experiment. If we orient our two dimensional sensor at a forty-five degree angle, we should have an equal amount of signal in both sensors. If we rotate the sensor on the x axis the magnitude of the *X* component will stay constant while the *Y* component will vary. This rotation will cause a significant directional error. To compensate for the error, most designs include a z-axis sensor. The z-axis sensor allows the magnetic sensor to accurately determine the direction of the magnetic field vector, so as to correct for this tilt. There is a single point in which the method breaks down. If all of the signal is in the z-axis sensor, then there is no projected component in either the x or y direction and the compass is indeterminate.

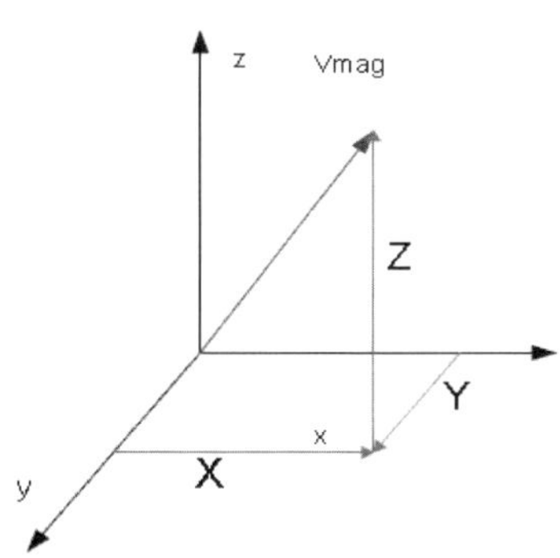

Fig. 5. Simple 3-d representation of a magnetic field. The projection of this vector onto the the relative position of a three axis magetometer is what is necessary to determine position.

The rotation of the reference plane requires the sensor manufacturer to supply a set of flattening equations to allow for easy implementation, by the customer, of the sensor. The flattening equations are based on the concept of roll, pitch and yaw. These equations are well known and can be found in numerous sources. The following equations [14] are matrices for rotation: α around the x-axis, β around the y-axis, and γ around the z-axis.

$$R_x(\alpha)=\begin{bmatrix} 1 & 0 & 0 \\ 0 & \cos\alpha & -\sin\alpha \\ 0 & \sin\alpha & \cos\alpha \end{bmatrix} \quad (4)$$

$$R_y(\beta)=\begin{bmatrix} \cos\beta & 0 & \sin\beta \\ 0 & 1 & 0 \\ -\sin\beta & 0 & \cos\beta \end{bmatrix} \quad (5)$$

$$R_z(\gamma)=\begin{bmatrix} \cos\gamma & -\sin\gamma & 0 \\ \sin\gamma & \cos\gamma & 0 \\ 0 & 0 & 1 \end{bmatrix} \quad (6)$$

The general rotation matrix is $R=R_xR_yR_z$ and by identifying the values of the rotational angles, the relationship of the compass plane can be referenced to the Earth's surface. As mentioned earlier, these values are described as roll, pitch, and yaw and are compared to a local axis reference. Even though the Earth is spherical, the observer sees the surface of the Earth as a flat plane. This is why we can use the Cartesian or orthogonal coordinate system. The transformation of the vectors onto a new coordinate system can done by using the following relationship

$$V_{mag}'=RV_{mag} \;. \quad (7)$$

The final result of this type of transformation is represented by Hong Wan [15] in the following equations

$$X'=Xcos\beta+Y\sin^2\beta-Z\cos\beta\sin\beta \;, \quad (8)$$

$$Y'=Ycos\gamma+Z\sin\gamma \quad (9)$$

where the heading is

$$\alpha=\tan^{-1}(Y'/X') \;. \quad (10)$$

Several cell phone manufacturers, such as Apple [16], have combined the results from similar equations with the results of the their accelerometers into fairly accurate global orientation methods. It is very important for a magnetometer to have a reasonably high signal to noise ratio for these methods to be used.

Another problem, when using a compass device, is that there are significant points in orientation on the Earth's surface in which there is no directional information. If the z sensor device is directly aimed in the z-direction, there will be no x and y axis information since there is no projection of the vector onto the x and y direction. This means also that near this point the signal is relatively small. The low value of this signal can create heading errors.

4 GMR Sensor Behavior

The use of GMR sensors requires a brief discussion of the general behavior of GMR material and how this affects compass behavior. Giant magnetoresistive sensors have generated a significant amount of interest as compass devices. This interest is due to the high magnetoresistive ratio i.e $\Delta R/R_0$. From Figure 6 [17] it is easy to see that a comparison of the key GMR technologies with the AMR devices show that the difference in sensitivity is greater than an order of magnitude. This difference requires additional handling to try to match the sensitivity. When GMR sensors were first invented, the most common type of sensor was the multilayer device. This GMR sensor usually consists of a structure similar to the one in Figure 1a, which is a layer of magnetoresistive material followed by a non-magnetic spacer then another magnetoresistor. To increase the $\Delta R/R_0$, the pattern is repeated. This type of magnetoresistor is not very useful for a compass since it behaves as an even function. The most common type of GMR sensor used for a compass device is the spin valve. Figure 1b shows a typical spin valve structure which has a pinned layer and a pinning layer to set the magnetic behavior to a reference direction. Figure 7 shows a typical spin-valve response from NVE Corporation [18]. A major characteristic of these type of magnetoresistors is the odd-function response and the hysteresis behavior of these spin valves. The odd-function response allows for a directional component from the output of the sensor. This directionality is key to the compass. The hysteresis behavior is driven by instabilities and generally is not something that can be changed.

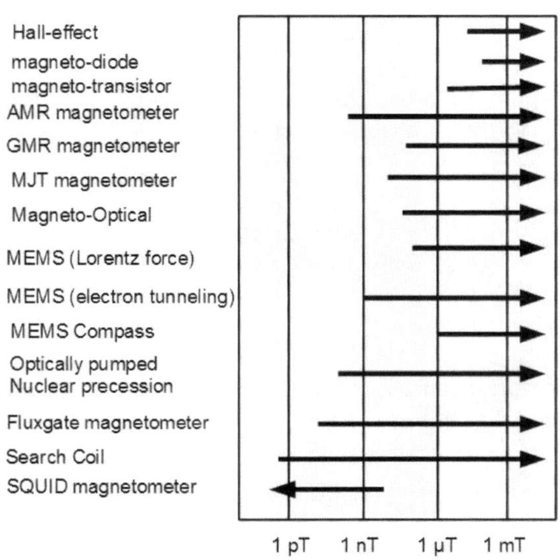

Fig. 6. Sensitivity range of different magnetic devices used for magnetometry from Lenz [17]

This hysteresis behavior requires that the compass designer put in some provisions to limit exposure of the sensor to large external magnetic fields. These provisions are often maximum field exposure values which limit the field exposure to less than 10 Gauss or 1 milli-Tesla. A common method to make a GMR sensor sensitive enough to be used as a compass device is the addition of pole pieces. This addition is needed to increase the directional bias due to the problem that the GMR sensor is sensitive to more than one field direction. Historically, anisotropic magnetoresistors, AMR, are used for thin film compasses due to their behavior which is directly coupled to current direction. In the barber-pole configuration, this device is naturally odd-functioned and a low cross-sensitivity. These AMR sensors have a naturally high sensitivity to low fields, but have a low dynamic range. The low dynamic range forces the user to make the magnetic environment of the sensor magnetically quiet.

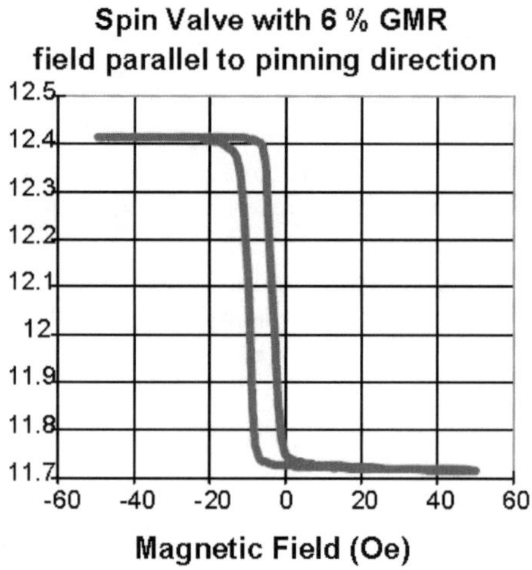

Fig. 7. Measured resistance *vs.* applied field for an antiferromagnetically pinned spin valve with the field applied parallel to the magnetization of the pinned layer with GMR = 6 % from Smith and Schneider[18]

To compete with this sensitivity, the soft magnetic pole pieces are also needed to intensify the field. These pole pieces require a significant amount of design, usually performed using some form of finite element analysis. The GMR compass unfortunately, has another disadvantage compared to the AMR compass. For the GMR sensor, unlike the AMR sensor, there is no equivalent structure to the barber-pole sensor. Figure 8 shows a typical schematic of a resistive bridge using GMR elements [19]. Figure 9 shows shows an actual sensor with two of these elements are used as reference devices while two are used to detect the field [19]. This cuts the effective signal in half as compared to the AMR device which can use all four elements. The advantage that the GMR devices have is that they are relatively small for the equivalent impedance. A typical AMR sensor has a sheet resistance of 6 Ω/□ to 11 Ω/□ where a typical spin-valve has a resistance of 16 Ω/□. The dynamic range of a GMR sensor is also greater than that of an AMR sensor. The best AMR sensors have a 3% $\Delta R/R_0$ and a 6% $\Delta R/R_0$ for the GMR spin-valve. Even with the addition of the pole pieces, these sensors can occupy less area than the equivalent AMR sensor. This size difference is due to the non-active elements being under the pole-pieces. As shown in Figure 6, the sensitivities of the GMR devices, even with the pole pieces, are going to be lower than the equivalent AMR sensor. Compass devices, for hand-held applications such as cell phones, only require between one to five degrees of accuracy. This range of directional accuracy is achievable using GMR compass devices. The specific GMR device that approaches the behavior of the AMR device better than the spin valve is the magnetic tunnel junction or MTJ.

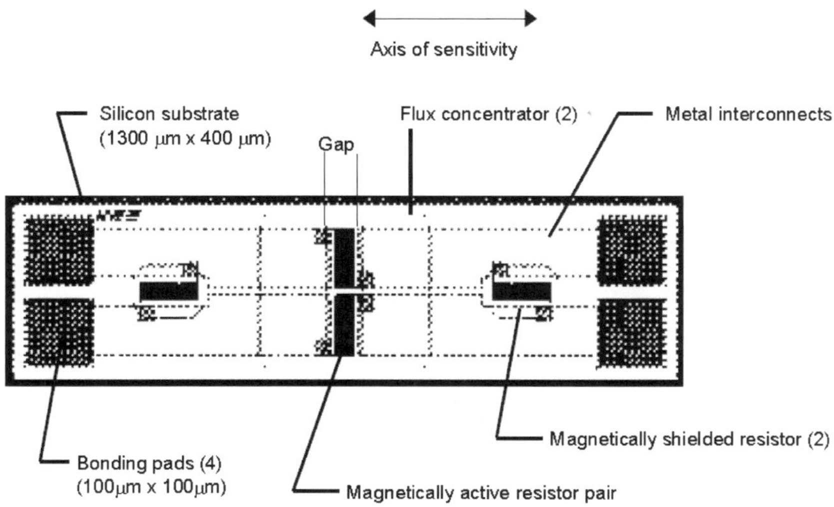

Fig. 8. Schematic representation of a typical GMR sensor [19]. The pole pieces also block the field from the two reference resistors, R_4 and R_3.

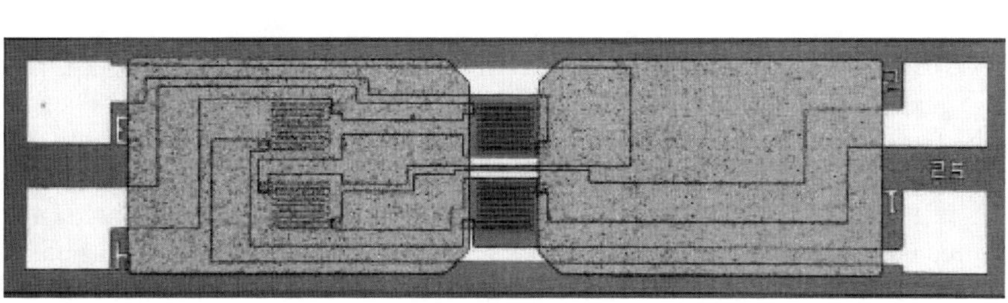

Fig. 9. GMR sensor from NVE [19] with integrated pole pieces. Note that the inactive sensors are shielded from the external field by the soft-magnetic pole pieces.

Magnetic tunnel junction devices, unlike spin-valve devices, modify the resistance through the thickness of the devices. The current, in essence, is perpendicular to the sense plane. Magnetic tunnel junctions have the same issues as the spin-valve devices and require pole pieces to form a directionally sensitive sensor. Figure 10 shows a comparison of detectivity versus frequency of several commercial sensors [20]. The MTJ and spin-valves have issues with 1/f noise on top of the hysteresis issues. Generally, a compass device is measuring a very low DC field which may or may not have noise depending on the last state of the sensor. Another condition which may impact the behavior of the compass will be the impact of environmental noise i.e. the effect of AC fields generated by AC electrical equipment.

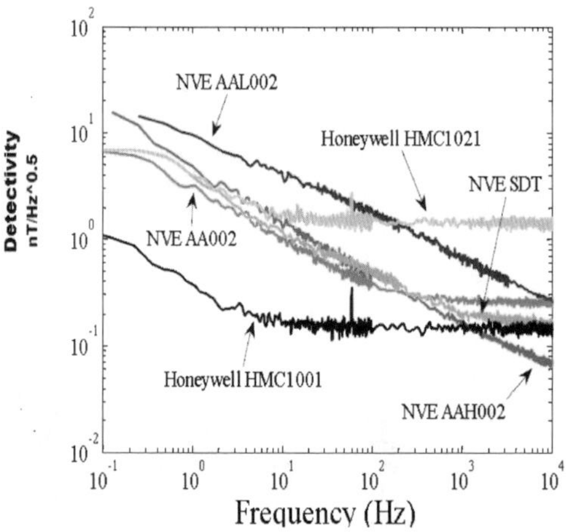

Fig. 10. MTJ sensor with flux concentrators from A. Jander et. al. [20]. Magnetic field detection in the low frequency range is shown.

Since background AC magnetic fields are ubiquitous, we will need to consider the effect of 1/f noise on the compass device. It turns out that the 1/f noise in GMR sensors is greater than the noise in AMR sensors by several orders of magnitude [21]. The source of magnetic noise in permalloy (AMR) is most likely lattice noise [22]. The 1/f noise in GMR is also dependent on the voltage bias of the sensor. Wan et. al. [21] also noted that the 1/f noise at low frequencies can change by a factor of 100 with a voltage bias change of a factor of 40 [23]. For MTJ sensors, the additional noise generated by tunneling electrons also affects the signal to noise ratio. The effect of the free layer thickness directly affects the 1/f noise and the sensitivity [24] of the MTJ devices.

As the free layer is reduced in thickness, the 1/f noise reduces but the MR ratio is also reduced. An additional design consideration for the use of GMR sensors as a compass device is the effect of temperature on the both the conductivity and on the sensitivity. Smits [25] shows that the conductivity of the MTJ, over temperature, is linear in the range of interest (233K-353K). Another parameter that can be used, is the current density and how it changes the sensor behavior. Russek [26] demonstrates the effect of current density on a spin valve as shown in Figure 11. This current bias effect can also be used to design the sensitivity of these compass.

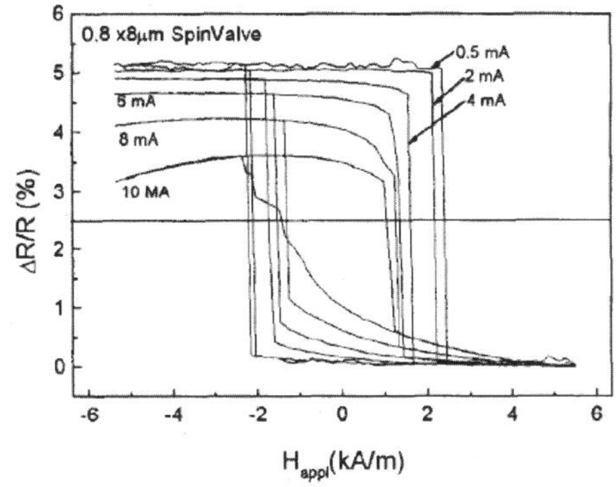

Fig. 11. Russek [26] et. al. plot of resistance change as the bias current is changed

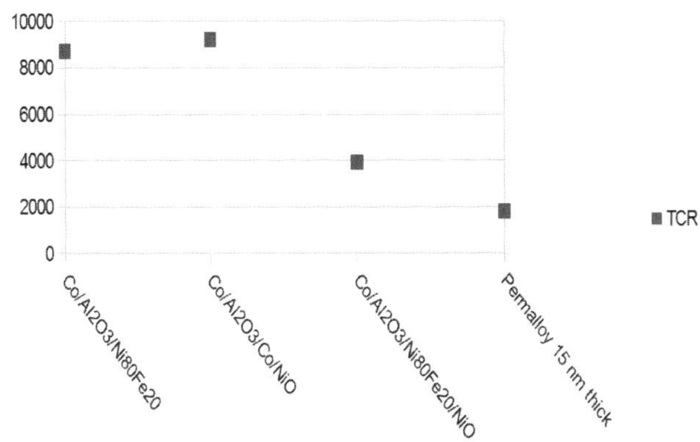

Fig. 12. Conductance versus temperature for three different MTJ devices and 15 nm permalloy. These films are $Co/Al_2O_3/Co/NiO$, $Co/Al_2O_3/Ni_{80}Fe_{20}/NiO$, $Co/Al_2O_3/Ni_{80}Fe_{20}$) and 80 Ni 20 Fe permalloy [25].

The importance, for a compass device, of the error sources is that the total signal necessary to find one degree of rotation is very small. If we take a simple Wheatstone bridge of approximately 1 kΩ with a 6% GMR over a 1mT range as shown in Figure 7 and use a pole piece construction as shown in Figure 8. This will increase the sensitivity by a factor of 2 to give us a sensitivity of 1.2%/ 10^{-4}T so we can calculate the bridge output with the following equation

$$V_{out} = V_s * \left(\frac{R_1}{(R_1 + R_4)} - \frac{R_3}{(R_3 + R_2)} \right) \tag{11}$$

where V_{out} is the output of the bridge and V_s is the supply voltage. With a supply voltage of 5V, the bridge output would be around 30 mV/10^{-4}T. If we then assume that the Earth's magnetic field in the area of Chicago, USA is around 16 mT then, small angular changes will give very small changes in the field. To measure a one degree change, (the difference from zero degrees to one degree) the field changes by $2.4x10^{-6}T$. This means that the voltage difference is 16 μV and the minimum requirement to guarantee a one degree angle change (due to Nyquest sampling rules) is 8 μV . The minimum noise level then must be less than the 8 μV. If the device is exposed to a field level that saturates the sensor, it is likely that the sensor may not continue to respond at the same level in which the sensor was calibrated. Hysteresis is, by its nature, an effect that is repeatable though difficult to deal with. For most compass schemes, it is important to reset the sensor back to its original state. In AMR sensors, the hysteresis is reset by means of set-reset field straps. This is the method that both help to set a reference direction and zero out any bridge offset. Another significant problem in compass measurement is the effect of temperature. This problem is clearly shown in Figure 12 [26] and in Chien-Tu Chao et. al [27]. This graph shows the variation of ambient temperature on the conductivity of MTJ sensors. This material changes in base resistivity, in sensitivity and hysteresis loop size. This temperature behavior requires, to improve repeatability and signal accuracy, a temperature sensor to either directly put into a feedback scheme or as a reference in for a digital circuit.

The third type of GMR that is used for compass devices is giant magneto-impedance or GMI. The GMI effect is an offshoot of the MI or magneto-impedance. Magneto-impedance relies on the frequency skin-effect which is modified by the magnetic field. It was initially documented in the 1990's [28] and detailed in multiple experiments over time. Initial structures tested were made from ribbons of amorphous materials [29].

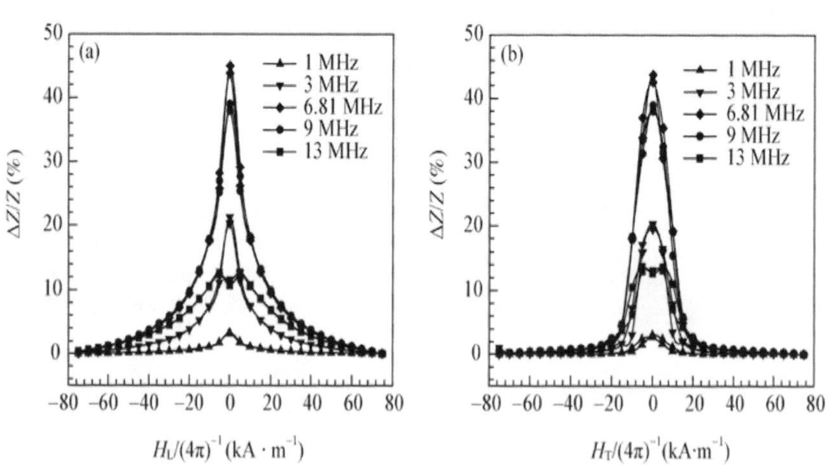

Fig. 13. Frequency behavior of GMI sensor between 1 MHz to 13 Mhz. This response is an even-function and not as suitable for compass application [29].

This sensitivity to the applied field, for the GMI, is frequency based, as shown in Figure 13 [29]. These ribbons originally were rather large and not suitable for compass devices but have benefited from miniturization methods used in the electronics industry. Morikawa et al. [30] developed a thin-film GMI sensor for Toyota research using thin films with a layered structure. The structures are Co-Si-B/Cu/CoSi-B,vCo-Si-B/Ag/Co-Si-B, and Fe-CoSi-B/Cu/Fe-Co-Si-B. Figure 14 [30] shows the sensor schematic used to test GMI by Morikawa et al [30]. Figure 15 [31] the results of testing the GMI thin-film sensor constructed by Maylin *et al.* showing the effect of frequency on the sensitivity. This device still produces an even function type of transfer function. An even function is a mathematical function that has a result that is the same in the left-half of the plane as it does in the right half of the plane. The behavior follows

$$Z = R\frac{\alpha}{[2\delta_0]}\left(\sqrt{\mu_R} - j\sqrt{\mu_L}\right) \tag{12}$$

$$\delta_0 = \sqrt{2\frac{\rho}{\omega}} \tag{13}$$

$$\mu_R = |\mu_t| + \mu_l'' \tag{14}$$

$$\mu_L = |\mu_t| - \mu_l'' \tag{15}$$

where α is the wire radius, R is the resistance, ρ is the resistivity, ω is the angular frequency, and μ_t is the circumferential permeability of the wire. The permeability μ_t is complex, and μ_r significantly modifies the impedance at frequencies above 100 MHz. Unfortunately these sensors, to work properly, have very low starting impedance which increases their power consumption. It also makes the pattern sizes quite small. Additional work on GMI devices have shown that a DC bias can create an asymmetrical response, as shown in Figure 16, which gives the odd-function response necessary to produce a compass [32]. To achieve an odd-function result, the top and bottom layers can have anisotropy angles of between 15° and 45°. Figure 17 [33] compares the GMI output to an single AMR barber-pole resistor. The function described still has an issue with the output of of the sensor pair having the same value at different fields.

This problem would only be an issue when the sensor is exposed to fields not normal in nature such as hard magnets and electric motors. To make a functional compass, the device made from these elements would need to be reduced in size so that all three axis could be measured.

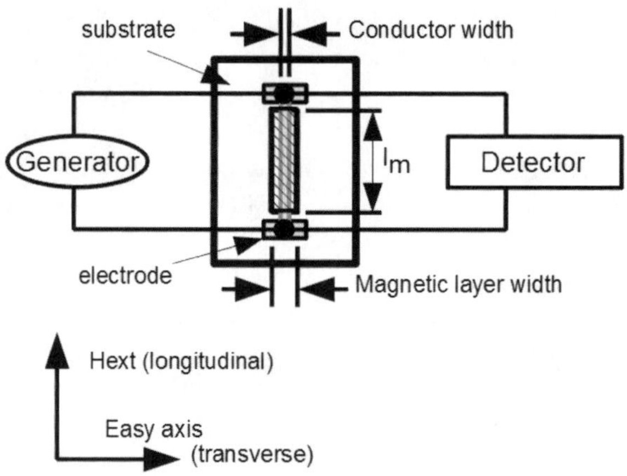

Fig. 14. Morikawa et al. [30] thin film GMI sensor. (a) Is the top view and (b) is the cross-sectional view.

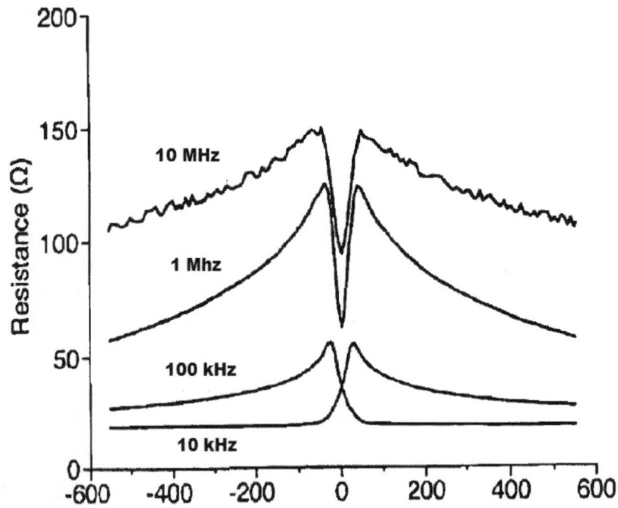

Fig. 15. Maylin et al. [31] thin film GMI sensor impedance results at 0.01, 0.1, 1, and 10 MHz. The output is an even function which is not ideal for a compass. This is because a vector pointing left has the same magnitude as a vector pointing right.

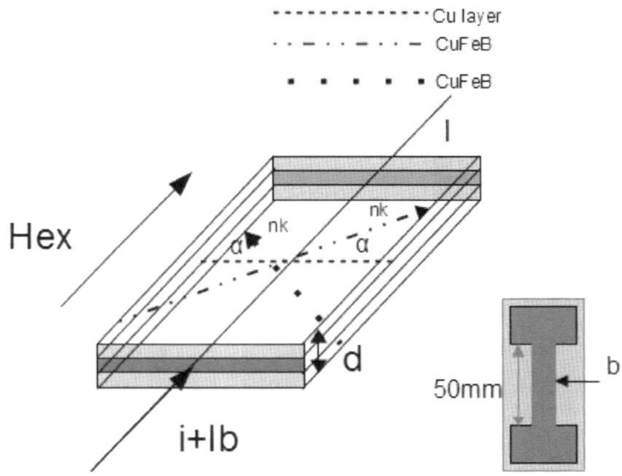

Fig. 16. Schematic of an experiment run by Delooze et. al. [32] (a) Film layer structure and principle quantities and directions. (b) In-plane view of the test structure.

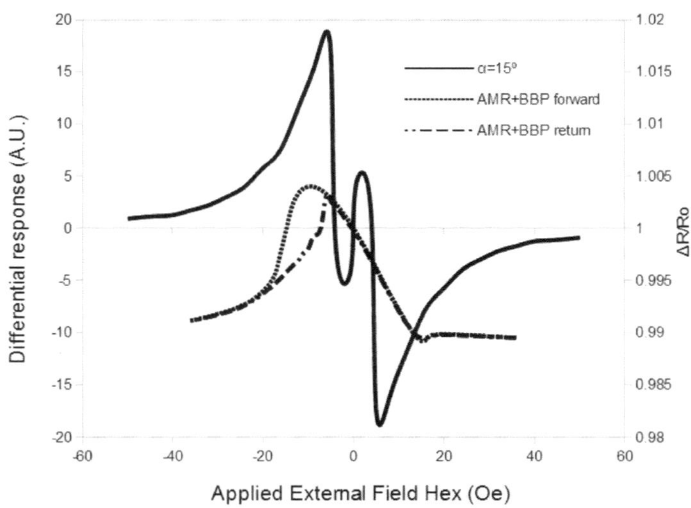

Fig. 17. Data taken from Delooze et. al. [32] and the Author. This device is a combination of a GMI devices using opposite DC current directions on two sensors compared with a typical barber-pole AMR sensing element. The excitation frequency used for the GMI is 90 MHz, b = 40 μm and I = 25 mA.

5 Commercial GMR Compasses

Table 2 shows a comparison of some selected commercially available GMR compass devices. Table 2 is constructed from the available data sheets from these selected manufacturers. Also included in this table is a comparison of two equivalent AMR devices. All of the devices shown have A/D converters to allow for compensation algorithms. The table shows four basic types of compass devices. These are, as have been discussed, AMR, Spin-valve, MTJ and GMI based sensors. Some of the issues, such as hysteresis, that were discussed in earlier sections show up in the commercial sensors.

The AMR sensors were included to act as a reference for the GMR sensors. It should be noted that the MEMSIC device in this table has a sensor licensed by Honeywell to MEMSIC. With this licensing, any variations in performance are probably due to the application specific integrated circuit or ASIC that was mated to the sensor. All five devices in this comparison have custom complimentary metal oxide semiconductor (CMOS) application specific integrated circuit (ASIC) devices to perform the analog to digital conversion and communication to an output device. Common output devices for these include automotive mirrors, cell phones and hand held GPS devices. Manufacturers over the years who have used such devices are Magna Donnelly, Garmin, Motorola, and Apple. The GMI sensor, from Aichi Steel, in this table, solves the dimensional issue and the vector issue by using fluxgate magnetometry methods. The size of the package, for the compass devices, is being driven by the needs of the digital phone market. The chief marketing handle for these communication devices is the size of the phone itself. Thin phones are important to the consumer base.

The thickness of the package, due to the z-axis sensor, is the biggest issue. In some devices, the z-axis device is a separate chip, mounted on its side. In other devices, the z-axis sensor is in the plane of the xy sensors but with z-axis pole-pieces. All the devices shown in Table 2 are between eight hundred microns to one millimeter in thickness, which seems to be a manufacturing limit. As expected, the largest devices are the AMR devices. These devices have the lowest impedance to device area. This then requires that the AMR sensor needs to grow quite large to achieve a reasonable resistance. The smallest devices are the MTJ sensors. The MTJ sensors have the highest impedance and therefore require less physical space on the chip.

An important consideration for the compass device is the input-output configuration. This means analog to digital conversion bits, current consumption, interface type and interrupts. The interface type that is used in these commercial devices is I^2C which is a two wire communication protocol developed by Philips Electronics. It can be used with either seven or ten bit words and is a very common interface type. This basic circuit type is readily available at numerous foundries internationally. The continuous current consumption also gives insight on how much time the device remains "on".

The Yamaha device has the highest current draw is either "on" a significant amount of time and or has a significantly lower resistance in the sensor. The Aichi Steel and the Freescale devices have interrupt lines to add additional handshaking. The lowest supply voltage is 1.7 volts and the highest voltage is 5.25 volts with the majority able to operate at or around 3 volts. The analog to digital converters, ADC, range from 10 bits to 15 bits. The ADC also can control the accuracy of the commercial sensors depending on the volts/bit. The sense range of these devices is one to twelve Gauss of external applied field.

The basic performance of these selected commercial devices over temperature and initial offset versus temperature induced offset is important. These parameters are listed in Table 2 and can be compensated for using a temperature sensor.

Table 2. Selected AMR and GMR devices and their specifications from their data sheets. All these devices have signal processing. At this time, some of these devices are only prototypes.

General	Technology	**GMI**	**AMR**		**GMR**	**MTJ**
	Company	***Aichi Steel***	***MEMSIC***	***Honeywell***	***Yamaha***	***Freescale***
	Product	AMI306	MMC314XMR	HMC5883L	YAS529	Mag3110
PKG	PKG	LGA 10	LGA 10	LGA 16	WLCSP 10	DFN 10
	Size (mm)	2 * 2 * 1	3 * 3 * 1	3 * 3 * 0.9	2 * 2 * 1	2 * 2 * 0.85
I/O	Voltage (V)	1.7 ~ 3.6	2.7 ~ 5.25	2.16 ~ 3.6	2.5 ~ 3.6	1.95 ~ 3.6
	Current_continuous working (mA)	>1	~2	~2	4	>1
	Current_samples per second (mA)	0.15 @ 20sps	0.55 @ 50sps	0.10 @ 7.5sps	4	0.14 @ 10sps
	Interface	IIC	IIC	IIC	IIC	IIC
	Interrupt	Y	/		/	Y
Maximum Ratings	Storage temp	-40 ~ 125	-55 ~ 125	-40 ~ 125	-50 ~ 125	-40 ~ 125
	Operating temp	-20 ~ 85	-40 ~ 85	-30 ~ 85	-40 ~ 95	-40 ~ 85
	Max exposed field		10000G		2000G	1000G
Performance	Range (+-Gauss)	12	4	1~8	3	10
	ADC (output bits)	12	12	12	10	15
	Sensitivity/Resolution (mGauss)	6	2	2	6 for XY / 12 for Z	1
	Offset (+-Gauss)		0.2			0.01
	Accuracy (deg)	1	2	2	5	
	Linearity (%FS)	0.5	1	0.1		1
	Hysteresis (%FS)		0.1	0.0025		1
	Repeatability (%FS)		0.1			
	Sensitivity TC	+-7% @0~60degC	0.11%/degC	+-5%		0.1%/degC
	Offset TC	+-3mG/degC @0~60degC	+-0.4mG/degC			+-0.1mG/degC
	Bandwidth (Hz)		40	75	40	40
	Noise (RMS)		0.6mG@25Hz			0.5mG
Features	Onchip temp sensor	Y	Y		Y	Y
	Single-chip-integration	/	/	/	Y	
	Offset removal	Y	Y	Y	Y?	
	Self test	/	/	Y	/	Y
	Others				3 AD for external	Oversampling configuration

If the application needs a wide temperature range of operation, these values must be accounted for. If the user combines other types of sensor information with the compass values it is still possible to compensate for this offset. Most hand held devices, such as cell phones, also have global positioning devices, gyroscopes, and accelerometers. The maximum external field that the GMR and MTJ sensors can be exposed to is also quite important since both sensors depend on a pinned layer. The maximum field for the MTJ and the spin-valve GMR devices are 1000 G to 2000 G. This seems high, but it is the level that a hard magnet in a circuit assembly line can produce. Often production assembly lines use magnets to activate proximity sensors. These sensors are used indicate the position of assembly tooling and can expose these types of devices to high fields. These compass devices are also sensitive to solder temperature and epoxy curing temperatures due to the thermodynamics of thin film diffusion. The maximum storage temperatures for these devices is 125°C. This generally means that the epoxy curing temperature should be 100°C or less. The wire bonding temperatures should also be limited to this range.

6 Discussion

GMR compass devices have expanded the market for electronic compasses due to their acceptable sensitivities and large dynamic ranges. These devices have significant issues, noise, hysteresis, and sensitivities. The initial issue is the cross-sensitivity of the typical GMR sensor and it bears restating how much this effects everything to do with this class of sensor. These sensors require magnetic pole pieces to enhance the directionality or the accuracy becomes poor. Figure 18 [18] shows a demonstration of the cross-axis behavior using pinned spin-valve devices. Hill [33] compared multi-layered GMR sensors versus spin-valve sensors for vector applications. The analysis in Hill [33] demonstrated that the more traditional multi-layered GMR devices are better vector devices. Unfortunately, multi-layered devices have sensitivity issues.

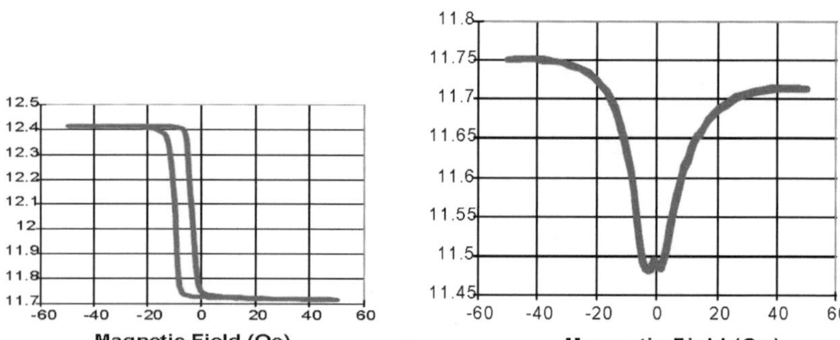

a. b.

Fig. 18. Typical behavior demonstrating of cross-axis sensitivity issues as related to an anti-ferromagnetic coupled spin-valve [18]. a) Field parallel to the pinning direction (NVE 6% device). b) Field perpendicular to the pinning direction (NVE 2.4% device).

The spin-valve device in this study had poor cross-axis sensitivity. This is not the complete picture, a spin-valve device does have cross-axis behavior in one direction and none in the other direction. This cross-axis behavior probably does not really effect the actual compass device since the pole-pieces do most of the selectivity. To contrast this with the AMR sensor, the cross-axis sensitivity of the standard AMR sensor is completely dependent on magnetization direction (which can be controlled as in the set reset scheme sold by Honeywell). In an AMR compass device the magnetization can be controlled by the set-reset coil. The introduction of commercial GMR sensors has proven that these type of devices are viable. The goal of increasing the sensitivity of GMR sensors has driven researchers to try variations of other sensing methods to create novel sensors. The methods used by Aichi Steel may in the end be the most sensitive and effective magnetometer, this is due to the incorporation of the fluxgate technique as shown in Figure 19 [34].

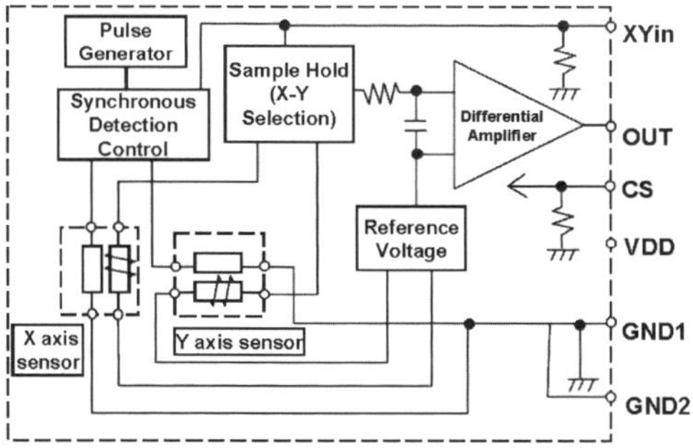

Fig. 19. Schematic of Aichi Steel's flux-gate GMI sensor [34]. This is for a 2-axis system.

The idea of adding magnetic structures to micro-electronic machines [35, 36] has expanded the possibilities for compassing. Figure 20 is a schematic of a proposed MEMS concentrator structure. The flux concentrator i.e. pole-pieces are placed on "springs" and "comb drives" to modulate the magnetic field. The magnetic sensor is placed on a spring structure [37] in-between the pole pieces. These type of compass devices may be accurate but have additional manufacturing issues, such as etch chemistry compatibility, which need to be answered. The additional complication of repeatability and calibration will still need to be answered. A new direction for compass devices is the use of nanotechnology to create a new category of vector sensor [38, 39].

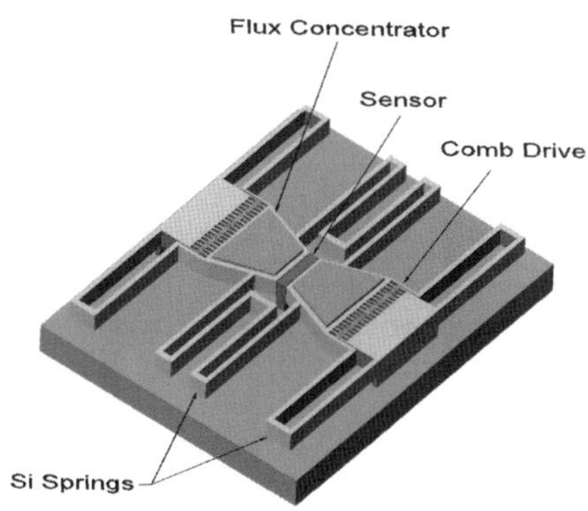

Fig. 20. Edelstein's [37] concept for a MEMS flux concentrator to reduce 1/f noise

Colossal magnetoresistors have issues still with operational temperature range which will in the short run impede implementation. These include nanowires made from variations on permalloy to colossal magnetoresistive materials.

7 Conclusions

Giant magnetoresistors are capable of being effective compass devices. Certain repeatability errors will have to be managed for the GMR devices to supplant AMR devices for high accuracy applications. As with all new technologies, it is now a race to come up with a method that can be accurate and manufactured in an inexpensive and reliable manner. The cost for implementation will always drive whether or not a new technology will be adopted. Compass devices are part of our history of exploration and discovery. From ancient times, where the compass made navigation across the Mediterranean possible, to modern times as an aid to personal navigation to and through the local shopping mall. We continue to find more ways to make compasses and new applications to use these devices.

References

[1] Morley, L.W.: Early Work Leading to the Explanation of the Geomagnetic Imprinting of the ocean Floor. EOS 67(36), 665–666 (1986)

[2] Gubbins, D., Herrero-Bervera, E.: Encyclopedia of Geomagnetism and Paleomagnetism, August 17. Springer (2007)

[3] Tumanski, S.: Thin Film Magnotoresistive Sensors (Series in Sensors), 1st edn. Taylor & Francis (June 8, 2001)

[4] Reig, C., Cubells-Beltrán, M.-D., Ramírez Muñoz, D.: Magnetic Field Sensors Based on Giant Magnetoresistance (GMR) Technology: Applications in Electrical Current Sensing. Sensors 9, 7919–7942 (2009)

[5] Robbes, D.: Highly Sensitive Magnetometers—A Review. Sensors and Actuators A 129, 86–93 (2006)

[6] Marchesi, M.: Fluxgate Magnetic Sensor System for Electronic Compass, Dissertation, Universita' Degli Studi di Pavia

[7] Glatzmaier, G.A., Roberts, P.H.: A Three-dimensional Convective Dynamo Solution with Rotating and Finitely Conducting Inner Core and Mantle. Phys. Earth Planet. Inter. 91, 63–75 (1995)

[8] Glatzmaier, G.A., Roberts, P.H.: A Three-dimensional Convective Dynamo Solution with Rotating and Finitely Conducting Inner Core and Mantle. Phys. Earth Planet. Inter. 91, 63–75 (1995)

[9] Glatzmaier, G.A., Roberts, P.H.: An Anelastic Evolutionary Geodynamo Simulation Driven by Compositional and Thermal Convection. Physica D 97, 81–94 (1996)

[10] Song, Richards: Seismological Evidence for Differential Rotation of the Earth's Inner Core. Nature 382, 221–224 (1996)

[11] `http://www.nasa.gov/vision/earth/lookingatearth/29dec_magneticfield.html`

[12] `http://education.usgs.gov/lessons/compass.html`

[13] `http://geomag.usgs.gov/`

[14] `http://mathworld.wolfram.com/RotationMatrix.html`

[15] Wan, H.: System for Using a 2-axis Magnetic Sensor for a 3-axis Compass Solution, United States Patent 6,836,971 B1

[16] Mayer, R., Piemonte, P., Huang, R., Patel, P.: Magnetometer Accuracy and Use, United States Patent 7891103 B2

[17] Lenz, J., Edelstein, S.: Magnetic Sensors and their Applications. IEEE Sensors Journal 6, 631–649 (2006)

[18] Smith, C.H., Schneider, R.W.: Low-Field Magnetic Sensing with GMR Sensors. Nonvolatile Electronics, Inc., http://www.nve.com/docs/doc430dcd6b2a5c4.pdf

[19] NVE Magnetic Sensors Catalog

[20] Jander, A., Smith, C., Schneider, R.: Magnetoresistive Sensors for Nondestructive Evaluation. Presented at the 10th SPIE International Symposium, Nondestructive Evaluation for Health Monitoring and Diagnostics, Conference 5770

[21] Wan, H., Bohlinger, M.M., Jenson, M., Hurst, A.: Comparison of Flicker Noise in Single Layer, AMR and GMR Sandwich Magnetic Film Devices. IEEE Transactions on Magnetics 33(5) (September 1997)

[22] Briaire, J.: 1/f Noise in Permalloy. Technische Universiteit Eindhoven (2000)

[23] Nor, A.F.M., Hill, E.W.: Noise Power Spectral Density in Single-Strip NiFeCo–Cu GMR Sensors. In: InterMag Europe 2002, April 28- May 2 (2002)

[24] Wisniowski, P., Almeida, J.M., Freitas, P.P.: 1/f Magnetic Noise Dependence on Free Layer Thickness in Hysteresis Free MgO Magnetic Tunnel Junctions. IEEE Transactions on Magnetics 44(11) (November 2008)

[25] Smits, A.A.: Tunnel Junctions Noise and Barrier Characterization. Eindhoven University of Technology (2001)

[26] Russek, S.E., Oti, J.O., Kim, Y.K., Cross, R.W.: Field Angle and Current Density Effects in Submicrometer Spin Valves for Digital Applications. IEEE Transactions on Magnetics 33(5) (September 1997)

[27] Chao, C.-T., Chen, C.-C., Kuo, C.-Y., Wu, C.-S., Horng, L., Isogami, S., Tsunoda, M., Takahashi, M., Wu, J.-C.: Temperature Dependence of Electrical Transport and Magnetization Reversal in Magnetic Tunnel Junction. IEEE Transactions on Magnetics 46(6) (June 2010)

[28] Panina, L.V., Mohri, K.: Magneto - impedance effect in amorphous wires. Applied Physics Letters 65, 1189 (1994)

[29] Le, A.-T., Tung, M.T., Phan, M.-H.: A Study of Giant Magnetoimpedance Effect and Magnetic Response in Micro-patterned F/Ag/F Magnetic Ribbon Structures (F=Co-rich Amorphous Ribbon). Journal of Superconductivity and Novel Magnetism 25(4), 1133–1138 (2012), doi:10.1007/s10948-011-1379-y

[30] Morikawa, T., Nishibe, Y., Yamadera, H., Nonomura, Y., Takeuchi, M., Taga, Y.: Giant Magneto-Impedance Effect in Layered Thin Films. IEEE Transactions on Magnetics 33(5) (September 1997)

[31] Maylin, M.G., Gore, J.G., Square, P.T., Atkinson, D.: Elongate GMI integrating magnetic sensor, US Patent 6747449 (June 8, 2004)

[32] Honkura, Y., Yamamoto, M., Mori, M., Koutani, Y.: Magnet with electromagnetic coil/impedance/sensor element, US Patent 7224161 (May 29, 2007)

[33] Delooze, P., Panina, L.V., Mapps, D.J., Ueno, K., Sano, H.: Sub-Nano Tesla Resolution Differential Magnetic Field Sensor Utilizing Asymmetrical Magnetoimpedance in Multilayer Films. IEEE Transactions on Magnetics 40(4) (July 2004)

[34] Hill, E.W.: A Comparison of GMR Multilayer and Spin-ValveSensors for Vector Field Sensing. IEEE Transactions on Magnetics 36(5) (September 2000)

[35] http://www.aichi-mi.com/3_products/catalog%20e.pdf
[36] Choi, S.: A Micromachined Magnetic Field Sensor for Low Power Electronic Compass Application. Dissertation, Georgia Institute of Technology (May 2007)
[37] Lenz, J., Edelstein, S.: Magnetic Sensors and Their Applications. IEEE Sensors Journal 6, 631–649 (2006)
[38] Edelstein, A.S., Fischer, G.A.: Minimizing 1/f Noise in Magnetic Sensors Using a Microelectromechanical System Flux Concentrator. J. Appl. Phys. 91, 7795–7797 (2002)
[39] Ziolo, R.F., Palacios, J.T., Zhang, X.: Magnetic Nanocompass Compass Compositions and Processes for Making and Using, U. S. Patent 5,889,091 (March 30, 1999)
[40] Peczalski, A.: Nanowire Magnetic Sensor, U. S. Patent 7,926,193 B2 (April 19, 2011)

Commercial Off-The-Shelf GMR Based Sensor on Board Optos Picosatellite

M.D. Michelena

Instituto Nacional de Técnica Aeroespacial (INTA),
28850 Madrid, Spain
diazma@inta.es

Abstract. Space is an environment of extreme parameters. Wide temperature swings, very low pressures (vacuum), moderate to high radiation, mechanical vibrations and impacts, etc. Thus, components for space applications, which need to stand these hard conditions, are normally very expensive and it often takes a while to include emerging technologies in the space market. This means that space components are not always that innovative.

The case of vector magnetometers is not an exception. Since the beginning of the space exploration mainly fluxgate magnetometers have been used for magnetic mapping [1]. Fluxgates are robust sensors and massive core fluxgates present very good performances for geomagnetic mapping and further exploration in the solar system. Besides, they are normally combined with a scalar absolute sensor for calibration of the vector magnetometer.

In an attempt to be able to get ready as fast as possible to use emerging magnetic sensing technologies for space applications, INTA has devoted some effort in the qualification for flight use, of Commercial Off-The-Shelf (COTS) solid state magnetic sensors, as AMR and GMR sensors [2-4].

In this chapter we describe the chain of testing and adaptation of the available commercial GMR sensors for an experimental payload in a picosatellite (OPTOS, 3 kg). We present the calibration tests results and the expectations we have for the in-orbit measurements.

1 Introduction – A Flight Opportunity

This chapter reports on the first GMR commercial sensor on board a satellite: from the conception to the preflight calibrations. The idea of such a challenge starts in 2005 encouraged by the increasing interest of GMR sensors mostly in recording magnetic heads [5-9]. The group of Space Magnetism of the Spanish National Institute of Aerospace Technology (INTA), born in the end of the nineties, defined its strategy with the upqualification of commercial off-the-shelf (COTS) magnetic sensors for space applications with the aim to be ready for the qualifications of the new emerging technologies [2]. In this line already in 2005 a miniaturized AMR-based sensor had been launched onboard NANOSAT-01, and the idea to broad this initiative to GMR

C. Reig et al.: *Giant Magnetoresistance (GMR) Sensors*, SSMI 6, pp. 181–210.
DOI: 10.1007/978-3-642-37172-1_8

sensors with half volume of an AMR, double response to magnetic field, and a potential lower power consumption, was very challenging.

At that time, three Spanish space missions started almost simultaneously:

1) An INTA picosatellite (mass lower than 3 kg) called OPTOS "A New Light in Space": a platform for technology experimentation [10]. Among the proposed payloads there were an optical camera APIS, an experiment to measure the temperature by means of a set of optical fibers (FIBOS), and we were asked to propose a magnetic experiment: a GMR-based experiment. Actually OPTOS had another AMR-based magnetometer onboard for the attitude and orbit control subsystem (AOCS), so the whole experiment consisted in the intercomparison of both vector sensors.

OPTOS, composed of three piled cubesats [11-13], is a real challenge of compactness. Imagine how to place the computer, the power unit, etc. and several payloads in a volume of 10 cm x 10 cm x 30 cm. In this compact scenario, the communications between the computer and the rest of the units are performed by means of optical wireless links (OWLS). The wires are completely removed fundamentally to make it easier the integration and assembly tasks.

2) A second nanosatellite in the frame of NANOSAT program: NANOSAT-1B. A more advanced and experimented platform for technology demonstration. Onboard this platform the initial experimental payloads were a couple of magnetic sensors: the GMR and a magnetoimpedance vector, and an experiment for the measurement of the total radiation flux by means of the effect the radiation has in the dark current of optoelectronic devices: *Las Dos Torres* (The Two Towers) [14].
3) A national satellite for Earth observation: SEOSAT. This was supposed to be the first Spanish satellite with the capability to take high resolution multispectral images for different purposes like cartography, agricultural and aquifers mapping, catastrophes and security management, etc. Again, in this platform, two very small payloads were proposed: the Two Towers and a GMR vector sensor called MAGNETITA.

In the three space mission the scientific objective for the GMR sensors was to measure the geomagnetic field (± 60 μT) with accuracy in the order of 10 nT, which is a typical resolution for attitude determination not very strict.

Figure 1 shows the block diagram of the different experiments on board the three platforms. It can be seen that the more complex experiment was proposed for SEOSAT, and consisted in a couple of vector sensors for intercomparison, a calibration coil and a temperature sensor, for thermal compensation due to the high temperature dependence of solid state sensors, within the same experiment. Experiment for

NANOSAT-1B followed this one in complexity, with calibration coil and temperature sensor in the experiment and the open possibility to compare GMR measurements with the AMR sensor. Finally the experiment onboard OPTOS was very limited as a whole but again there were coils in the satellite for GMR calibration and the above mentioned AMR sensor for the AOCS. It can be seen how the complexity of the experiment was reduced as mass was more limited: SEOSAT mass is above 700 kg, NANOSAT-1B mass is 20 kg and OPTOS is less than 3 kg.

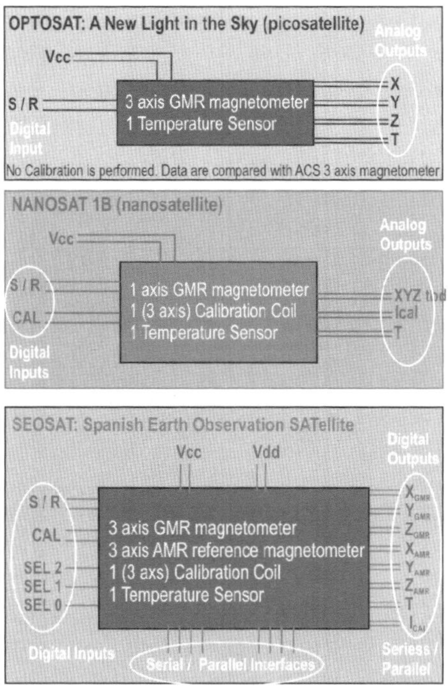

Fig. 1. Blocks diagram of the GMR experiments proposed for the three platforms

In 2006 NANOSAT-1B restricted the magnetic experimental payloads to one and then the magnetoimpedance experiment continued alone onboard this platform. In 2008 MAGNETITA also got off of SEOSAT so the GMR experiment onboard OPTOS is the only one that has survived up today and the protagonist of this story.

2 Selection of an Appropriate Commercial Off-The-Shelf (COTS) GMR Sensor

As it has been described previously in this book, GMR effect is the change in electrical resistance of a multilayer of ferromagneticaly coupled magnetic layers, separated by non magnetic layers, when it is exposed to a magnetic field. In this chapter it is only considered the group of sensors consisting of a multilayer of two ferromagnetic layers separated by a metallic layer (GMR). Other types of sensors like the Magnetic Tunnel Junctions (MTJs) consisting of a trilayer of ferromagnetic material with an inner insulator are not taken into consideration [15-16].

In the absence of a magnetic field the magnetization of the GMR above mentioned sensors, is perpendicular to the measuring axis and alternate among the layers. When a magnetic field is applied, the magnetization of the magnetic layers rotates towards the direction of the external magnetic field (Figure 2).

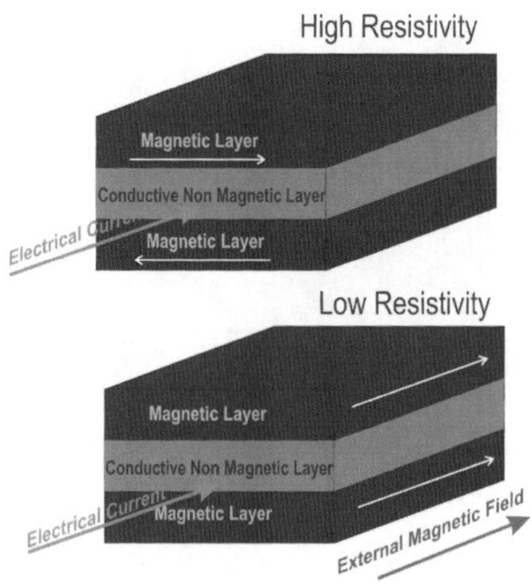

Fig. 2. GMR working principle scheme

The effect was first described in 1988 but in the time until 2005, the GMR commercial devices have been used mostly as magnetic encoders and switches, and very much related to the field of magnetic recording. There was not a significant manufacture of sensors. Among the 47 companies we bench-marked manufacturing magnetic devices: sensors, reed switches, encoders, etc (Table 1), it was only possible to find suitable magnetic sensors for geomagnetic mapping in NVE and Hitachi Metals.

Table 1. Companies manufacturing magnetic sensors

Companies commercializing magnetic sensors (2005)				
N.	ENTERPRISE	WEB	PRODUCTS	
			SENSOR TYPE	APPLICATION FIELD
1	Allegro	www.allegromicro.com/Products.aspx	HALL EFFECT	Currrent Sensors Magnetic speed sensors Magnetic Digital Position Sensors Engine speed sensing Electronic Power Steering (EPS) Fan and pump systems Infotainment Lighting systems Hybrid/Electric vehicles
2	Analog Devices	www.analog.com/en/index.html	HALL EFFECT	Automotive Throttle Position Sensing Pedal Position Sensing Suspension Position Sensing Valve Position Sensing Industrial Absolute Position Sensing Proximity Sensing
3	Asahi Kasei Electronics	www.asahi-kasei.co.jp/ake/en/	HALL EFFECT MR	One chip pulse encoder IC Programmable linear Hall effect IC for one axis position detection Non-contact rotation angle sensor
4	Ascension Technology	www.ascension-tech.com/	Third-generation pulsed-DC magnetic tracking technology	Surgical Navigation and Image Guided Procedures Biomechanics & Ergonomics Simulation & Training Real-time Visualization & Virtual Reality 3D Animation & Live Performance
5	Asonik	www.asonik.eu/	HALL EFFECT	Electro Cardio Recorder Computerized Cavendish Balance Multiresonance minispectrometer type NEQ-5

Table 1. *(continued)*

Companies commercializing magnetic sensors (2005)				
N.	ENTERPRISE	WEB	PRODUCTS	
			SENSOR TYPE	APPLICATION FIELD
6	Bartington	www.bartington.com/	Fluxgate	Defence: Marine Defence: Land Aerospace Geosciences Near Surface Geophysics Physics Medical Industrial
7	Baumer Electric	www.baumer.com/int-en/homepage/	HALL EFFECT MR	Monitoring of rotational speed Detection of final positions Object detection Indirect fill level detection
8	EMI Technology Center	www.slb.com/about/rd/technology/emi.aspx	Induction	High frequency measurements Marine sensors Magnetotellurics
9	EMX Industries	www.emxinc.com/	HALL EFFECT	Proximity Sensors
10	ETS lindgren	www.ets-lindgren.com/	Magnetic Resonance Imaging Shielding	Magnetic Resonance Imaging Equipment
11	F.W. Bell	fwbell.com/default.aspx	HALL EFFECT MR	Laboratory Instrumentation
12	Fischer Custom Communicatio ns	www.fischercc.com/	Induction	Emissions detection
13	Group3 Technology	www.group3technology.com/	HALL EFFECT	Control Systems
14	Hitachi Metals America	www.hitachimetals.com/	GMR	Wind Power Energy

Table 1. *(continued)*

Companies commercializing magnetic sensors (2005)				
N.	ENTERPRISE	WEB	PRODUCTS	
			SENSOR TYPE	APPLICATION FIELD
15	HL Planartechnik	www.meas-spec.com/hl-planartechnik.aspx	MR	Position sensors Rotary displacement sensors
16	Honeywell	honeywell.com/Pages/Home.aspx	MR	Navigation systems Aerospace
17	IC Haus	www.ichaus.de/	HALL EFFECT	Encoder
18	IFM efector	www.ifm.com/ifmus/web/home.htm	Induction	Food and Beverage Applications
19	Infineon	www.infineon.com/cms/en/product/index.html	HALL EFFECT	Automotive Industry Consumer applications
20	Infra	infra-electronic-sensors.blogspot.com.es/	Induction	Position sensors Rotary sensors
21	Lake Shore Cryotronics	www.lakeshore.com/Pages/Home.aspx	HALL EFFECT Induction Fluxmeter	VSM, Cryogenic applications
22	Less EMF	www.lessemf.com/	HALL EFFECT	AC and DC meteres
23	Magnetic Instrumentation	www.maginst.com/	Induction HALL EFFECT Fluxgate	
24	Magnetic Physics	www.magnet-physics.com/	Induction HALL EFFECT	Laboratory Instrumentation
25	Magnetic Sensor Corporation	www.magsensors.com/	Induction	VR Speed Sensor Electromagnets Solenoids Motors IR Steering Mirrors
26	Micronas	www.micronas.com/en/home/index.html	HALL EFFECT	Automotive

Table 1. *(continued)*

Companies commercializing magnetic sensors (2005)				
N.	ENTERPRISE	WEB	PRODUCTS	
			SENSOR TYPE	APPLICATION FIELD
27	Nanjing New Zhongxu Microelectronics	www.zhongxu.com/english/index.asp	HALL EFFECT	
28	NVE	www.nve.com/index.php	GMR	Medical Sensors Gear Tooth Sensors
29	Ohio Semitronic	https://www.ohiosemitronics.com/	HALL EFFECT	Magnetic Field Sensing applications
30	Philips	www.philips.com	AMR	Compas Position sensor
31	PNI	www.pnicorp.com/	Induction	Military applications Scientific applications Robotic Mobile & Tablet Gaming
32	Polatomic Inc	polatomic.com/	Laser-pumped helium	UAVs and Helicopters Underwater Space / Planetary
33	Prodyn Technologies	prodyntech.com	Induction	
34	Quasar Federal Systems	www.quasarfs.com/	Induction	Geophysical Exploration Defense
35	Rikei	www.rikei.com/en/index.html	MR	Mobile Phone PDA
36	RMS Instruments	www.rmsinst.com/	VLF (Very Low Frequency)	Geophysical
37	Samsung Electro Mechanics	www.samsungsem.com/index.jsp	HALL EFFECT	Mobile communication devices

Table 1. *(continued)*

Companies commercializing magnetic sensors (2005)				
N.	ENTERPRISE	WEB	PRODUCTS	
			SENSOR TYPE	APPLICATION FIELD
38	Sensitec	www.sensitec.com/en/?s=	MR	Angle, Length, Current and Magnetic field measurement
39	Senstronic	www.senstronic.com/	Induction MR HALL EFFECT	Proximity sensors
40	Tel Atomic	www.telatomic.com/	NMR HALL EFFECT	NMR equipment Magnetic Field Measurement
41	Ultra Electronis PMES	www.ultra-electronics.com/	Induction	AC magnetic signals Magnetic Signatures Measuring and degaussing systems
42	Xensor Inc	www.xensor.com/	Induction MR HALL EFFECT	Magnetic sensors
43	Xushi	www.xushi-sensor.com/	HALL EFFECT Reed switches	Position sensors and valves for industrial equipments and automated pipelines in plastic, light industry, shoemaking, textile, chemical industry, petroleum, tobacco, food, metallurgy
44	Zetex	www.diodes.com/	MR HALL EFFECT	Switches (Unipolar and Omnipolar) Bipolar latches (Single Output and Complementary Output) Smart Fan Driver
45	AnaChip	www.chipdocs.com/manufacturers/ANACH.html	HALL EFFECT AMR Non Volatil Memory	Distributor
46	GMW Associates	www.gmw.com/	HALL EFFECT Induction Fluxgate	Distributor and integrator
47	Sensor Solutions	www.sensorsolutions.biz/	Sensors technologies (not magnetic)	Capable to work in high magnetic fields (7T)

The candidate COTS for the mission are limited to those listed in Table 2. Actually the gradiometers are purchased to complement the measurement of the magnetic field with a measurement of the gradient.

Table 2. List of initial candidates for GMR COTS sensors

Item	Reference	Characteristics	Manufacturer
1	HM55B	GMR Magnetic Sensor	Hitachi
2	AA002-02	GMR Magnetic Sensor	NVE
3	AA005-02	GMR Magnetic Sensor	NVE
4	AAH002-02	GMR Magnetic Sensor	NVE
5	AAL002-02	GMR Magnetic Sensor	NVE
6	AB001-02	GMR Gradiometer	NVE
7	ABH001-00	GMR Gradiometer	NVE
8	AAV001-11	Spin Valve Magnetic Sensor	NVE
9	AAV002-11	Spin Valve Magnetic Sensor	NVE

Among the COTS firstly taken into consideration: sensors by Hitachi and NVE, it was decided to focus on the NVE sensors because the Hitachi sensors are explicitly not recommended for radiation proof [15].

The former devices under test were the magnetic sensors AA002, AAH002 and AAL002 and the gradiometers AB00102 and ABH001-00 of NVE [16]. The selection was based on their higher sensitivity compared to other sensors of the same families and the appropriateness of the dynamic range. For instance, AA005 has a significant higher dynamic range between 1 and 7 mT, above the specifications of the mission.

GMR magnetic sensors have an active area of approximately 350 microns by 1400 microns. The sensors are configured as a Wheatstone bridge with resistors of 5 kΩ made of GMR material. The sensors are provided with flux concentrators for two purposes: to provide the Wheatstone bridge with an asymmetry that unbalances the output of the Wheatstone bridge due to the different shielding of the resistors and to tailor the sensitivity.

The formula NVE provides for the calculation of the field in the position of the sensors due to the effect of the flux concentrators (FC) is:

$$B_{sensor} = 60\% \frac{l_{FC}}{g_{FC}} B_{applied} \qquad (1)$$

where l_{FC} (~ 400 μm) is the length of the flux concentrators and g_{FC} (~ 100 μm) the gap between them.

As a result these sensors create an artificial field gradient in the chip by means of which the magnetic field is measured.

In NVE magnetic gradiometers the flux concentrators are removed so the Wheatstone bridge has zero output in the presence of a uniform magnetic field all along the dimensions of the chip. In contrast the output is unbalanced in the presence of a gradient field being the shift in voltage proportional to the field and to the spacing between pairs of resistors (0.3 mm, 0.5 mm, or 1.0 mm depending on the sensor product). As it has been mentioned, these sensors were purchased in case that a complementary measurement could be performed, mostly motivated by the presence of a nearby dipole, since OPTOS has not a magnetic cleanliness program. This option was later discarded.

Finally a couple of spin valve sensors were purchased for comparison but they were not used for this project, so from now on this work will focus on the magnetic sensors.

Characteristics of the purchased sensors can be seen in Table 3.

When comparing with AMR sensors, GMR are more efficient in terms of sensitivity per voltage. Also the power consumption of GMRs is lower due to the higher resistance (5 kΩ compared to a 1 kΩ typical value of AMR.

Table 3. Characteristics of AA002, AAH002 and AAL002 sensors [16]

Parameter	Sensor type			Unit
	AA002	AAH002	AAL002	
Electrical resistance	5	2	5.5	kΩ
Sensitivity	[30, 42]	[110, 180]	[30, 42]	mV / V mT
Saturation Field	1.5	0.6	1.5	mT
Bridge Voltage Range	[<1, 24]	[<1, 24]	[<1, ±25]	V
Operating Frequency	[DC, > 1]	[DC, > 1]	[DC, > 1]	MHz
Operating Temperature Range	[-50, 125]	[-50, 150]	[-50, 150]	ºC
Bridge Electrical Offset	[-4, +4]	[-5, +5]	[-4, +4]	mV / V
Signal Output at Maximum Field	60	40	45	mV / V
Linear Range	[0.15, 1.05]	[0.06, 0.30]	[0.15, 1.05]	mT
Nonlinearity	2	4	2	% (unipolar)
Hysteresis	4	15	4	% (unipolar)
Tempco of Resistance	0.14	0.11	0.11	% / ºC
Tempco of Output at Constant supply Current	0.03	0.1	-0,28	% / ºC
Tempco of Output at Constant supply Voltage	-0,1	0	-0,4	% / ºC
Off Axis Characteristic	Cos β	Cos β	Cos β	β angle between field and sensitivity axis
ESD Tolerance	400	400	400	V pin-to-pin HBM

Figure 3 shows the response curves of the candidate sensors. Two characteristics should be remarked: one is that GMR is an even effect and thus their response is the same no matter the direction of the field and the other is that response curves present a not negligible hysteresis.

The transition from the ordered saturated magnetized state up to the high resistance state goes through a line with higher sensitivity while the transition from the spontaneous alternated magnetization towards the saturation magnetized state goes through a lower sensitivity curve. In the low field region, the repeatability is really poor. However the sensitivity of any of the branches is very good and appropriate for the magnitude of the Earth magnetic field intensity.

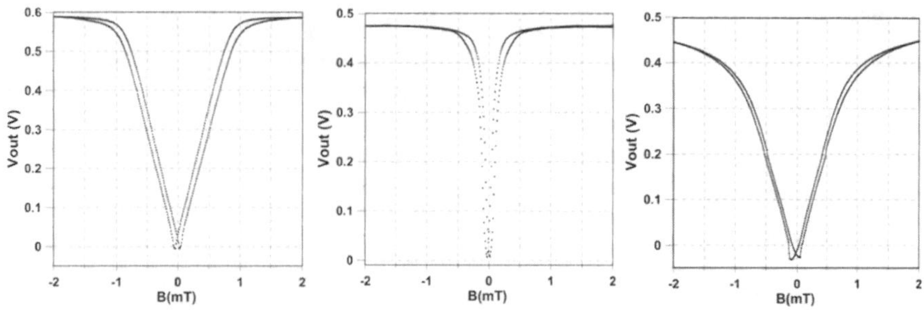

Fig. 3. AA002, AAH002 and AAL002 response curves for a 2 mT range

Another issue (mentioned in) is the high dependence of GMR sensors with temperature. Figure 4 corresponds to a sequence of measurements of the AAL002 device performed at a fast rate, while the sensor is being warmed up by means of a close electrical resistor (Joule effect). It can be seen how the sensitivity of the sensor decreases monotonically with the increase of temperature. This characteristic will have to be taken into account in two aspects: on the first hand a thermal compensation will be essential, and on the other hand, any coil for the correction of the previously mentioned odd response will have to be placed at a certain distance from the sensor to avoid direct heating.

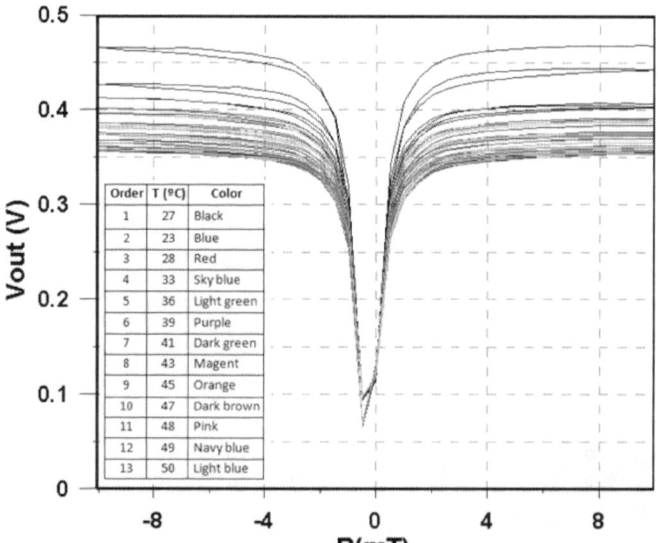

Fig. 4. Degradation of the response of the AAL002 magnetic sensor as the temperature increases

The next step is to study the linearity of the different candidates to discern which the most suitable working region is and with which sensor.

The response of the GMR sensors under study is not linear in the whole range. Though this actually can be compensated calibrating the curve, for basic applications with simple proximity electronic and data processing it is much more convenient to have a linear response.

To study where the sensor is linear attention is paid to the first derivative of the output response respect the external magnetic field (denoted by dVout/dB in the graphs). When the value of the derivative is constant, i.e. the sensitivity is constant, the sensor is considered to have a linear response. The ideal situation is that both branches of the hysteresis loop had the same slope and thus, the same sensitivity and that the linearity is a maximum.

To study this, a simple set up is developed (Figure 5). A MATLAB program controls the intensity of a power source (E3631E by Agilent Technologies) that supplies with current a coil setting the external magnetic field. The direction of the current is changed by means of two relays connected to the switches of the Data Acquisition / Switch Unit (HP 34970). The value of the electrical current is requested to calculate the magnetic field. The output signal of the sensor is acquired by the Data Acquisition / Switch Unit.

Each measurement consists of the average of only two samples. These first tests are performed at very low frequencies to get the external field completely stabilized.

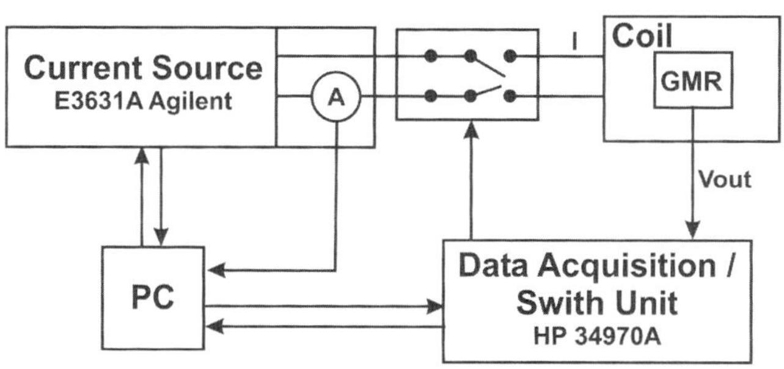

Fig. 5. Block diagram of the set up for the measurements

The coil used is not a perfect Helmholtz one and the area of homogeneous magnetic field is confined in a region of 5 mm with an homogeneity better than $9 \cdot 10^{-2}$ %. A mechanical fixture is manufactured to guarantee the repeatability in the positioning of the sensors.

Figures 6, 7 and 8 show the response curve (dark blue referenced to the left hand side Y axis) and the derivative / sensitivity (light blue referenced to the right hand Y axis) of the AA002 , AAH002 and AAL002 respectively when a ramp of ± 2 mT is applied.

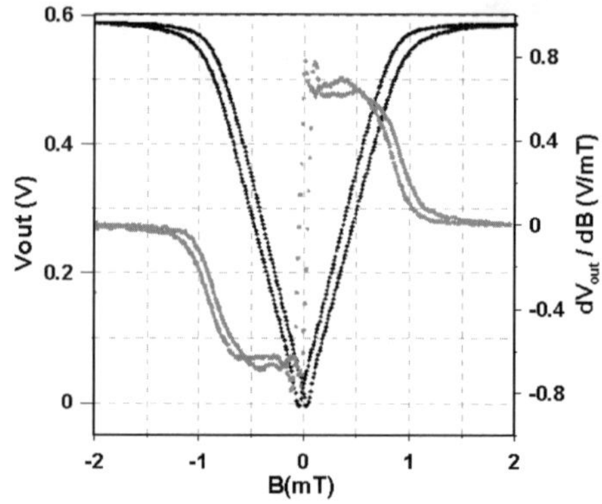

Fig. 6. Response curve and derivative of the AA002 sensor vs. the external magnetic field

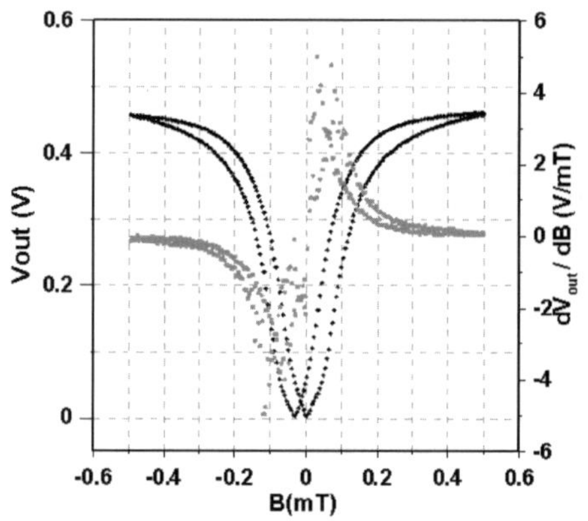

Fig. 7. Response curve and derivative of the AAH002 sensor vs. the external magnetic field

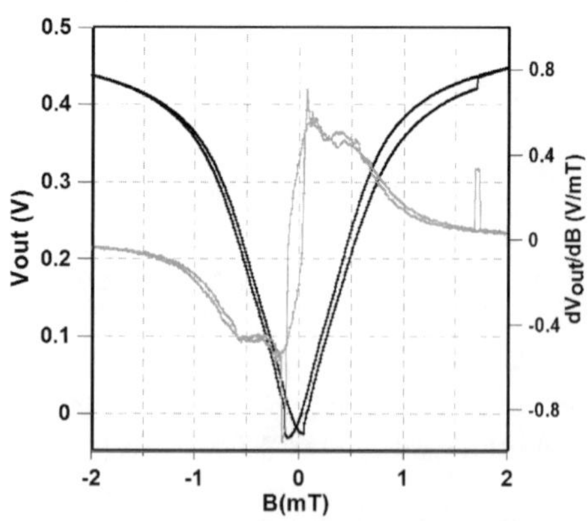

Fig. 8. Response curve and derivative of the AAH002 sensor vs. the external magnetic field

At a glance it can be seen that AA002 and AAH002 show higher hysteresis than AAL002, and the linearity is better in AAH002 and AAL002.

Regarding the derivative of the field (sensitivity) AAH002 sensor (Figure 7) does not even present a linear interval in the field range. This sensor is discarded for the in flight experiment.

However, AA002 and AAL002 derivatives are constant for a certain interval of the field (~0.5 mT). In the sensor AA002 the derivative is split into two branches due to the hysteretic behaviour of this sensor while in the AAL002 the split is minimal. Actually, the dispersion of the derivative in the two branches is higher in the AA002 than in the AAL002.

Following this argument of a less hysteretic and a more linear response, AAL002 sensor is selected for the OPTOS experiment.

3 Biasing Mechanism

Previously it has been discussed the even nature of the GMR effect, which for practical purposes limits the capability of the sensors to measure the magnitude of the field but not the direction. Other magnetoresistive sensors have this problem as well but even at commercial level, the devices are provided with a biasing mechanism called Barber pole. In these sensors the magnetic material (a monolayer) has a set of copper microstraps inserted at an angle of 45° of the magnetic strap making the current pass through the material at an angle of -45° following the shortest distance between better conductive layers. Also the devices have a microcoil assembled so a magnetization in the easy axis can be performed. In such conditions, when a magnetic field with a component perpendicular to the easy axis is applied, the magnetization rotates to an equilibrium position between the action of the field and the previous state magnetized in the easy axis [17]. With this mechanism, AMR devices are sensitive to field direction since the projection of the magnetization on the direction of the current is different for both signs of the field.

More recently, it has been reported the use of crossed axis GMR for odd response as a function of the field [18], which has been recommended for linear applications without excitation. However, available COTS GMR sensors in 2005 are not provided with such a mechanism and thus, external biasing needs to be performed to measure the vector field [20-24]. In this subchapter this mechanism is introduced.

The first step for biasing is to find an optimum working point. It has been described that the zero field area is a very poor area for measuring so biasing fields are applied to find a better working area. Figure 9 shows biased curves for AAL002.

The mechanism proposed to solve the hysteresis problem is the following. If both branches are characterized it is possible to measure with the sensor in any of them but what is more, if it is possible to locate the magnetic sensor in one of the branches, the measuring procedure is easier and the repeatability higher.

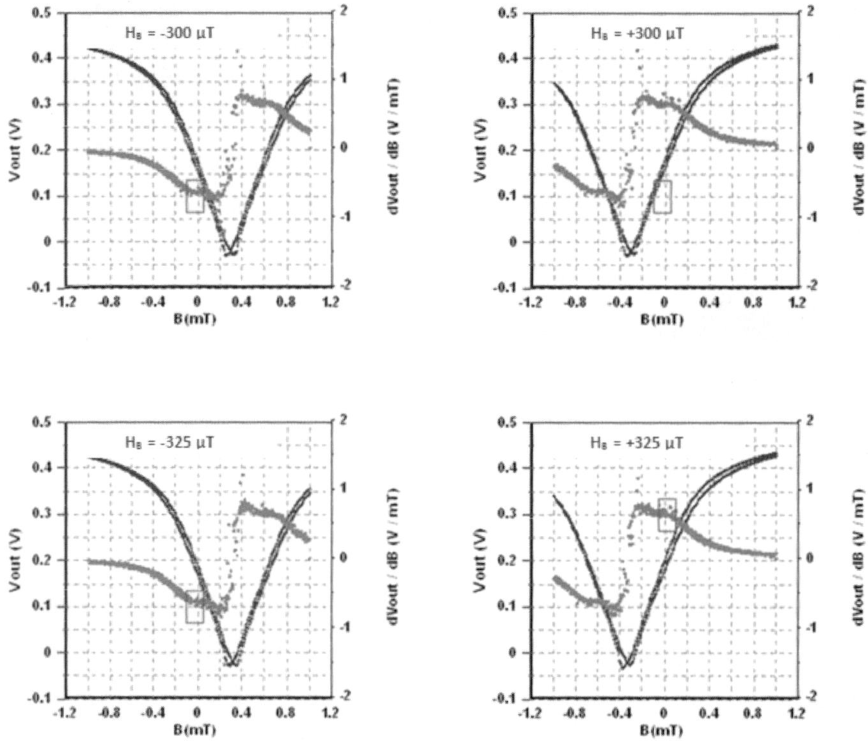

Fig. 9. Response curve (dark blue referred to the left hand side Y axis) and derivative (light blue referred to the right hand side Y axis) of the AAL002 sensor vs. the external magnetic field. For different bias field in both senses.

This is the goal of this mechanism: to apply a magnetic field high enough so as to guarantee that the sensor is following a certain branch when this field ceases.

In Figure 10 it is shown a graph of the response (dark blue) of the AA002 and its derivative (light blue) in a range of field of ± 2 mT. It has been used the AA002 curve because it is the one which shows the higher hysteresis and it is easier to distinguish between the branches.

In the absence of an external bias and S / R magnetic field (State 1), the response of the sensor is very uncertain and highly dependent of the previous states.

If a bias field is applied (State 2) there are two possible states depending on the previous state of the sensor. If the environment is relatively clean, the most probable branch is the one on the left in the upper part of Figure 10, this is, the one with less sensitivity. To avoid this uncertainty around the working point a Set field, with a higher value of that of the biasing field is applied (State 3) followed by a bias field centred in the range of constant derivative (State 4). In this way the sensor goes to this 4th state from state 3 through the most sensitive path.

In the negative part of the measurement, the procedure is exactly the same. The State 5 is uncertain but after the Reset pulse (State 6), the State 7 is already defined and the corresponding sensitivity characterized.

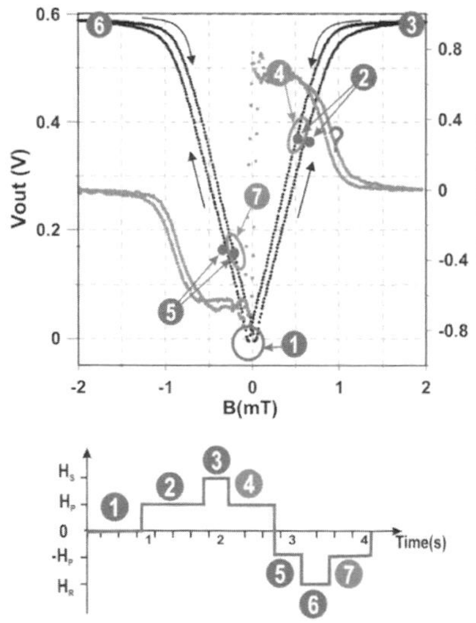

Fig. 10. Set Reset mechanism. Correspondence between the field applied and the magnetic state of the GMR multilayer

The result of this procedure for different values of the S / R field is shown in Figures 11 and 12.

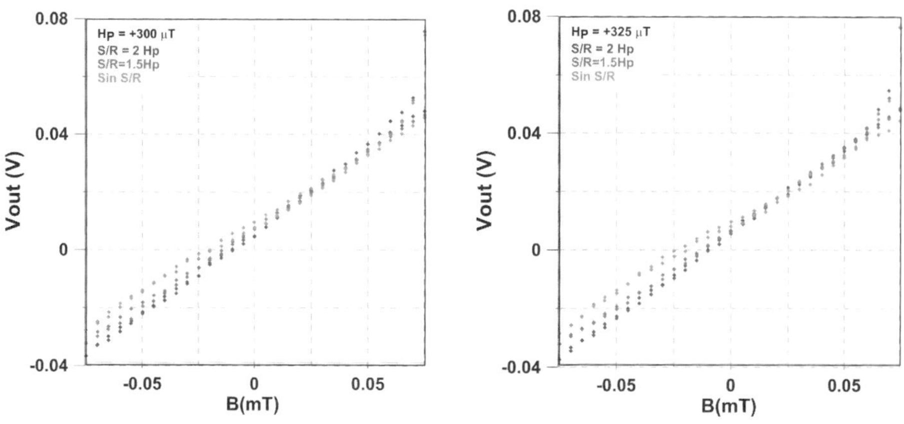

Fig. 11. Response of the AA002 sensor with the bias field mechanism (green), with a 1.5 times the bias field Set / Reset Field (red) and with two times the bias field Set / Reset field (blue)

It can be seen in these figures that the higher the S / R field the higher sensitivity. Also at the appropriate value of the bias field the derivative has the less dispersion.

The Set Reset mechanism together with the alternating bias improves notably the behaviour of the sensors being the best choice the AAL002, which presents the less hysteretic behaviour and thus the higher repeatability. The value of the bias field is around 300 and 325 μT and the minimum detectable field is now limited by the power sources used.

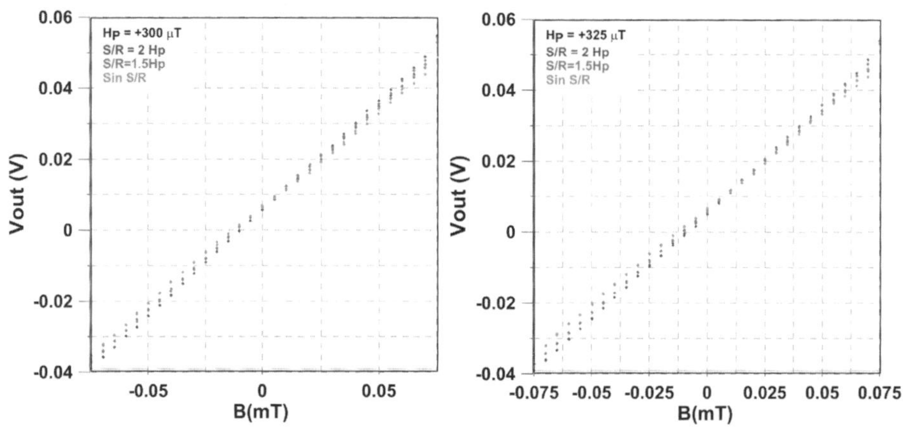

Fig. 12. Response of the AAL002 sensor with the bias field mechanism (green), with a 1.5 times the bias field Set / Reset Field (red) and with two times the bias field Set / Reset field (blue)

4 Qualification of the Sensor

As it has been explained, GMR sensors were in a commercial stage of development, thus the technology readiness level (TRL) attributable to the COTS sensors was around 4 following the NASA and ESA standards.

For OPTOS mission the qualification of the sensor for flight use was performed by means of three different tests:

- An outgassing test to check the suitability of the plastic package
- An irradiation test with protons to discard a malfunction of the sensors under radiation (potential creation of defects of the material, and thus in the magnetic response)
- Tailored upscreening for the mission

OUTGASSING

Space environment is often thought as vacuum. In this atmosphere, components can outgas. This implies mainly three concerns:

- Their operational properties can change giving rise to dimensional stability and lubrication problems for instance
- They can generate a “gas cloud”, which can perturb measurements specially in scientific missions
- The “gas cloud” can condensate on surfaces modifying their operational properties: thermo-optical, radiation effects and electrical with the corresponding thermal and electrical problems

Thus, an outgassing test was performed following the ECSS-Q-70-02 norm from the European Space Agency “Thermal vacuum outgassing test for the screening of space materials”. The purpose of this test is to ascertain if the package of the sensor outgases by means of the measurement of the:

- Total mass loss (TML)
- Returned mass loss (RML)
- Collected volatile condensable material (CVCM)

This test was performed at INTA. GMR sensors showed a negligible outgassing, and thus, no expected change in operational properties of the plastic package, neither do “gas cloud” emission or condensation, add derived problems. As a result, the package was considered apt for flight in terms of outgassing.

IRRADIATION

Satellites orbiting the Earth have a very different weather than equipment on the surface. “Space vacuum” involves plasmas, all range of energies electrons and ions, which potentially can damage the equipment of the spacecrafts. High energy particles (10 MeV) have the potential to ionize atoms in the materials through which they propagate and low enery particles (< 100 eV) they can produce accumulation of charge or other types of materials degradation.

At this point it is very difficult not only to simulate the interaction between these particles and the materials with experiments but also it is difficult to estimate the radiation environment of the missions because of the changing nature of the space weather. But one normal way of operation in space missions is on the first hand to generate an envelope of radiation for the mission life and on the other hand, to perform radiation tests generally with gamma and with protons to simulate the different damages as a function of their energy.

In our case an exhaustive campaign of gamma irradiation for anisotropic magnetoresistive (AMR) sensors had been performed for the sensors of NANOSAT-01, launched in 2004. Even though GMR and AMR technologies are not the same, it was assumed that the magnetic material was immune to the total irradiation dose up to tens of krad Si and consequently, a campaign of irradiation with protons was performed, to study the effects the high energy particles could have on the sensing properties of the devices.

At the time of the irradiation campaign the number of sensors for flight was limited and some of them had been reserved for the rest of the qualification process, so a further assumption was made: that all NVE magnetic sensors have similar materials and the possible damages will be homogeneous in all of them. Under this assumption, sensors of the type AA003 underwent the proton irradiation campaign.

The tests were performed in the RADEF facility, located in the Accelerator Laboratory at the University of Jyväskylä, Finland (JYFL). The facility includes a beam line dedicated to proton irradiation studies of semiconductor materials and devices.

The real fluence received by the GMR sensors is listed in Table 4.

Table 4. Mean Fluence and standard deviation (protons / cm^2) for the irradiated devices

Step	Fluence (p+/cm^2)	Std. Dev. (p+/cm^2)
0	0	0
1	8,00E+10	4,27E+09
2	4,80E+11	2,56E+10
3	8,80E+11	4,70E+10
4	1,28E+12	6,83E+10
5	1,68E+12	8,97E+10
6	2,08E+12	1,11E+11

The fluence is calculated for a typical satellite in a Low Earth Orbit (LEO).

Four AA003 GMR sensors were exposed to this radiation. The response of the sensors for a magnetic field ramp was measured after each step starting with a saturating magnetic field in one direction and returning to this point to measure changes in sensitivity and hysteresis.

A mechanical set up (Figure 13) was performed to reproduce the position of the sensor in every check.

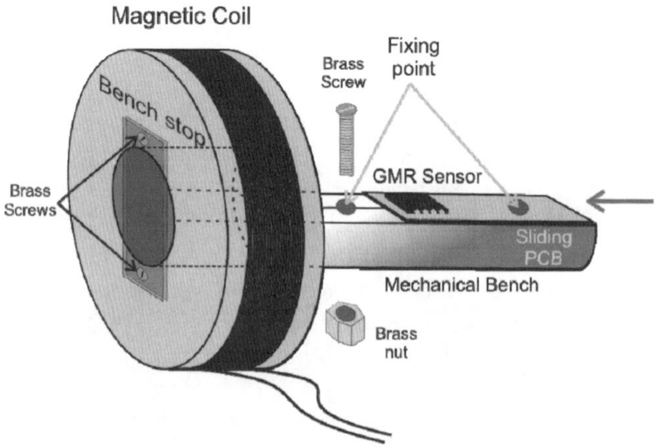

Fig. 13. Mechanical bench for repetitive magnetic characterization of GMRs

The sensors did not show appreciable changes in sensitivity and hysteresis along the irradiation, so they are considered as radiation robust for the fluence and rates used.

TAILORED UPSCREENING

OPTOS is a mission with an expected life time of three years. The components of the missions need to have proven that they can stand the conditions of the life time. To do so, an accelerated ageing of the devices is foreseen. This is called upscreening.

During the upscreening process a representative number of AAL002 GMR devices of the same badge are tested. The sequence of the upscreening, designed by INTA, and the number of components for each step are shown in Figure 14.

After a serialization a visual inspection of the 30 units of the lot is made to ascertain that there are no irregularities in the devices. This step is performed by means of an optical microscope at 10x. Purchased sensors did not present any visible non conformity.

For the following steps, the sensors are placed in a PCB with appropriate sockets inside a magnetic shielding chamber with a coil (Figure 15) to permit the checking of the magnetic characteristics of the sensors in the different conditions. The whole set up is placed in the thermal chambers for the qualification process.

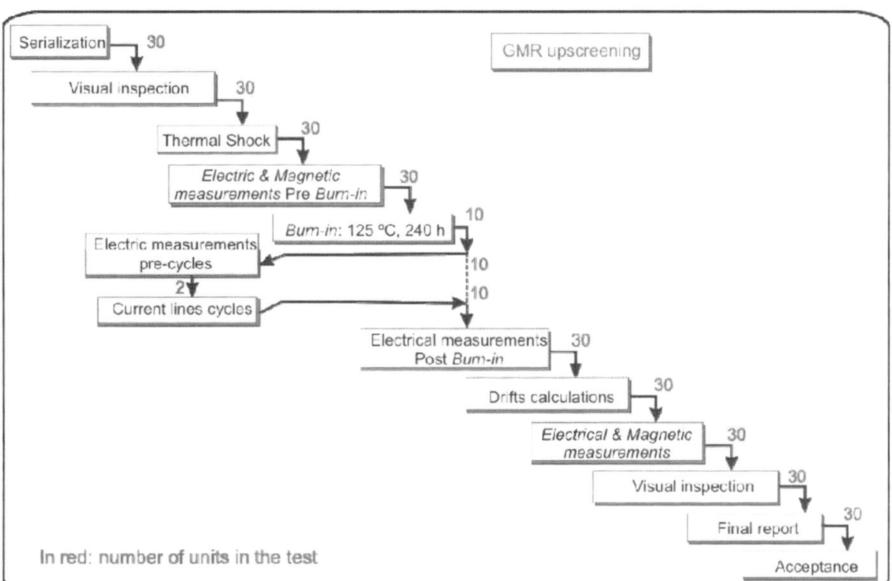

Fig. 14. Tailored upscreening for GMR sensors

Fig. 15. PCB with sockets inside the one-layer mumetal shielding chamber

The following step is a thermal shock which is performed to determine the resistance of the part to sudden changes in temperature. The parts undergo a specified number of cycles, which start at room temperature. The parts are then exposed to the highest qualification temperature (+70ºC) and, within a short period of time, exposed to the lowest qualification temperature (-20ºC), before going back to room temperature. After the final cycle, external visual examination of the package, pins, and seals is performed at 10x. The marking is also inspected (magnification lower than 3x). As no illegible mark and/or or evidence of damage to the package, pins, or seals after the stress test was noticed, the devices are apt to continue the burn-in test, in which the temperature is risen to 125ºC for a long time. In contrast to the specified time of 240 hours the devices remained at 125ºC for 92 hours, which was agreed by the system engineer.

At this point it is checked the electrical and magnetic response of the devices, i.e. the measurement of the Wheatstone bridge resistance (not powered) and a five points ramp of magnetic field generated by means of an external coil.

Devices showed a negligible deviation of their electrical Wheatstone bridges resistors (lower than 100 ppm) and the sensitivity measured does not change appreciably (~1 % outside the magnetic shielding chamber).

The next step is the whole life simulation of the devices. This consists in a non stop operation of the sensors at an accelerated rate of excitation to age the components.

Since the simulation of the mission life cycle is potentially stressing for the components, only a small amount of the devices undergo this proof.

After this test the resistance of the components is measured with no observed deviation respect to the former measurements.

The calculation of the drifts resulted in a negligible deviation of the response of the sensors with magnetic field (~1 % outside the magnetic shielding chamber) and no observable variation of the Wheatstone bridges electrical resistances (lower than 100 ppm). Thus, AAL002 GMR sensors are considered qualified for OPTOS mission.

5 Design of the Device

In this section it is explained the mechanical design of the sensor. The description of the front end electronics are out of the scope of this chapter.

AAL002 GMR sensors are one axis sensor. To develop a vector magnetometer they need to be located in perpendicular planes. A cube of FR4 material was used for this purpose (Figure 16).

The cube was mechanized with some slots for the biasing mechanism, which consists in magnetic coils wounded around each magnetometer producing a biasing field in the sensing direction. The field needs to be high enough to saturate the GMR sensor (in the order of 600 μT).

The biasing mechanism permits the measurement of the field with its direction. To do so, attention needs to be paid in the winding of the coils around each sensor for a right handed system aligned with the reference system of the spacecraft.

Since only four of the pins of each magnetometer need to be connected, no printed circuit board was performed. The connection of the pins was wired.

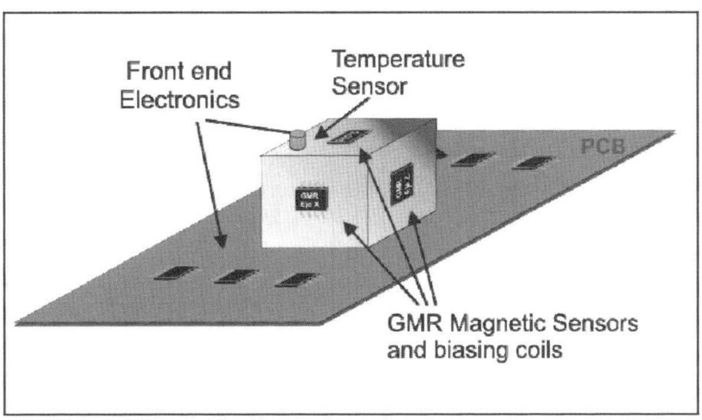

Fig. 16. Scheme of the FR4 cube with three orthogonal GMR devices

Figure 17 shows a picture of the GMR flight model. Towards the centre it can be seen the cube with the three GMR sensors and their biasing windings.

The whole PCB consists in the GMR sensor, the optical link and the corresponding adapter. The dimensions of the PCB are 89 mm, 78.8 mm ad 19 mm and the nominal power is lower than 500 mW.

Fig. 17. Flight model of the GMR based vector magnetometer

6 Performance

GMR sensor for OPTOS is a vector magnetometer with a dynamic range of ± 150 μT in every component. The sensitivity of the different axes is: 28 mV/ μT – X axis, 24 mV/ μT – Y axis ad 17 mV/ μT - Z axis.

The offsets in the different axes: -115 mV (X axis) , -105 mV (Y axis) and -124 mV (Z axis), though this offset will have to be in flight calibrated due to the proximity to magnetic materials.

The stability of the sensor is fairly good (better than 500 ppm in one hour) and the cross axis is less than 1 %. Sensors still need a final calibration with temperature.

GMR sensors have higher sensitivity than other magnetoresistive sensors (AMR) but their intrinsic noise, likely due to their more complicated structure, is higher and implies a higher low-field detectivity limit. Due to the ultimate performance-limiting factor of the noise in GMR sensors, it is worthy to devote some effort in the determination of it and in the understanding of the noise sources. Besides, the measurement of the power spectral density – PSD, can give any clue of the better operation frequency of the sensor.

In the low-frequencies range, the two sources of noise are the white noise or thermal noise and the 1/f noise [25-26].

The former is related with spontaneous fluctuations induced by the thermal excitations and its PSD is given by the Nyquist equation:

$$PSD(f) = 4k_B TR \tag{2}$$

Where "k_B" is the Boltzmann constant, "T" the absolute temperature and "R" the electrical resistance. In our case this noise is lower than 2 pT /√Hz.

This noise is a function exclusively of the resistance (not of the magnetic field), and cannot be suppressed nor modified.

The latter source of noise, the 1/f noise, appears in magnetic and non magnetic systems and it is related to fluctuations of energy around equilibrium. Its PSD, given by Hooge in 1969, corresponds to:

$$PSD(f) = \frac{\gamma_H V^2}{N_C f} \tag{3}$$

Where "γ_H" is the Hooge´s constant, "V" the applied voltage and "N_C" the number of charge carriers in the active volume.

This noise is inversely proportional to the volume so it becomes dominant in small structures.

The set up for the measurement of the PSD in the GMR AAL002 sensor is depicted in Figure 18.

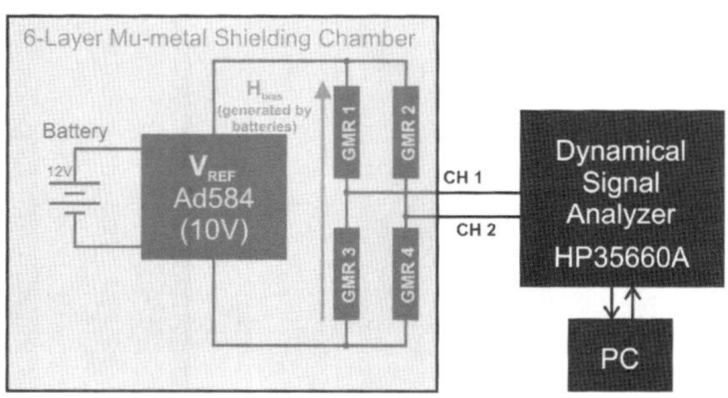

Fig. 18. Set up for the measurement of the PSD of the GMR AAL002 magnetic sensor

The sensor is supplied with 10 V by means of a voltage reference AD584 supplied with batteries. The coils to provide the bias field are also supplied with batteries. The bridge configuration has the advantage that voltage fluctuations will appear on both signal lines and thus, they will be able to be subtracted. To reduce as much as possible the unwanted pickup of external magnetic field fluctuations due to both the noise in the bias field or the environmental noise, the Helmholtz coils generating the bias field are high enough so as to avoid magnetic gradients in the region of the magnetic sensor and the battery voltage is monitored not permitting a voltage drop higher than a 0.1 %. Finally, the whole system is isolated from external fluctuations of the magnetic field by means of a 6 layer mu–metal shielding structure.

There is no pre-amplification of the magnetic sensor signals because the signal analyzer used has input impedance higher than 1 MΩ.

The dynamical signal analyzer is PC controlled to measure the PSD of the two individual channels from DC up to 10 kHz in intervals of 200 Hz with a LabView programme.

Afterwards the two channels are cross-correlated (Figure 19).

The graphs of the noise density are in Volts. The sensitivity is in the order of 300 – 450 mV/mT, so the magnetic noise density at 1 Hz is in the order of tens of nT√Hz and at 1 kHz in the order of nT√Hz.

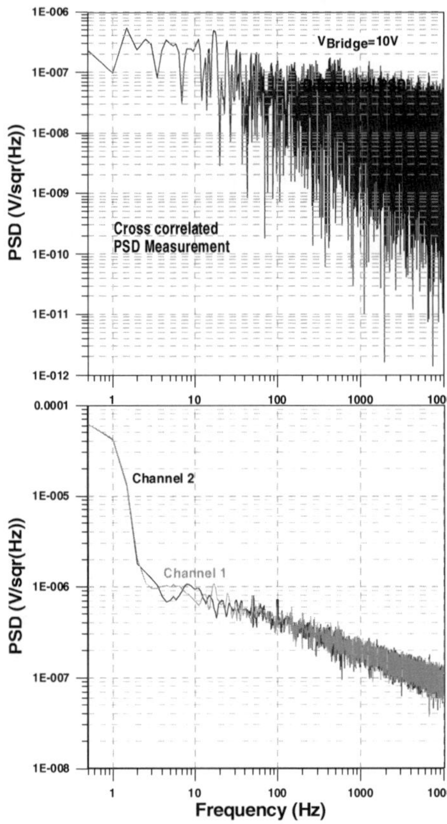

Fig. 19. Cross-correlation (upper graph) of the measurement of the PSD in the two branches of an AAL002 magnetic sensor (graph in the bottom)

The fast decay of the PSD with the frequency together with the suitability of the S / R and biasing mechanism suggest an operation in the range of the kHz instead of a DC operation in which the external magnetic field supposes a modulating signal to the excitation alternating bias field and measure by means of a lock in amplifier.

7 Summary and Conclusions

Working in space conditions is always a challenge; due to the extreme conditions of temperature, radiation and vacuum. This gets even more complicated when one tries to use commercial devices, not thought from scratch to work in space.

The process overviewed in this chapter shows how it was faced the use of commercial giant magnetoresistive devices for flight conditions. The procedure for the qualification of a GMR COTS based vector magnetometer from the definition of the mission and payloads on board to the pre-flight calibrations prior to final assembly and launch is a critical procedure.

The first part of the procedure consists in a bench mark of the enterprises manufacturing magnetic sensors in search of GMR candidates. As a result, it was concluded that the GMR COTS sensor candidate for the fight demonstration is AAL002 by NVE to be onboard OPTOS picosatellite.

A biasing mechanism was designed to the measurement of the field direction. To do so, extra magnetic coils had to be foreseen for the sensors.

The whole qualification process of the COTS sensors has been described: outgassing, irradiation with protons and upscreening. AAL002 sensors seem not to be damaged in the qualification process so they are selected for flight.

Finally the outcome sensor has been fully characterized and characteristics are explained.

In conclusion, GMR COTS sensors present a high dynamical range and high sensitivity (up to 40 mV / V / mT) but they present some disadvantages compared to other magnetic sensors as for instance the lack of linearity, thermal and time stability, repeatability and noise. At this point white noise is negligible and the dominant source of noise is 1/f noise.

This fact makes us think that an excitation of the sensor at frequencies in the order of 1 kHz can lead to a more stable and lower noise operation.

Finally, after all these years of work, OPTOS flight model (Figure 20) is on the bench about to be ready for the launch.

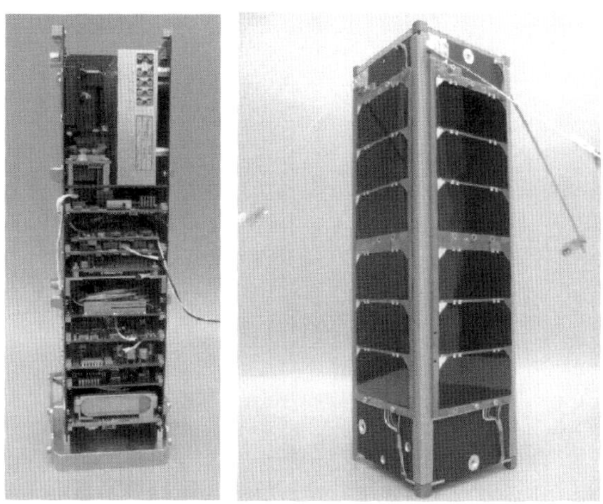

Fig. 20. Current status of OPTOS spacecraft flight model (FM)

GMR COTS roadmap to space can be the fastest ever (Figure 21). Our fingers are crossed for a successful launch and future operation of OPTOS spacecraft, our laboratory is waiting expectantly for the new emerging sensor to qualify. Our hopes are that we have the opportunity to qualify a disruptive technology of magnetic sensing that revolutionizes the space magnetometry.

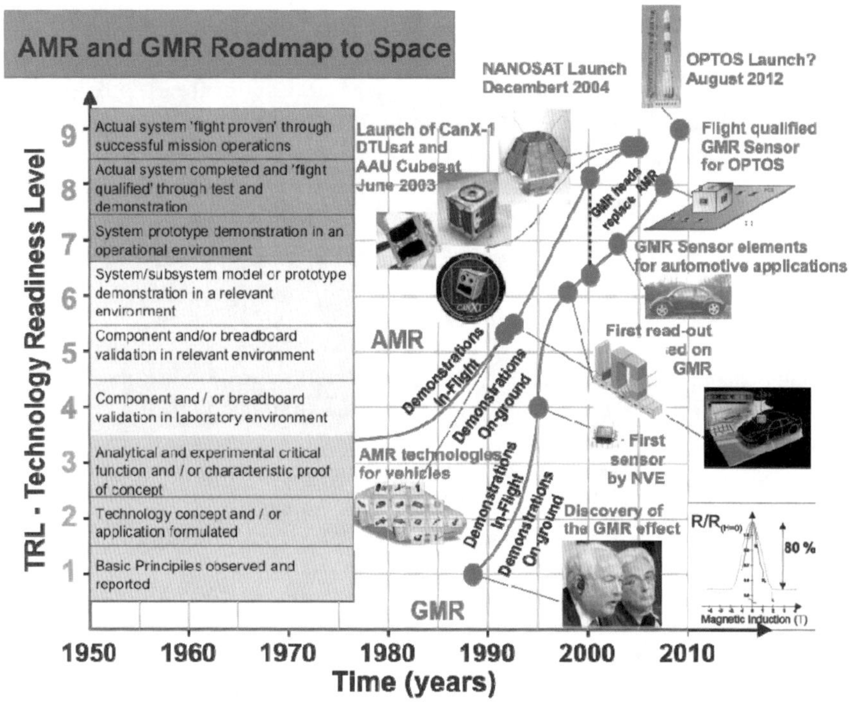

Fig. 21. AMR and GMR roadmaps

Acknowledgements. Author acknowledges OPTOS team for all the support. In particular, OPTOS/GMR team in part supported by MEIGA PROJECT AYA2009 - 14212-C05 -01 / MICINN & AYA2011 -29967-C05 -01 / MINECO and TD02/08 INTA Calvo Rodés grant.

References

[1] Acuña, M.H.: Space-based magnetometers. Review of Scientific Instruments 73(11), 3717 (2002)

[2] Michelena, M.D., Arruego, I., Oter, J.M., Guerrero, H.: COTS-Based Wireless Magnetic Sensor for Small Satellites. IEEE Transactions on Aerospace and Electronic 46, 542–5572 (2010)

[3] Magnes, W., Diaz-Michelena, M.: Future Directions for Magnetic Sensors for Space Applications. IEEE Transactions on Magnetics 45, 4493–4498 (2009)
[4] Michelena, M.D.: Small magnetometers for Space Applications. Sensors 9, 2271–2288 (2009)
[5] Klaassen, K.B., Contreras, J.T.: Thermal transient response of electrical self heating in current-in-plane magnetoresistive heads. IEEE Transactions on Magnetics 42(10), 2438–2440 (2006)
[6] van Peppen, J.C.L., Klaassen, K.B.: A new approach to micromagnetic simulation of thermal magnetic fluctuation noise in magnetoresistive read sensors. IEEE Transactions on Magnetics 42(1), 56–69 (2006)
[7] Maqableh Mazin, M., Liwen, T., Huang, X., et al.: CPP GMR Through Nanowires. IEEE Transactions on Magnetics 48(5), 1744–1750 (2012)
[8] Yupin, S., Pornchai, S., Yoshihiro, O., et al.: Performance Improvement System for Perpendicular Magnetic Recording with Thermal Asperity. IEICE Transactions on Electronics E94C (9), 1472–1478 (2011)
[9] Jia, Z., Misra, R.D.K.: Magnetic sensors for data storage: perspective and future outlook. Materials Technology 26(4), 191–199 (2011)
[10] Aragon, V., Garcia, A., Amaro, R., et al.: OPTOS communications: A high-performance solution. Microwave and Optical Technology Letters 54(4), 943–946 (2012)
[11] Daniel, S., David, K.: A survey and assessment of the capabilities of Cubesats for Earth observation. Acta Astronautica 74, 50–68 (2012)
[12] Tsitas, S.R., Kingston, J.: 6U CubeSat commercial applications. Aeronautical Journal 116(1176), 189–198 (2012)
[13] Najam, A.N., Han, X., Yan, J.L., et al.: An Architecture Analysis of ADCS for Cube Sat: A Recipe for ADCS Design of ICUBE. In: Mechanical and Aerospace Engineering 2012. Applied Mechanics and Materials, vol. 110-116(pts 1-7), pp. 5397–5404 (2012)
[14] Jiménez, J.J., Oter, J.M., Apestigue, V., Hernando, C., Ibarmia, S., Hajdas, W., Sánchez-Páramo, J., Álvarez, M.T., Arruego, I., Guerrero, H.: Proton monitor Las Dos Torres: first intercomparison of in-orbit results. Accepted to IEEE Transaction Nuclear Science, doi:10.1109/TNS.2012.2198241
[15] Magnetic Compass Sensor HM55B Datasheet 19-SM-001-2 p.9 Precaution for use 9.2
[16] NVE Corporation GMR Sensor Catalog, pp.14–19
[17] Caruso, M.J.: Set/Reset Pulse Circuits for Magnetic Sensors. Honeywell Applications Note, AN-201
[18] Freitas, P.P., Ferreira, R., Cardoso, S., Cardoso, F.: Magnetoresistive sensors. J. Phys.: Condens. Matter 19, 165221–165241 (2007)
[19] Reig, C., Cubells-Beltrán, M.-D., Ramírez Muñoz, D.: Magnetic Field Sensors Based onGiant Magnetoresistance (GMR) Tecnology: Applications in Electrical Current Sensing. Sensors 9, 7919–7942 (2009)
[20] Kools, J.C.S., Ruigrok, J.J.M., Postma, B., et al.: Fabrication and characterization of giant magnetoresistive elements with an integrated test coil. IEEE Transactions on Magnetics 33(6), 4513–4521 (1997)
[21] Ripka, P., Tondra, M., Stokes, J., Beech, R.: AC-driven AMR and GMR magnetoresistors. Sens. Actuators A 76, 227–232 (1999)
[22] Vopálenský, M., Ripka, P., Platil, A.: Precise magnetic sensors. Sens. Acuators A 106, 38–42 (2003)

[23] Vopálenský, M., Ripka, P., Kubík, J., Tondra, M.: Improved GMR sensor biasing design. Sens. Acuators A 110, 254–258 (2004)
[24] Dimitropoulos, P.D., Avaristsiotis, J.N.: Integrating the Fluxgate principle in the Spin-Valve and AMR sensor Technologies. Sens. Actuators A 106, 43–47 (2003)
[25] Pannetier, M., Fermon, C., Le Goff, G., Simola, J., Kerr, E., Coey, J.M.D.: Noise in small magnetic systems-applications to very sensitive magnetoresistive sensors2. J. Mag. Mag. Mat. 290-291, 1158–1160 (2005)
[26] Stutzke, N.A., Russek, S.E., Pappas, D.P., Tondra, M.: Low-frequency noise measurements on commercial magnetoresistive magnetic field sensors2. J. Appl. Phys. 97, 10Q107 (2005)

High-Spatial Resolution Giant Magnetoresistive Sensors – Part I: Application in Non-Destructive Evaluation

K. Chomsuwan[1], T. Somsak[2], C.P. Gooneratne[3], and S. Yamada[4]

[1] King Mongkut's University of Technology Thonburi,
Bangkok, Thailand
[2] Rajamangala University of Technology Lanna, Chiang-Mai, Thailand
[3] King Abdullah University of Science and Technology,
Thuwal, Saudi Arabia
Chinthaka.Gooneratne@kaust.edu.sa
[4] Kanazawa University, Kanazawa, Japan

Abstract. In this chapter, we report the utilization of spin-valve type giant magnetoresistance (GMR) sensors in non-destructive evaluation (NDE). The NDE application is the inspection of high-density printed circuit boards (PCBs) based on the eddy-current testing (ECT) technique. An ECT probe with a GMR sensor is presented for the inspection of high-density double-layer PCB models. The utilization of a GMR sensor as a magnetic sensor showed that PCB inspection could be performed with high-spatial resolution and sensitivity, over a large frequency range.

1 Introduction

Giant magnetoresistance (GMR) sensors have great potential to be used as magnetic field detectors because they are sensitive to low-magnetic fields with high spatial resolution and can be easily integrated with existing semiconductor electronics [1-13]. Due to advancements in micro/nano technology they can be expanded to compact array structures and fabricated in a large scale [14-18]. GMR sensors are energized by applying a constant current and the output voltage is an indication of the change in resistance due to the change in applied magnetic field. High spatial resolution GMR probes are presented in this chapter for inspection of printed circuit boards (PCBs).

Electrical contact tests, and/or non-electrical, non-contact methods such as automatic visual/optical inspection, have been routinely used for the inspection of PCBs for defects for many years [19-25]. While electrical tests can be used to obtain information about many types of defects it cannot detect potential defects such as line width or spacing reductions. Automatic visual/optical inspection procedures are inexpensive and provides high throughput but can only inspect the outer surface of a PCB. Eddy-current testing (ECT) is a popular non-contact, non-destructive evaluation (NDE) method that is usually applied to evaluate material flaws without changing or altering the material under test [26, 27]. Defects on a PCB conductor can be

C. Reig et al.: *Giant Magnetoresistance (GMR) Sensors*, SSMI 6, pp. 211–241.
DOI: 10.1007/978-3-642-37172-1_9

investigated by eddy-current flow. The ECT probe employed in this research for the inspection of PCBs consists of a planar meander coil for excitation and a GMR sensor for detection. The aims of this research is to use the ECT probe to analyze and enhance the ECT signal for the easy identification of any defect points on the PCB conductor, to improve the scanning speed and resolution for detecting imperfections of the PCB conductor, and also to obtain PCB conductor dimensions and alignment.

2 ECT Technique Based Application

2.1 Introduction

NDE is an examination, test, or evaluation performed on any type of tested object without changing or altering that object in any way, in order to determine the absence or presence of conditions or discontinuities that may have an effect on the usefulness or serviceability of that object [28, 29]. NDE may also be conducted to measure other test object characteristics, such as size; dimension; configuration; or structure, including alloy content, hardness, grain size, etc. The most common NDE methods are visual inspection, magnetic particle testing and ultrasonic testing. ECT is an NDE technique which is sensitive to very small cracks or flaws on a test specimen surface and subsurface. Several high-performance ECT probes have been developed based on exciting coils and magnetic sensors for, detecting dangerous cracks around fastener hole and engine components in aircrafts [30-33], and inspection of cracks, flaws as well as corrosion in power plant equipment, reactor, turbine, thick structures and pipeline systems [34-37]. Magnetic sensors play a key role in ECT-based inspection systems [38]. Several kinds of magnetic sensors such as Hall, GMR, SQUID, etc. have been successfully used as ECT probes for non-destructive detection of material cracks [39]. Moreover, inspection techniques such as multi frequency ECT and pulse ECT have been developed for thin metal tubes and sheets, as well as metal cladding for measuring thickness and for the location and sizing of internal defects [40].

Eddy-currents are closed loops of induced current circulating in a plane perpendicular to the direction of a time varying magnetic flux density $\boldsymbol{B}$. The variation of $\boldsymbol{B}$, generates an electric field intensity $\boldsymbol{E}$, in the loop as expressed by Maxwell's equation in equation (1).

$$\nabla \times \boldsymbol{E} = -\frac{\partial \boldsymbol{B}}{\partial t}. \tag{1}$$

Therefore, the current density $\boldsymbol{J}$, in a material with conductivity σ, also circulates in the loop because:

$$\boldsymbol{J} = \sigma \boldsymbol{E}, \tag{2}$$

where $\boldsymbol{B}$, $\boldsymbol{E}$, $\boldsymbol{J}$ and σ are in T, V/m, A, and S/m respectively.

Eddy-currents are referred to as "eddy-current losses" when they occur in electrical machines such as motors and transformers because they degrade the performance of such devices. On the contrary, eddy-currents are useful when applied to detect flaws or cracks on metallic materials; this forms the basis for the ECT technique.

The frequency of the applied magnetic flux has an effect on how deep the electric field intensity penetrates into the conductive material. Its amplitude is attenuated exponentially with depth. The distance through which the amplitude decreases by a factor e^{-1} is known as the "skin depth" or the "penetration depth" of the material and it can be expressed as

$$\delta = \frac{1}{\sqrt{\pi f \mu \sigma}}, \tag{3}$$

where δ is the skin depth (m), f is the frequency of B (Hz) and μ is the permeability.

Skin effect arises when the eddy-currents flowing in the test object at any depth produce a magnetic flux which opposes the primary flux, thus reducing the total magnetic flux and causing a decrease in current flow as depth increases. Alternatively, eddy-currents near the surface can be viewed as shielding the coil's magnetic flux, thereby weakening the magnetic flux at greater depths and reducing induced currents. From equation (3), increasing of frequency of the applied B, σ and μ of the specimen are the cause of decreasing of skin depth. This effect is very useful when it is applied to ECT technique for detection of cracks or flaws on conductive material at difference depths.

The proposed ECT probe consists of an exciting coil and a magnetic sensor as shown in Fig. 1. A sinusoidal current is fed to the exciting coil to generate a B over the conductive material. Eddy-currents are induced and circulate in the conductive material due to the B from the exciting coil; the eddy-currents in turn also generate its own B. B generated from eddy-currents are normally uniform. Therefore, the output signal that is obtained from the pick-up coil is constant. Whenever there is a defect on the conductive material, eddy-currents change its path leading to a non-uniform B, resulting in a change in the output signal at the pick-up coil; thus defect points can be identified on a given material under test.

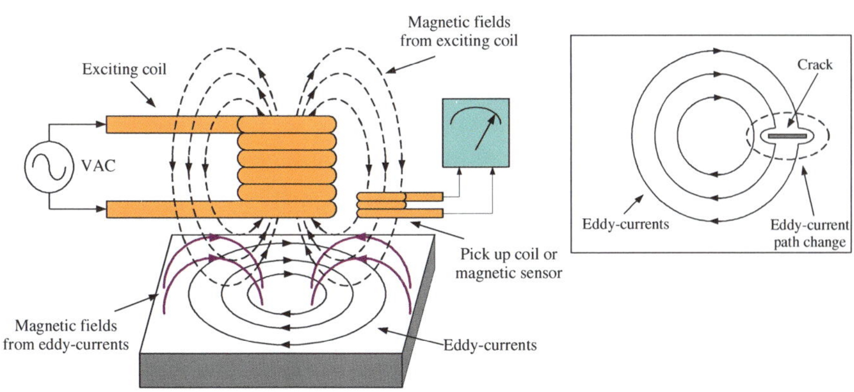

Fig. 1. Basic principle of eddy-current testing technique on crack detection

Eddy-currents can be applied to test not only cracks or flaws but also the following variables:

- Conductivity variations
- Spacing between probe and specimen
- Material thickness
- Thickness of plating or cladding on a base material
- Spacing between conductive layers
- Permeability variations

Although the ECT technique can be used for many applications, many factors such as conductivity, permeability, lift-off height and coil design have an influence on test performance.

2.2 PCB Inspection Based on ECT Technique

2.2.1 Design and Construction of the ECT Probe

The proposed ECT probe, which consists of a long planar meander coil serving as an exciting coil, and a magnetic sensor, was fabricated for PCB inspection as shown in Fig. 2. The long planar meander coil was used as an exciting coil because it provides the advantages of easily developing the matrix sensor to improve the scanning speed [41, 42] and of providing a short distance between the sensor and tested object. The long planar meander coil was made from copper with 35 μm thickness. Two films made from Polyimide are needed to separate the planar meander coil from the magnetic sensor and the PCB conductor. Thickness of the film is 50 μm therefore; the total thickness of the planar meander coil is around 135 μm. The two-dimensional (2-D) *B* distribution of the planar meander coil is shown in Fig. 3. This figure shows that the *B* generated by the planar meander coil are distributed only in the *x* and *y* direction.

Magnetic sensors have to be set up in the sensing direction in order to detect the *B* only in the scanning direction. As shown in Fig. 4, the spin-valve type GMR used as the magnetic sensor consists of 4 strips and each strip has dimensions of 100 μm × 18 μm. Therefore, the total effective area of the GMR sensor is 100 μm × 93 μm with a 7 μm gap between the strips. In comparison of sensor structure, it can be said that distance between the GMR sensor surface and PCB conductor or lift-off height is at least 135 μm, whereas other sensors such as solenoid coils have higher lift-off than the GMR sensor.

2.2.2 PCB Defect Detection Using the ECT Probe

High-frequency exciting currents are fed into the planar meander coil to generate a *B* distributed over the PCB conductor as shown in Fig. 5. The exciting currents normally flow in z axis or scanning direction. The eddy-currents flowing in the PCB conductor are induced by the applied B and also flow in z axis or scanning direction. Because of the skin depth effect, the eddy-currents flow very close to surface or boundary of the PCB conductor. Whenever a defect or the PCB conductor boundary that is perpendicular to scanning direction is found, the eddy-currents will change its path and flow in x direction generating a magnetic flux density B_z, flowing in the z direction. Therefore, defects on the PCB conductor or the conductor boundary can be identified if B_z is detected. Partial defects, occurring on both PCB conductor width and PCB conductor thickness, also have an effect on the eddy-current path. As show in Fig. 6, low eddy-currents flowing in the x axis generate a weak B_z.

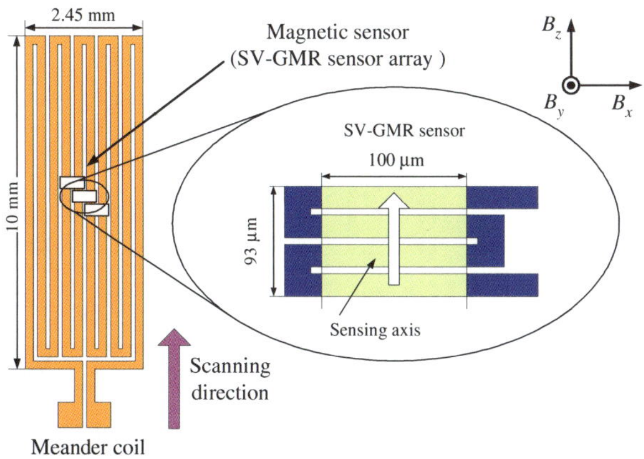

(a) Top view

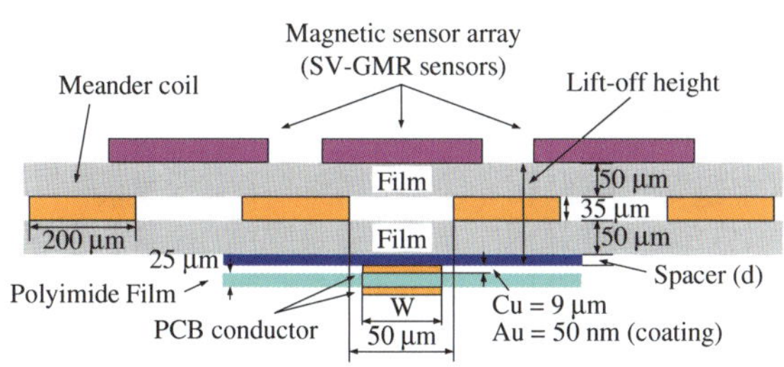

(b) Cross section

Fig. 2. Proposed ECT probe for printed circuit board inspection

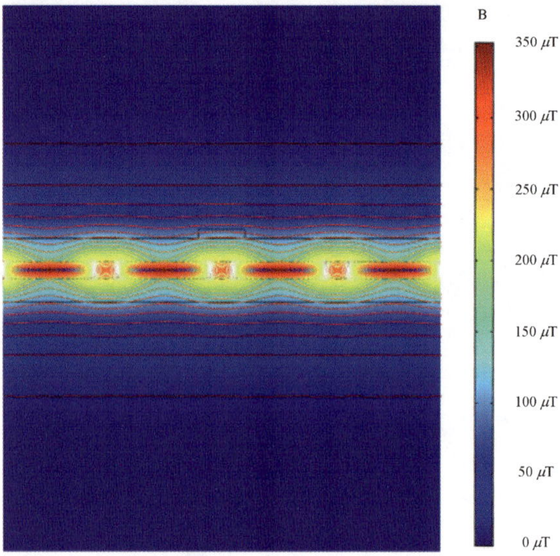

Fig. 3. Magnetic flux distribution obtained from a meander coil with 4 turns

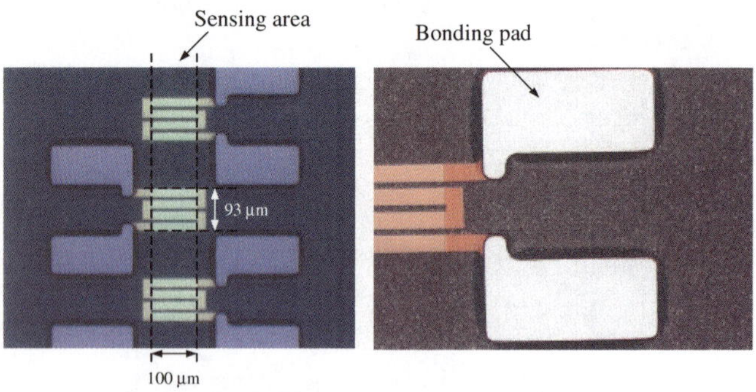

Fig. 4. The GMR sensors used in experiments

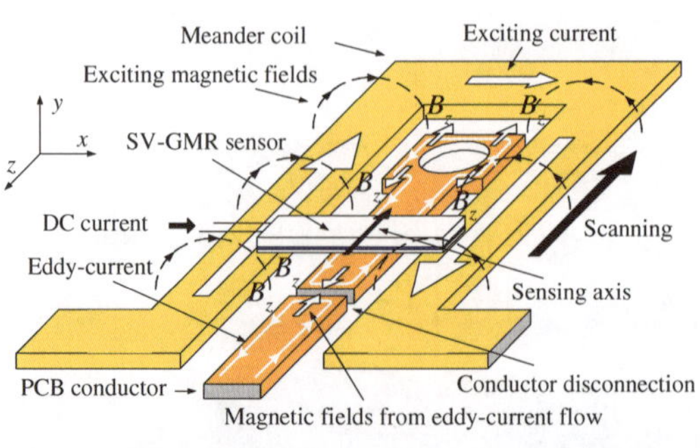

Fig. 5. Basic principle of ECT technique for printed circuit board inspection

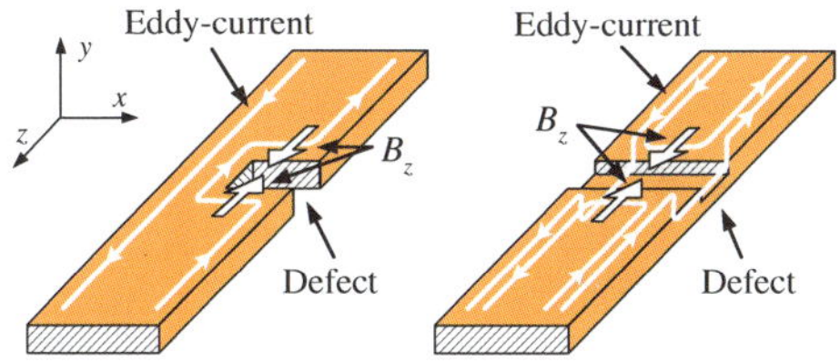

Fig. 6. Eddy-current paths for partial defects occurring on a PCB conductor, in width (left) and thickness (right)

It is more difficult to measure B_z than B_y, because B_z appears at a very short distance from the test PCB conductor and its value is not as high as B_y. However, B_z only appears when there are defects or at the boundaries of a PCB conductor. The output signal does not have to be extracted from other signals as in case of the inspection utilizing B_y. The magnetic sensor moving above the test PCB conductor is not penetrated by B_z until it encounters a defect; therefore, the magnetic sensor is less susceptible to noise in the output signal. This is a big advantage of the proposed ECT probe and the main reason for its high sensitivity. To detect the defect occurring on the PCB conductor, the magnetic sensor, therefore, has to detect only B_z or the magnetic flux density that is parallel to the scanning direction.

2.2.3 Finite Element Analysis of Eddy-Current Flow

Eddy-currents flowing in the PCB conductor and B distribution generated by the eddy-currents were studied based on the Finite Element Method (FEM). Three types of defects on the PCB conductor, namely conductor disconnections, partial defects on

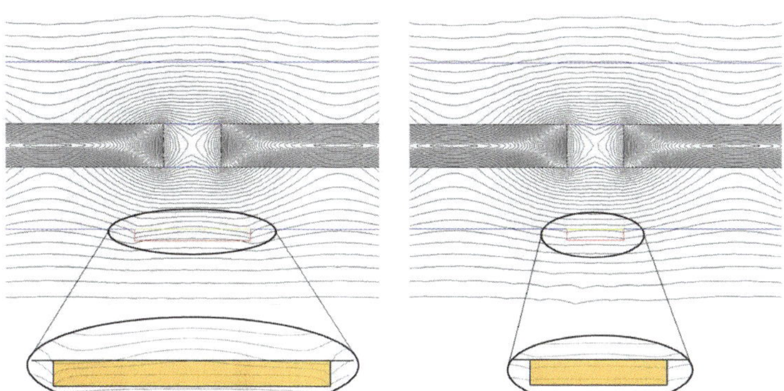

(a) 100 µm PCB conductor width. 50 µm PCB conductor width.

Fig. 7. Magnetic flux distribution over a PCB conductor at an exciting frequency of 5 MHz

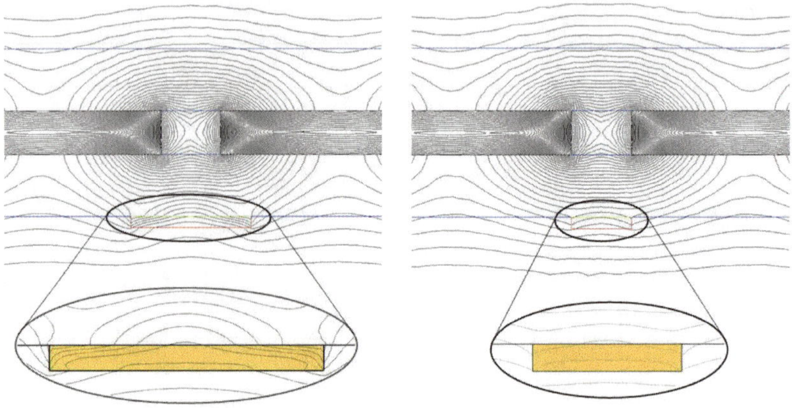

(a) 100 μm PCB conductor width. (b) 50 μm PCB conductor width.

Fig. 8. Magnetic flux distribution over a PCB conductor at an exciting frequency of 10 MHz

PCB conductor width, and partial defects on PCB conductor thickness, are analyzed. Disconnection length of 50 μm is allocated on the conductor disconnection model. For partial defects, the disconnection region also is set to 50 μm where the disconnection region is only 50 % of the PCB conductor. The skin depth or depth of penetration is an important parameter that must be considered. The skin depth of copper at a frequency of 5 MHz can be calculated by referring to equation (3). Sinusoidal current of 200 mA at a frequency of 5 MHz and 10 MHz was fed to the planar meander coil model to generate a *B* over the PCB conductor. As shown in Fig. 7, because of flux penetration effects, *B* distribution over the PCB conductor at a frequency of 5 MHz differs from one without a PCB conductor, as was seen in Fig. 3. A narrow PCB conductor has less effect than a wide PCB conductor. In case of high frequency (10 MHz) excitation, the flux penetration effect is stronger than that at a frequency of 5 MHz as shown in Fig. 8. Eddy-currents flow on a PCB conductor and the resulting magnetic flux distribution over the PCB conductor when a defect is found on the PCB conductor are shown in Fig. 9. Eddy-currents usually flow along the scanning direction, but whenever there is a defect or soldering point on the PCB conductor the eddy-currents change its path resulting in a measurable B_z. Moreover, the distribution of eddy-currents is mainly along the PCB conductor boundary because of skin depth effect. Therefore, peak values of B_z appear at the defect point region or the PCB conductor boundaries.

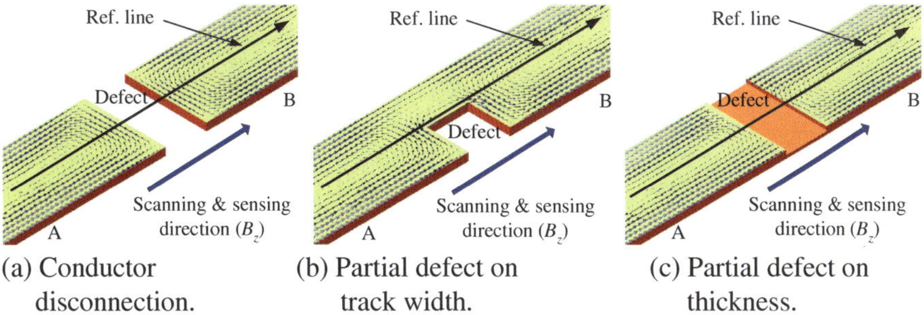

Fig. 9. Eddy-current flow in a PCB conductor for different types of defects

Based on FEM analysis, the magnetic flux distribution over a PCB conductor with defects is shown in Fig. 10. B_z fluctuates at the defect point or displacement distance at 0 mm. The wider PCB conductor generates a higher B_z variation than the narrow PCB conductor because of high-density eddy-current flow. For a partial defect on the PCB conductor track width, magnitude of magnetic flux density variation at partial defect on PCB track width does not differ from that of a conductor disconnection point. In addition, the effect of exciting frequency and PCB conductor width influences the B variation at a defect point. A partial defect on PCB thickness generates the lowest B variation because of low eddy-current flow. However, the B variation is high enough to be detected by the magnetic sensor.

2.2.4 Solder Microbead Detection by the ECT Probe

The proposed ECT probe is also applied to detect solder microbeads that are placed on the PCB for assembly; Ball Grid Array (BGA) package based on surface mounting technology. As shown in Fig. 11, the principle of solder microbead detection is similar to the inspection of PCB defects. Eddy-currents are induced by alternating exciting B which flow on the surface of the microbead. These eddy-currents generate a B_z that usually occur at the microbead boundary. Therefore, the ECT probe can obtain information about the position of the microbead by detecting B_z as in the case of PCB defect detection.

The same technique as mentioned above was applied to analyse the magnetic flux distribution obtained from eddy-current flow inside a solder microbead. The microbead size of 125 μm radius (250 μm diameter) was used in the analysis model. Eddy-currents are induced and flow in the microbead as shown in Fig. 12 and this eddy-current flow generates a B_z as shown in Fig. 13. The shape of the signal obtained in the solder ball region is very similar to the signal obtained at the disconnection point of the PCB conductor. At the microbead center, the signal changes from positive to negative at a displacement distance of 0 mm. Therefore, the FEM analysis results confirm that the proposed ECT probe can also be applied for the detection of solder microbeads.

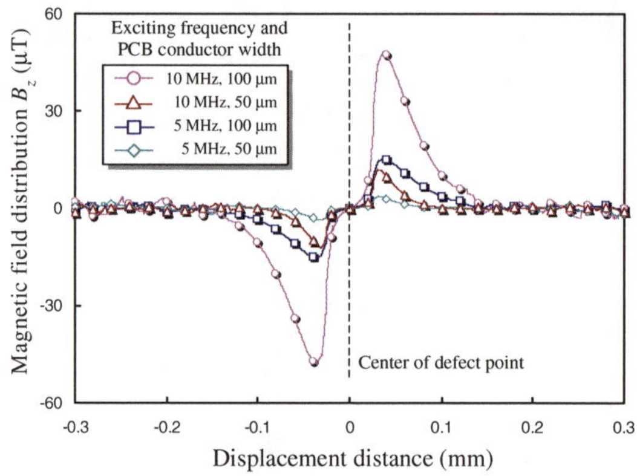

(a) Conductor disconnection

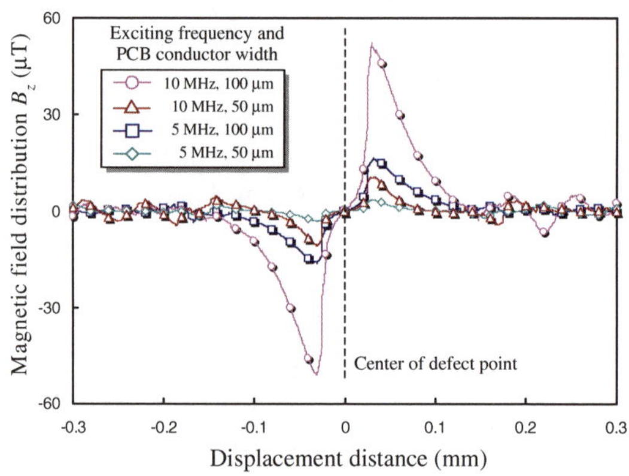

(b) Partial defect on track width

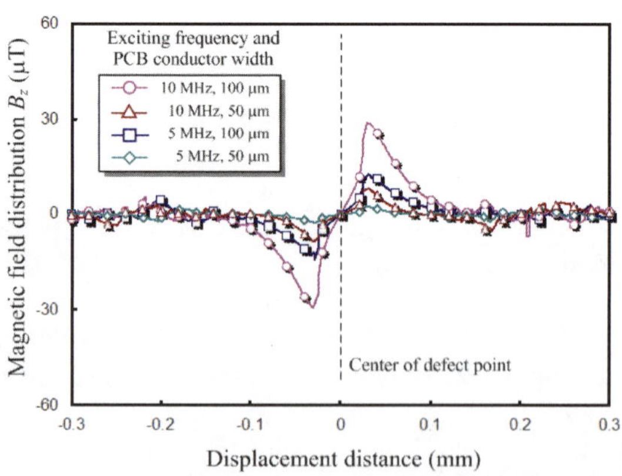

(c) Partial defect on thickness

Fig. 10. Magnetic flux density in scanning direction (B_z) for different defect types over the reference line, as in Fig. 9

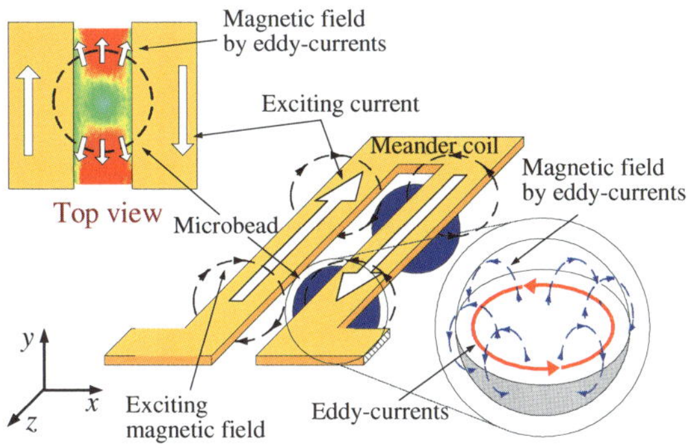

Fig. 11. Detection of solder microbeads by the ECT probe

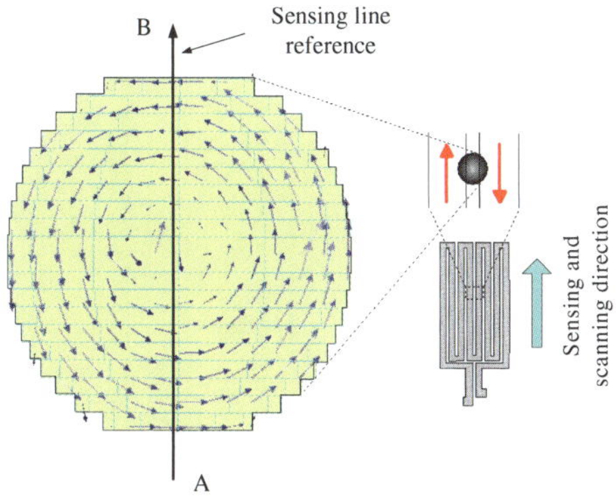

Fig. 12. Eddy-current flow inside a solder microbead of 125 μm radius when placed under a meander coil generating a magnetic flux density at a frequency of 5 MHz

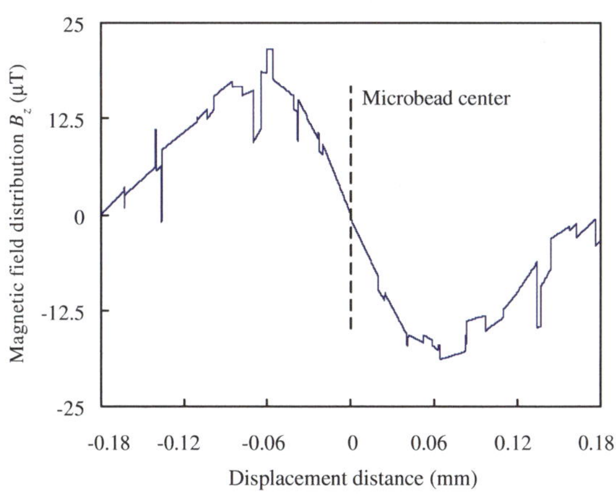

Fig. 13. Magnetic flux distribution B_z over the solder microbead obtained from FEM analysis; exciting frequency is 5 MHz

2.3 ECT Probe Characteristics and Inspection System

2.3.1 Characteristics of the GMR Sensor

A magnetic flux density ranging from -4 to 4 mT are generated by a DC exciting current that is fed to the Helmholtz coil. The nominal resistance of the GMR sensor is 400 Ω. The DC characteristic of the GMR sensor in its sensing axis is shown in Fig. 14. From the characteristics, the proposed GMR sensor has a maximum magnetoresistance ratio of approximately 12 % of its normal resistance, or has a resistance variation between 370 and 420 Ω. The linear region sensitivity of the proposed GMR is approximately 12 %/mT or 48 Ω/mT with a low hysteresis loop. Moreover, the GMR sensor has a high-sensitivity in the applied *B* range from -1 to 1 mT.

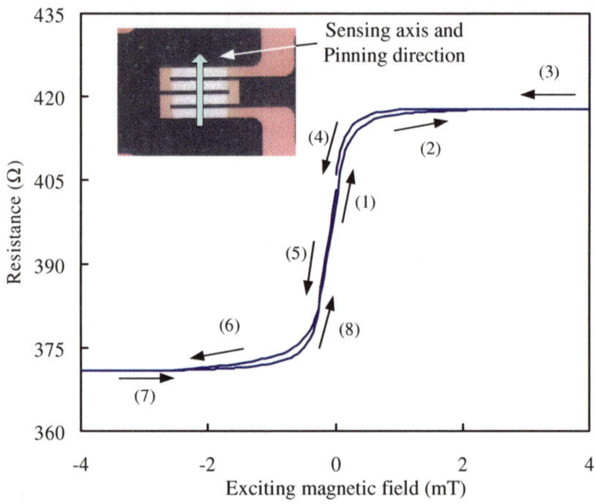

Fig. 14. DC characteristics of the GMR sensor

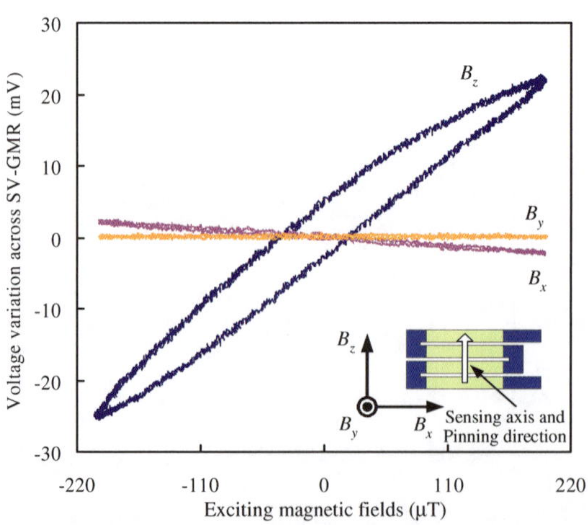

Fig. 15. Small signal AC characteristics of the GMR sensor at an exciting frequency of 100 kHz

Small signal AC characteristics of the GMR sensor in the *x*, *y* and *z* axes were tested with a *B* in the range of -200 to 200 μT; the detected *B* generated from eddy-current, for the purpose of PCB inspection, is smaller than ±200 μT. A DC bias current of 4 mA, the usual bias current used in the experiment, is applied to the GMR sensor. The small signal characteristics of the proposed GMR sensor are tested at a frequency of 100 kHz as shown in Fig. 15. From the characteristics it can be seen that the GMR sensor is most sensitive in the *z* (sensing) axis. The sensitivity of the proposed GMR sensor in the sensing axis is around 150 μV/μT.

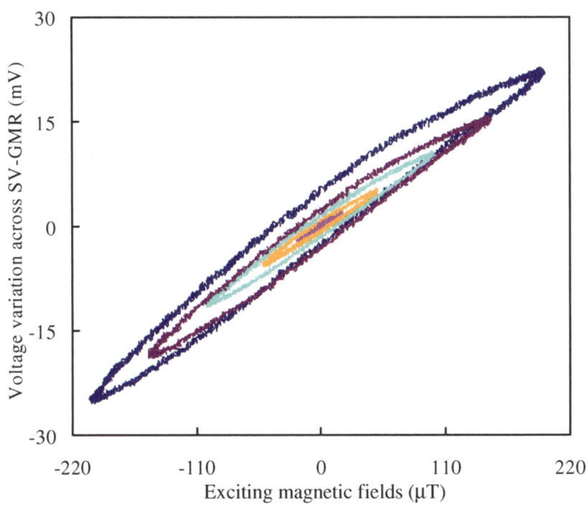

(a) Exciting frequency of 100 kHz.

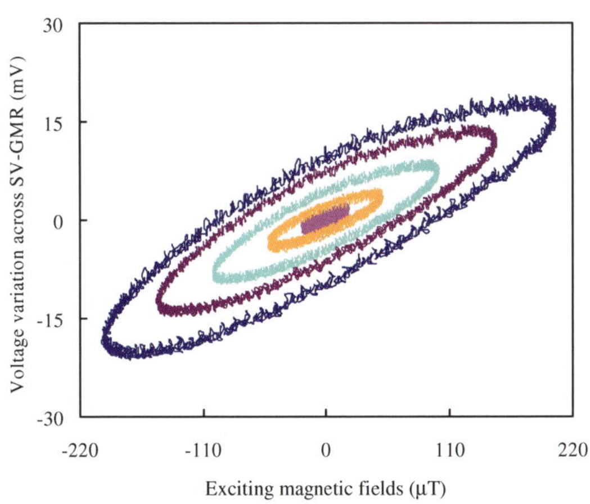

(b) Exciting frequency of 1 MHz.

Fig. 16. Effect of external magnetic flux amplitude on the GMR sensor output

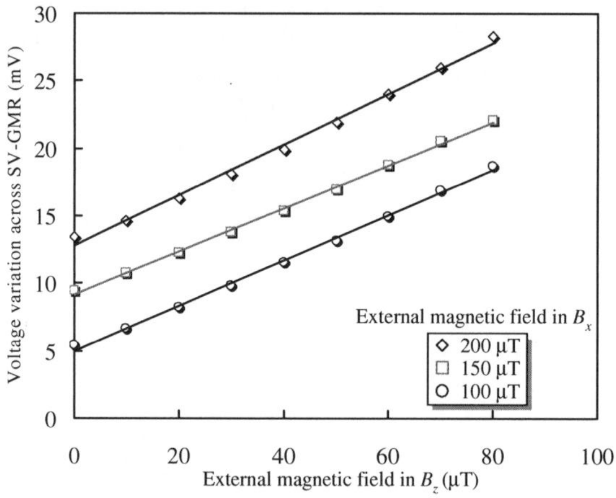

Fig. 17. Effect of an external magnetic flux density B_x on B_z and GMR sensor output

The sensitivity of the GMR sensor in the linear region is independent of the applied B amplitude as shown in Fig. 16. However, there is a marked increase in hysteresis with increasing B and frequency. Therefore, the GMR sensor used in the high-spatial ECT probe will be operated at magnetic flux densities lower than 20 μT; so that hysteresis effects will be at a minimum when the probe is used for defect detection. Planar meander coils normally generate a B in the x and y direction; therefore, experiments were done to observe the effect B_x has on the detection of B_z. Figure 17 shows the effect of constant external magnetic flux density B_x at the same frequency of magnetic flux density B_z and its effect on the GMR sensor output. It can be seen that the B_x results in a constant resistance variation, whereas resistance variation depends on the external magnetic flux density B_z. This shows that the constant magnetic flux densities in other axes have no effect on the detection of B_z.

2.3.2 Characteristics of the ECT Probe for PCB Inspection

Three types of defects on a PCB conductor are experimentally analyzed. 9 μm thick PCB conductors made from Cu coated with 0.05 μm Au were used as a PCB model with conductor disconnections and partial defects on the PCB track width as shown in Fig. 18. Conductor disconnections ranging from 50 to 500 μm were allocated on the PCB conductor as shown in Fig. 18 (a). Distance between the defects is 10 mm. For partial defects on PCB conductor track width, the conductor disconnection region was fixed at 100 μm as show in Fig. 18 (b). The conductor disconnection regions vary from 25 % of PCB track width to a total disconnection. For partial defects on PCB thickness, PCB conductors with thickness of 35 μm are used. Two chipping defects, as shown in Fig. 18 (c), were allocated on the model. The disconnection region is fixed at 100 μm as in the case of partial defect on PCB track width.

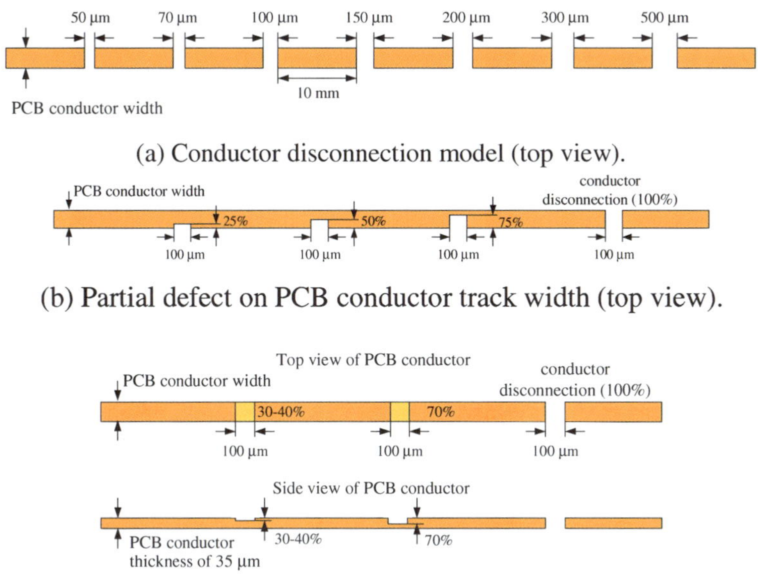

(c) Partial defect on PCB thickness.

Fig. 18. Model of defect on PCB conductor

The B generated by eddy-current flow at defect points on a PCB conductor can be detected by the ECT probe as shown in Fig. 19; the signal obtained from the ECT probe agrees with the aforementioned FEM analytical results. The signal has the same pattern and its amplitude depends on defect size; the phase of the ECT signal also varies at defect points. As shown in Figs. 20 and 21, although the ECT signal variations are very small, the utilization of the GMR sensor as a magnetic sensor provides the possibility of defect detection on a narrow, 50 μm wide, PCB conductor and a conductor disconnection length of 70 μm. The noise level is constant at around 0.7 μV. As shown in Fig. 21, signal variation depends on PCB conductor size and decreases almost linearly. Imperfections on a PCB conductor track width and thickness can also be inspected by the ECT probe. Figures 22 and 23 show the signal variations at partial defects on a PCB conductor track width and thickness, respectively. The signal variations at partial defects are lower than the signal variations at conductor disconnections. The signal variations at thickness defects are very small but large enough to identify the defect points.

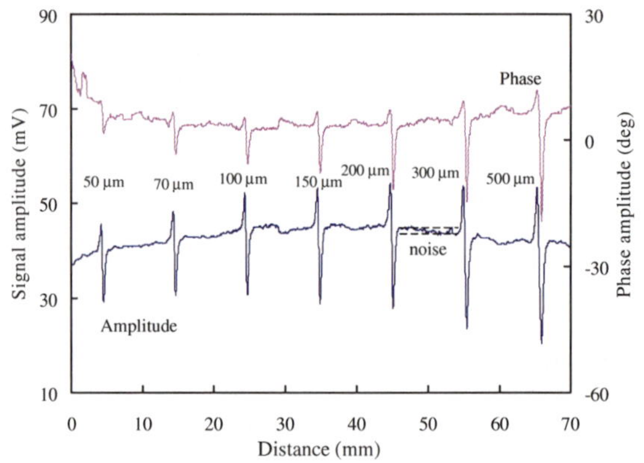

Fig. 19. ECT signal and its phase obtained from scanning over a 200 µm wide PCB conductor using the ECT probe

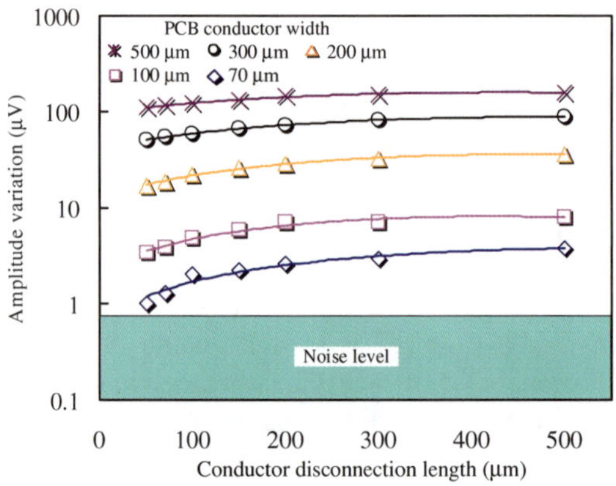

Fig. 20. Signal amplitude variation for different conductor disconnection lengths

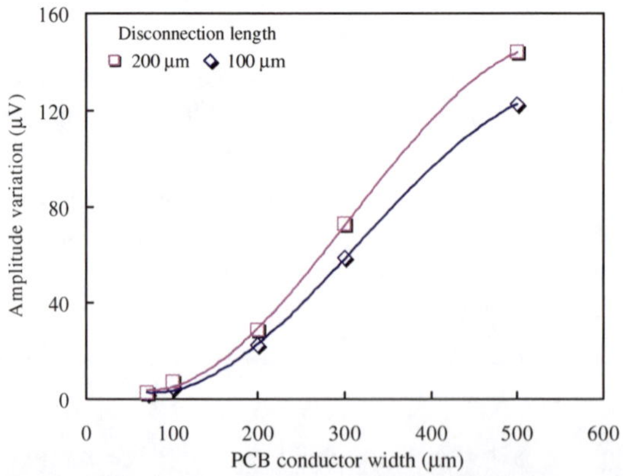

Fig. 21. Signal amplitude variation for different PCB conductor widths

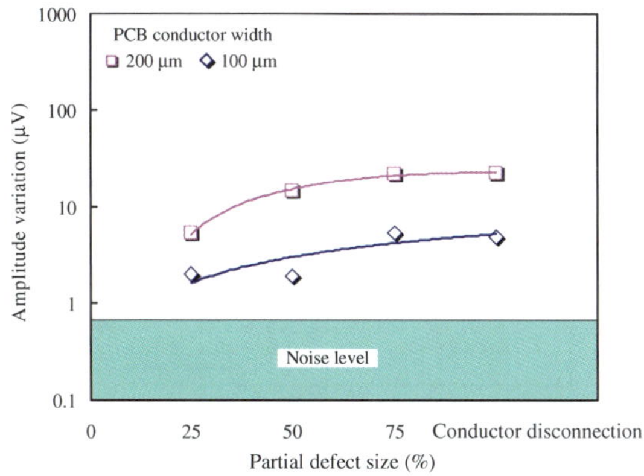

Fig. 22. Signal amplitude variation for partial defects on PCB conductor track widths

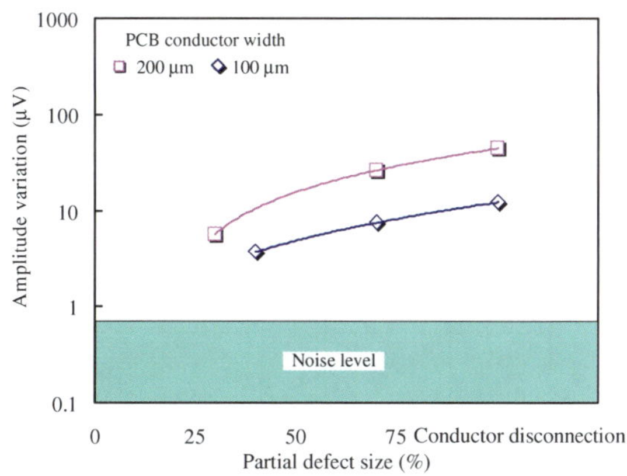

Fig. 23. Signal amplitude variation for partial defects on PCB conductor track thicknesses

The ECT signal waveforms in Fig. 24 obtained from the detection of a solder microbead with 125 μm radius at the frequency of 5 MHz and 10 MHz agree with the ECT signal waveforms obtained from FEM analysis. The determination of the microbead position is done by considering the peak of signal gradient. Figure 25 shows the maximum variation of the ECT signal versus the radius of the solder microbead, ranging from 125 to 300 μm. The maximum signal variation at an exciting frequency of 10 MHz decreases with the solder microbead radius and it is lower than the signal variation at exciting frequency of 5 MHz when the solder microbead radius is bigger than 200 μm. This is because the planar meander coil cannot generate a uniform *B* distribution. The experimental results also shows that signal variations at the solder microbead depend on the frequency of the exciting *B* and the solder microbead radius.

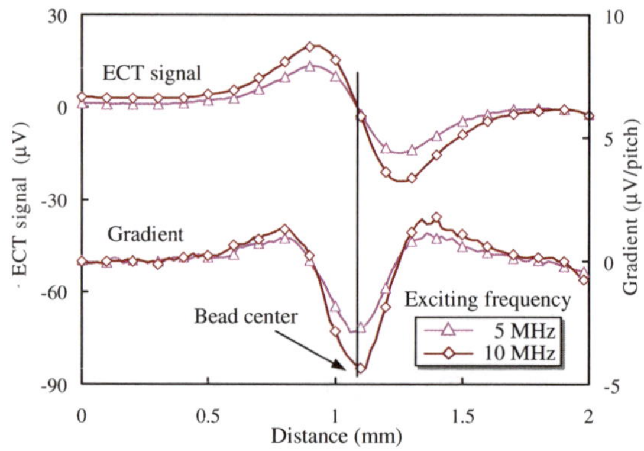

Fig. 24. ECT signal obtained from the detection of a solder microbead (PbSn) with 125 μm radius and its signal gradient

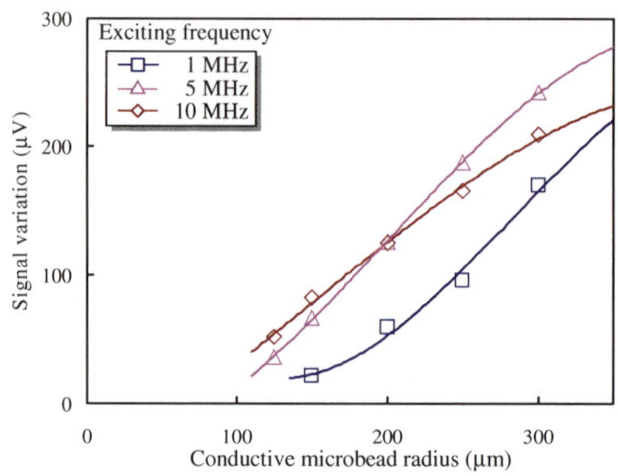

Fig. 25. Maximum signal variation vs. conductive microbead (PbSn) radius

2.3.3 PCB Inspection System

The PCB inspection system consists of two parts as shown in Fig. 26; the data acquisition process by the ECT technique and the image processing technique used to identify defect points on a PCB conductor. PCBs are scanned and the collected data is sent for identification of defects. The data acquisition apparatus consists of the ECT probe, a PCB position controller, exciting system, and measurement system, as shown in Fig. 27. The position controller is controlled by the computer for scanning regions on PCBs. High-frequency excitation at a frequency of 5 MHz is generated by a function generator and is fed to the power amplifier before feeding to the meander coil. The output signals from the ECT probe are very weak and comprise many unwanted signals; the harmonics and high-frequency noise. Therefore, the signal-to-noise ratio must be improved.

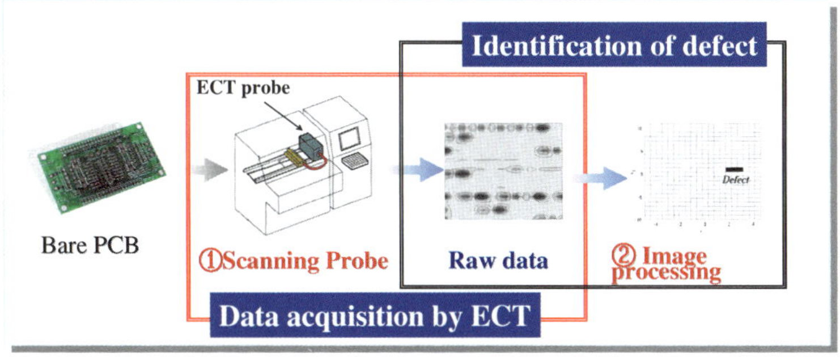

Fig. 26. PCB inspection system

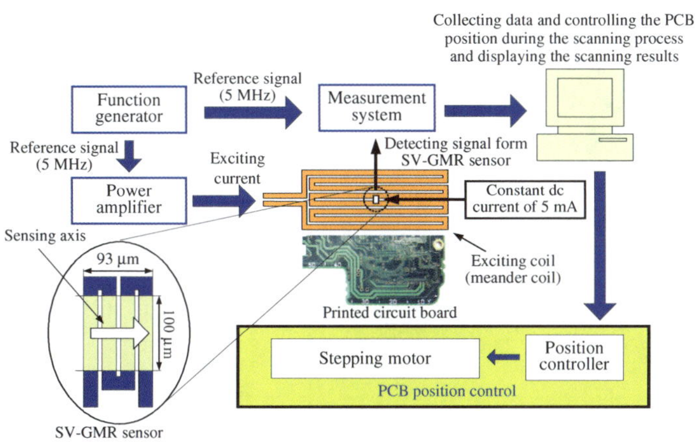

Fig. 27. Data acquisition system based on ECT technique

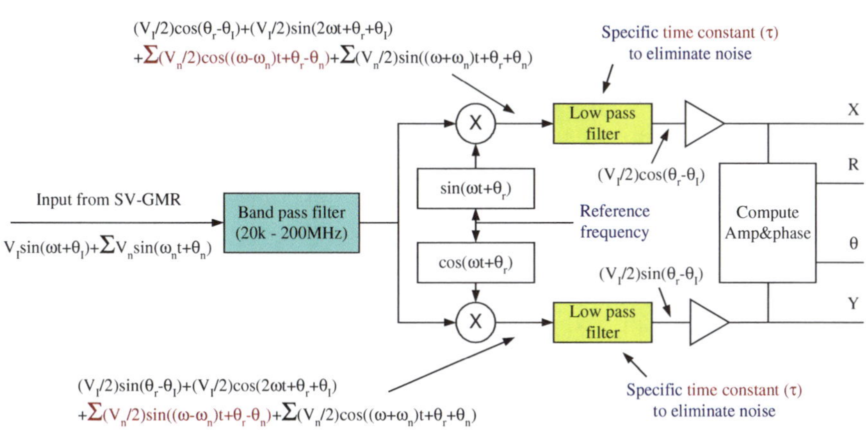

Fig. 28. Block diagram showing the basic principle of a lock-in amplifier

A lock-in amplifier based on phase-sensitive detection can be used to measure very small signals embedded in a lot of noise. This is because phase-sensitive detection can improve signal-to-noise ratio of noisy signals, which is usually higher than 60 dB [43]. A simple block diagram representing the basic principle of a lock-in amplifier is shown in Fig. 28. The input signal $v_{in}(t) = V_s \sin(\omega t + \theta_s) + \sum V_{noise} \sin(\omega_{noise} t + \theta_{noise})$, which consists of a signal and noise term, (where V_{noise}, ω_{noise}, and θ_{noise} are amplitude, frequency, and phase shift of the noise, respectively), is multiplied by the sine and cosine function at the fundamental frequency.

Therefore, the noisy signal can be obtained as following:

$$\begin{aligned} V_x/2 &= (V_s/2)\sin(\theta_r - \theta_s) + (V_s/2)\sin(2\omega t + \theta_r + \theta_s) \\ &+ \sum (V_{noise}/2)\cos((\omega - \omega_{noise})t + \theta_r - \theta_{noise}) \\ &+ \sum (V_{noise}/2)\sin((\omega + \omega_{noise})t + \theta_r + \theta_{noise}). \end{aligned} \tag{4}$$

$$\begin{aligned} V_y/2 &= (V_s/2)\sin(\theta_r - \theta_s) + (V_s/2)\cos(2\omega t + \theta_r + \theta_s) \\ &+ \sum (V_{noise}/2)\sin((\omega - \omega_{noise})t + \theta_r - \theta_{noise}) \\ &+ \sum (V_{noise}/2)\cos((\omega + \omega_{noise})t + \theta_r + \theta_{noise}). \end{aligned} \tag{5}$$

It can be seen from equations (4) and (5) that there are 3 components; DC, low-frequency, and high-frequency. Due to the relationship between time constant τ, and cut-off frequency f_c, the AC component can be rejected by the low pass filter resulting in only a DC component, by increasing the time constant τ. If the low-frequency component is very close to DC, time constant of the low pass filter needs to be increased in order to eliminate the low-frequency component. This means that the ECT probe have to wait until the lock-in amplifier acquires a steady state value before the ECT probe can be moved to measure the signal at the next position. This is a cause of scanning speed restriction if a lock-in amplifier is used to measure the ECT signal.

Fourier analysis can also be applied to measure low, noisy signals. By applying Fourier analysis on the measured signal, the signal amplitude at the required frequency can be extracted in a shorter time than by using a low pass filter. Amplitude of the signal at the fundamental frequency can be obtained within a few cycles of the measured frequency (5 MHz) [44]. Therefore, the PCB inspection with high-speed scanning can produce inspection results without or minimum distortion. A Fourier series is an expansion of a periodic function *f(t)* in terms of an infinite sum of sines and cosines. The computation and study of Fourier's series is known as harmonic analysis and is extremely useful as a way to break up an arbitrary periodic function into a set of simple terms that can be plugged in, solved individually, and then recombined to obtain the solution to the original problem or an approximation to it to whatever accuracy is desired or practical.

A signal *f(t)* can be expressed by a Fourier's series in the form:

$$f(t) = a_0/2 + \sum_{n=1}^{\infty}\{a_n \cos(n\omega t) + b_n \sin(n\omega t)\}, \tag{6}$$

where n represents the rank of the harmonics (n = 1 corresponds to the fundamental component). The term $a_0/2$ represents the mean value or DC component of waveform and a_n and b_n are Fourier coefficients.

2.4 High-Density PCB Inspection

2.4.1 PCB Defect Inspection

As shown in Fig. 29, ten PCB conductors with different conductor widths, W, and gaps, G, was used as a model. Two PCB models were used:

1) 100 μm PCB conductor width, W, with 200 μm gap, G
2) 100 μm PCB conductor width, W, with 100 μm gap, G

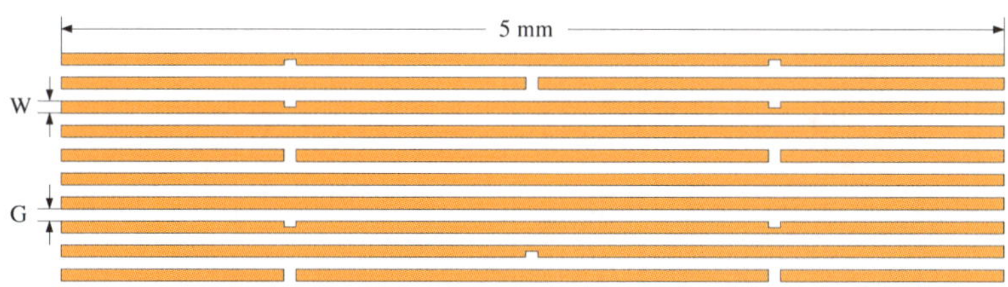

Fig. 29. Single layer PCB model used in an experiment

The PCB images and scanning results after image processing was applied, as shown in Fig. 30. The same defects were allocated on both the PCB models. The smallest conductor disconnection was only 20 μm. Furthermore, different kinds of partial defects were located on this model. The scanning results show that the ECT probe can inspect defects on a PCB conductor. From the inspection results of a bare PCB with a 100 μm track width, the defects on PCB conductor are not difficult to identify although the images are not clear. PCB conductor gap is one of the parameters that affect the signal variations. Signal variations at defect points decrease when the gap between PCB conductors is small and if the spatial resolution of the GMR sensor is not high enough. Hence, the spatial resolution needs to be increase for the inspection of high density PCBs that with less than 100 μm gap.

Distance between PCB conductors and sensing level is very important for inspection of the bottom-layer of double-layer PCBs. Figure 31 shows the inspection results obtained from scanning over a 100 μm wide PCB conductor. The GMR sensor can detect the B variation at defect points and provides variation of both signal amplitude and phase. The complex plane is convenient to represent the variation of both signal amplitude and phase at defect points as real and imaginary components. The results

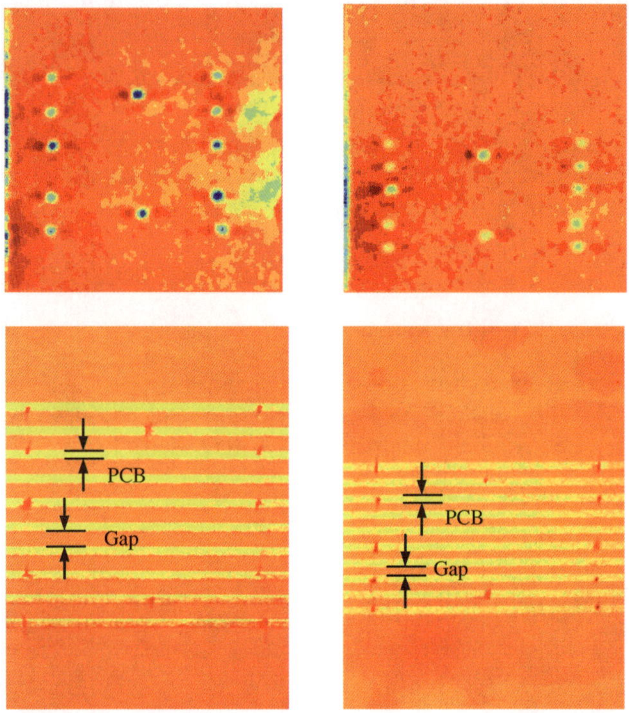

Fig. 30. Inspection results of the PCB model with 100 μm PCB conductor width (W) and gap (G) of 200 μm (left) and 100 μm (right)

show that the proposed ECT probe is able to inspect the defects on the PCB conductor with 235 μm lift-off height although the signal variations are very small. This means that the probe is capable of inspecting the defects at the bottom layer if the distance between PCB conductor and sensing level is less than 200 μm. Two models of high-density double-layer PCBs with a dimension of 5 mm × 5 mm as shown in Fig. 32 was used for experiments. Both models have a conductor width of 100 μm with 100 and 200 μm gaps between its conductors. The top layer has PCB conductors parallel to the *x* direction while the conductors are in the bottom layer. The disconnection and partial defects are also allocated on both the top and the bottom layer of the PCB model. 2-D images reconstructed from the ECT signals obtained from scanning over the top layer of the both PCB models in *x* and *y* directions are shown in Figs. 33 and Fig. 34. Numerical gradient technique is a simple image processing technique that is used to eliminate signal offset and to enhance the signal at the defect points. The 2-D images show that the proposed ECT probe is capable of inspecting the defect clearly although the defect points are also allocated on bottom layer of the test PCB. Moreover, not only conductor disconnections but also imperfections on PCB conductor can be detected.

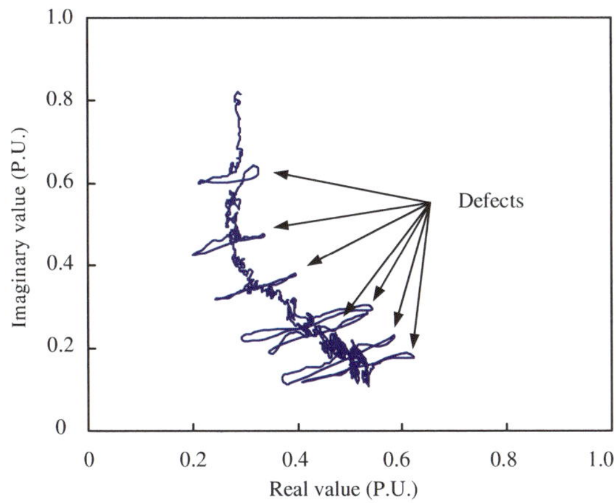

(a) Lift-off height = 185 μm

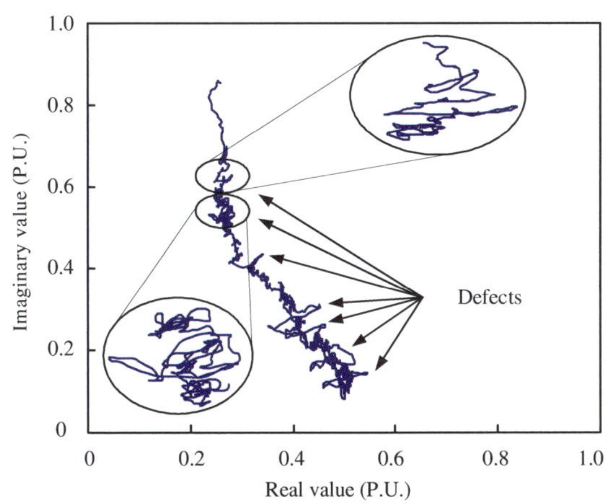

(b) Lift-off height = 235 μm

Fig. 31. Complex plane of ECT signal obtained from scanning over a 100 μm wide PCB conductor, at different distances from sensing level to the PCB conductor

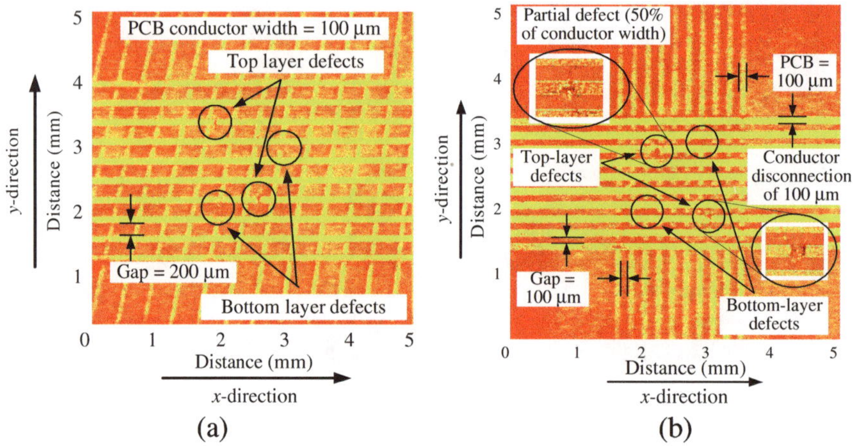

Fig. 32. Double layer PCB models. (a) 100 μm PCB conductor model with gap of 200 μm, and (b) 100 μm PCB conductor model with gap of 100 μm

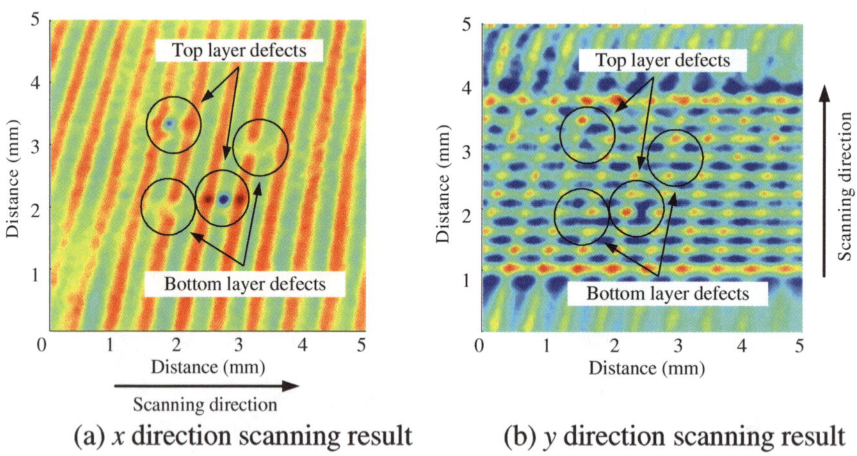

Fig. 33. PCB inspection results for a 100 μm conductor width with a gap of 200 μm

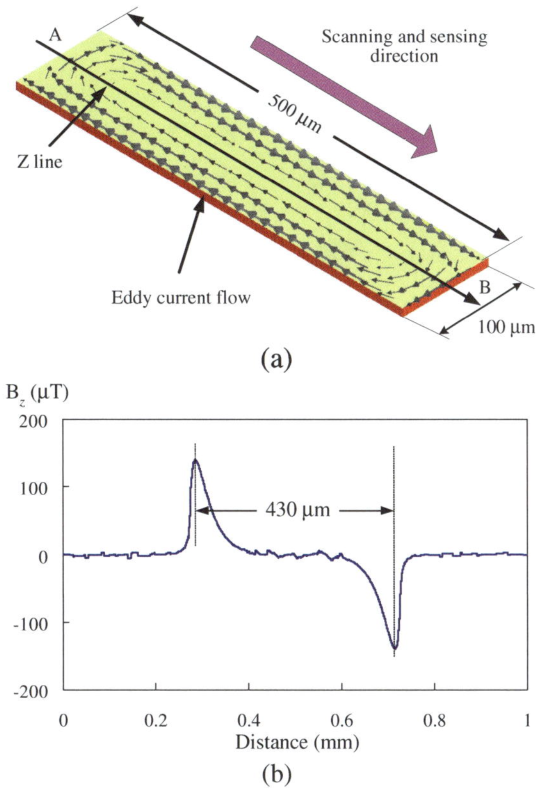

Fig. 34. Eddy-current flow (a) magnetic flux density over a PCB conductor in the sensing direction on the Z line from A to B; (b) when the probe scans along the PCB conductor

2.4.2 PCB Dimension and Alignment Inspection

Eddy-currents are usually distributed near the PCB conductor boundaries, as in Fig. 34 (a). Therefore, peak values of *B* in sensing direction occur near PCB conductor boundaries that are perpendicular to the scanning direction. These characteristics indicate that both length and width of the PCB conductor can specified by considering the peak values of *B* with measurement error less than 70 μm and 30 μm, for measurement of PCB conductor length and width, respectively. For precise measurement, exciting frequency should be higher than the frequency in simulation to generate eddy-currents flowing as close as possible to the PCB conductor boundary. Therefore, peak values of *B* in sensing axis will occur very close to PCB conductor boundary that is perpendicular to scanning direction. Absolute measurement errors, when the probe has been applied to measuring PCB conductor width and disconnection length, are shown in Fig. 35 (a) and (b) respectively. From these results, absolute measurement errors are lower than 100 μm for measurement of PCB conductor width. In case of conductor disconnection and PCB conductor length measurements, the absolute measurement errors are lower than 400 μm. However, it is not more than 200 μm when disconnection or gap between PCB conductors that are longer than 200 μm is measured.

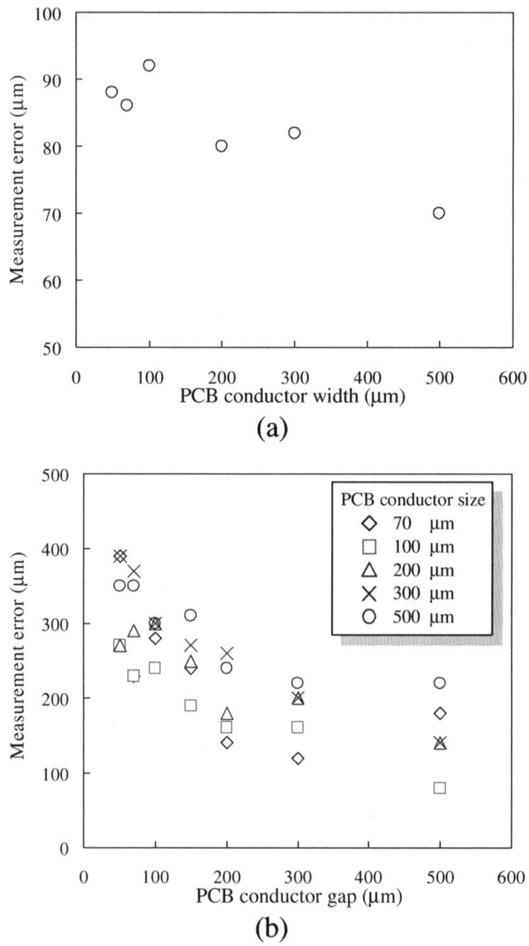

Fig. 35. Absolute measurement errors obtained when the ECT probe is used to measure PCB conductor width (a) and disconnection length (b)

High-density PCB models with 100 μm PCB conductor width were used as a model in the experiment and gap between the conductors is 100 μm, as shown in Fig. 36 (a). The high-density PCB model images and its reconstructed 2-D images from ECT signals without offset are shown in Fig. 36 (b). Conductor disconnection and partial defects were allocated on the conductor. Basic filtering techniques were applied to eliminate noise before the reconstruction of the 2-D image. From the reconstruction of the 2-D image, the defect points on the conductor were clearly identified. The PCB conductors are able to be identified by considering the peak of ECT signal as shown in the strip chart. Inspection of the larger gap model provided not only clearer details and easier identification of PCB conductor size but also gap between conductors (with error less than 100 μm) than the smaller gap model. In addition, the distance between defect points, both the larger and smaller gap model, is also able to be specified accurately with an error less than 200 μm. However, more effective image processing techniques should be applied to improve the information [45, 46].

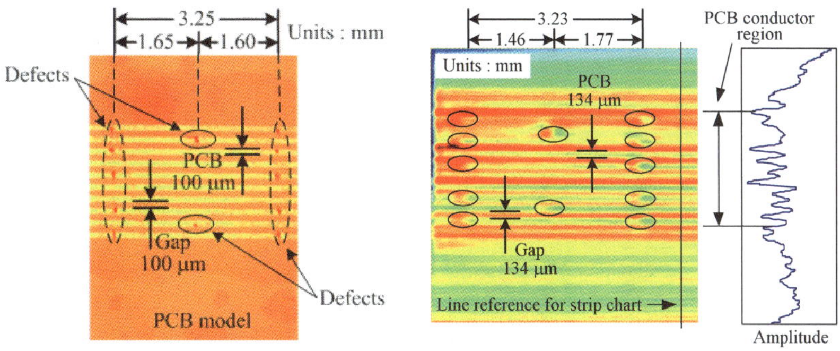

Fig. 36. High-density PCB images of a 100 μm wide PCB conductor; (a) inspection results (b) with gaps of 100 μm. The strip charts obtained from the line reference show the high signal variation occurring at the PCB conductor region.

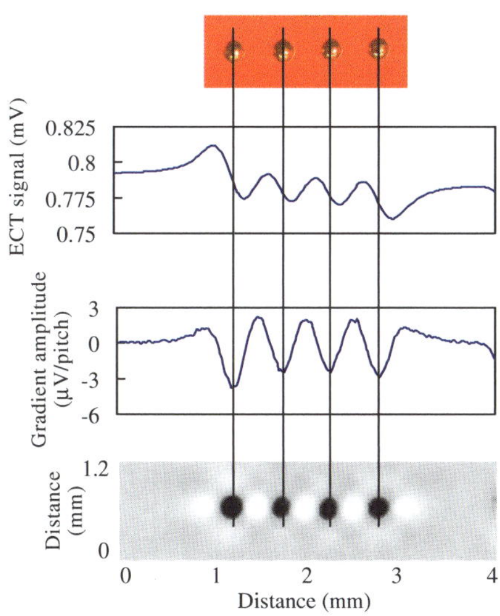

Fig. 37. Identification of the solder microbeads by the application of numerical gradient to the ECT signals

2.4.3 Ball Grid Array Detection

The numerical gradient technique is applied to ECT signals obtained from the detection of solder microbeads [47, 48]. The results show that the numerical gradient is able to enhance the signal at the microbead as shown in Fig. 37. As a result, the microbead position is easily identified by considering the peak of the signal gradient.

In addition, the pitches of the conductive microbead are also measured by considering the peak of the signal gradient. The solder microbead array model with 250 μm diameter and 410 – 460 μm microbead pitches and its detection results are shown in Figs. 38 (a) and (b) respectively. The proposed ECT probe scans over the solder microbead with areas of 4 mm × 4 mm. Form the detection results, the solder microbeads are clearly recognized and the pitches of the conductive microbead are also accurately specified with errors within 50 μm.

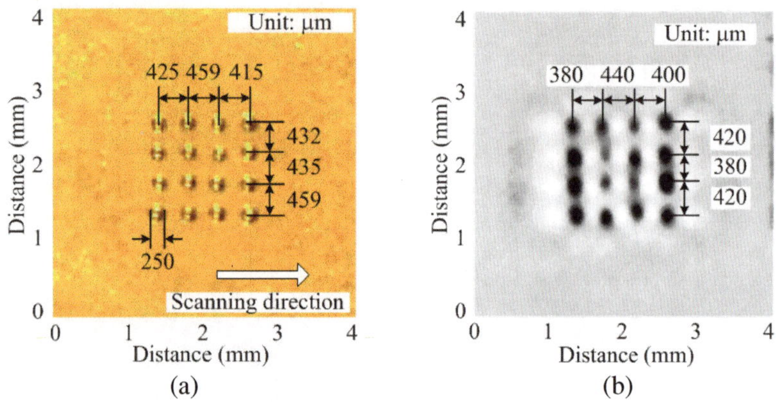

Fig. 38. Solder microbead array model (a) and its detection results (b)

3 Conclusions

The design and development of an ECT probe for high-density PCB inspection was reported in this section. The ECT probe consists of a magnetic sensor and a planar meander coil functioning as an exciting coil. A GMR was chosen as the magnetic sensor because it provides the advantages of high-spatial resolution due to small dimensions, high sensitivity to low magnetic fields over a broad range of frequencies. Moreover, it is inexpensive and is able to operate at room temperature. Behavior of eddy-current flow and magnetic flux distribution was studied using FEM and the results confirmed that the proposed ECT probe can be applied to high-density PCB inspection.

The GMR sensor was characterized in order to determine its capability of magnetic field detection and sensitivity; these results were used to develop a high performance ECT probe. Characterization of the probe with GMR sensor showed that measurement could be performed with high signal-to-noise ratio. Furthermore, the probe with GMR sensor is capable of defect detection on PCB conductor with width of 70 μm and thickness of 9.05 μm. Three kinds of defect were tested in the experiment. The first was conductor disconnection, the second was partial defect on PCB track width, and the third was partial defect on PCB thickness. The results show that the inspection results can be performed by the proposed ECT probe. Inspection of high-density single and double layer PCB models were demonstrated. Inspection of the PCB model with conductor width and gap of 100 μm was performed by the proposed ECT probe. In case of double layer PCB inspection, the probe is capable of inspecting the defects at the bottom layer if the distance between PCB conductor and sensing level is not over 200 μm. Moreover, dimension and alignment of PCB conductor can also be examined. Consideration of the peak of magnetic flux density that usually occurs at boundary of PCB conductor is useful for investigation of PCB dimensions and alignment. The inspection results showed that the proposed ECT probe is able to examine PCB conductor dimensions and alignment with an error less than 200 μm. Furthermore, the ECT probe was also successfully utilized to detect the position of 125 μm radius solder microbeads on a BGA.

Acknowledgment. The authors would like to thank Professor Masayoshi Iwahara (Retired) from the Institute of Nature and Environmental Technology, Kanazawa University, for his help in this research. Thank you to Professor Hiroyuki Wakiwaka from Shinshu University for the advice and support about GMR sensors. The authors gratefully acknowledge research support from the Ministry of Education, Culture, Sports, Science and Technology (MEXT) and the Japan Society for Promotion of Science (JSPS).

References

[1] Smith, C.H., Schneider, R.W.: Low-Field Magnetic Sensing with GMR Sensors Part 1: The Theory of Solid-State Sensing. Sensors 16, 76–83 (1999)
[2] Schneider, R.W., Smith, C.H.: Low Magnetic Field Sensing with GMR Sensors Part 2: GMR Sensors and Their Applications. Sensors 16, 84–91 (1999)
[3] Wakiwaka, H.: Characteristics of Giant Magnetoresistance Sensor and Their Applications. Journal of the Magnetics Society of Japan 28, 825–833 (2004)
[4] Schewe, H., Schelter, W.: Industrial applications of magnetoresistive sensors. Sensors and Actuators A 59, 165–167 (1997)
[5] Tumanski, S.: Thin Film Magnetoresistive Sensors. Institute of Physics, London (2001)
[6] Tsymbal, E.Y., Pettifor, D.G.: Perspectives of Giant Magnetoresistance. In: Ehrenreich, H., Spaepen, F. (eds.) Solid State Physics, pp. 113–237. Academic Press, California (2001)
[7] Daughton, J.M.: GMR applications. Journal of Magnetism and Magnetic Materials 192, 334–342 (1999)
[8] Lenz, J., Edelstein, A.S.: Magnetic Sensors and Their Applications. IEEE Sensors Journal 6, 631–649 (2006)
[9] Dieny, B.: Giant magnetoresistance in spin-valve multilayers. Journal of Magnetism and Magnetic Materials 136, 335–359 (1994)
[10] Dogaru, T., Smith, S.T.: Giant Magnetoresistive-Based Eddy Current Sensor. IEEE Transactions on Magnetics 37, 3836–3838 (2001)
[11] Childress, J.R., Fontana Jr, R.E.: Magnetic recording read head sensor technology. Comptes Rendus Physique 6, 997–1012 (2005)
[12] Gooneratne, C.P., Liang, C., Giouroudi, I., Kosel, J.: An integrated micro-chip for rapid detection of magnetic particles. Journal of Applied Physics 111, 07B327 (2012)

[13] Gooneratne, C.P., Liang, C., Giouroudi, I., Kosel, J.: A giant magnetoresistance ring-sensor based microsystem for magnetic bead manipulation and detection. Journal of Applied Physics 109, 07E517 (2011)
[14] Smith, C.H., Schneider, R.W., Dogaru, T., Smith, S.T.: Eddy-current testing system with GMR magnetic sensor arrays. Review of Progress in Quantitative Nondestructive Evaluation 2323, 406–413 (2003)
[15] Reig, C., Cubells-Beltrán, M.D., Muñoz, D.R.: Magnetic Field Sensors Based on Giant Magnetoresistance (GMR) Technology: Applications in Electrical Current Sensing. Sensors 9, 7919–7942 (2009)
[16] Freitas, P.P., Ferreira, H.A., Graham, D.L., Clarke, L.A., Amaral, M.D., Martins, V., Fonseca, L., Cabral, J.M.S.: Magnetoresistive DNA Chips. In: Johnson, M. (ed.) Magnetoelectronics, pp. 331–396. Academic, New York (2004)
[17] Miller, M.M., Sheehan, P.E., Edelstein, R.L., Tamanaha, C.R., Zhong, L., Bounnak, S., Whitman, L.J., Colton, R.J.: A DNA array sensor utilizing magnetic microbeads and magnetoelectronic detection. Journal of Magnetism and Magnetic Materials 225, 138–144 (2001)
[18] Gaster, R.S., Xu, L., Han, S.J., Wilson, R.J., Hall, D.A., Osterfeld, S.J., Yu, H., Wang, S.X.: Quantification of protein interactions and solution transport using high-density GMR sensor arrays. Nature Nanotechnology 6, 314–320 (2011)
[19] Schwartz, W.H.: Vision Systems for PC Board Inspection. Assembly Engineering 29, 18–21 (1986)
[20] Barnard, S.T.: Automatic Visual Inspection of Printed Circuit Boards. In: Advanced Systems for Manufacturing: Conference on Production Research and Technology, pp. 423–429 (1985)
[21] Mukai, S.: PCB Continuous Line System Proceeds from Manufacturing to Inpsection. Journal of Electronic Engineering 29, 34–39 (1992)
[22] Foster, J.W., Griffin, P.M., Messimer, S.L., Villalobos, J.R.: Automated Visual Inspection: A Tutorial. Computers in Industrial Engineering 18, 493–504 (1990)
[23] Goodall, A.J.E., Lo, E.K.: A Review of Inspection Techniques Applicable to PCB Manufacturing and Assembly particularly with respect to SMT. Advanced Manufacturing Engineering 3 (1991)
[24] Hroundas, G.: Importance of Electrical Inspection for the Printed Circuit Board. In: Proceedings of the Circuit Expo West, pp. 53–60 (1986)
[25] Jacob, G.: Advances in Board Inspection. Evaluation Engineering, 126–133 (1992)
[26] García-Martin, J., Gómez-Gil, J., Vázquez-Sánchez, E.: Non-Destructive Techniques Based on Eddy Current Testing. Sensors 11, 2525–2565 (2011)
[27] Le Bihan, Y., Pávó, Marchand, C.: Characteristics of small cracks in eddy current testing. The European Physics Journal Applied Phyics 43, 231–237 (2008)
[28] Hellier, C.J.: Handbook of Nondestructive Evaluation. McGraw-Hill (2001)
[29] Shull, P.J.: Nondestructive Evaluation Theory, Technique, and Applications. Marcel Dekker Inc. (2002)
[30] Burkhardt, G.L., Fisher, J.L.: Nondestructive Testing of Aging Aircraft. Southwest Research Institute Technology Today (2001)
[31] Dogaru, T., Smith, S.T., Smith, C.H., Schneider, R.W.: Eddy Current Inspection for Deep Crack Detection Around Fastener Holes in Airplane Multi-Layered Structures. In: 3rd Pan American Conference for Nondestructive Testing, Rio de Janeiro, Brazil, (2003)
[32] Dogaru, T., Smith, S.T.: Novel Eddy Current Probes for Detection of Deep Cracks Around Fastener Holes. In: Fifth Joint NASA/DOD/FAA Aging Aircraft Conference, Orlando, USA (2001)

[33] Ko, R.T., Petricola, D.L., Berens, A.P., Rambo, M.: A Study of Index Step Sizes for Varing Coil Diameters and Their Effects on Throughput and Probability of Detection (POD) for Improved Automated Eddy Current Inspections on Areospece Engine-Components. In: 16th World Conference on Nondestructive Testing, Montreal, Canada (2004)
[34] Stanic, D., Elsing, B.: Eddy Current Inspection of Cladding Material on VVER440 Reactor Pressure Vessels. In: 16th World Conference on Nondestructive Testing, Montreal, Canada (2004)
[35] Muzhitsky, V.F., Karabchevsky, V.A., Karpov, S.V.: Ten Years of Experience of Stress-Corrosion and Other Pipelines' Surface Flaws Testing by Means of Eddy Current Technique. In: 16th World Conference on Nondestructive Testing, Montreal, Canada (2004)
[36] Crowther, P.: Non Destructive Evaluation of Coatings for Land Based Gas Turbines Using a Multi-Frequency Eddy Current Technique. In: 16th World Conference on Nondestructive Testing, Montreal, Canada (2004)
[37] Tian, H., Yamada, S., Iwahara, M., Toyama, H., Miya, K.: Eddy-Current Model and Detection in a Thick Stainless Steel Plate. Transaction on Magnetics Society of Japan 5, 39–42 (2005)
[38] Rao, B.P.C., Jayakumar, T., Kalyanasundaram, P., Raj, B.: An Insight into Design and Development of Eddy Current Sensors for Non-Destructive Evaluation of Materials. In: Proceedings of the National Workshop on Sensors, Kalpakkam, India (2000)
[39] Dogaru, T., Smith, C.H., Smith, S.T.: New Directions in Eddy Current Sensing. Sensors Magazine 18 (2001)
[40] Mulligan, C., Lee, C., Danon, Y.: Characterization of Magnetron Sputtered Coating by Pulsed Eddy Current Techniques. Review of Progress in Quantitative Nondestructive Evaluation 24, 1721–1728 (2005)
[41] Nakamura, K., Yamada, S., Iwahara, M.: Inspection of Printed Circuit board Using Multi-ECT Probe with Solenoid Pickup Coils. Journal of the Magnetics Society of Japan 26, 555–558 (2002)
[42] Yamada, S., Nakamura, K., Iwahara, M., Taniguchi, T., Wakiwaka, H.: Application of ECT Technique fro Inspection of Bare PCB. IEEE Transactions on Magnetics 39, 3325–3327 (2003)
[43] Standford Research Systems, User's Manual: Model SR844 RF Lock-In Amplifier, Standford Research Systems. Rev. 2.5, USA, 2-3–2-18 (1997)
[44] Chomsuwan, K., Yamada, S., Iwahara, M.: Bare PCB Inspection System with SV_GMR Sensor Eddy-Current Testing Probe. IEEE Sensors Journal 7, 890–896 (2007)
[45] Taniguchi, T., Kacprzak, D., Yamada, S., Iwahara, M.: Wavelet-Based Processing of ECT Images for Inspection of Printed Circuit Boards. IEEE Transactions on Magnetics 37, 2790–2793 (2001)
[46] Kacprzak, D., Taniguchi, T., Nakamura, K., Yamada, S., Iwahara, M.: Superposition of Signal Components During Inspection of Printed Circuit Boards by an Eddy Current Testing Probe With a Solenoid Pick-Up Coil. IEEE Transactions on Magnetics 37, 2794–2796 (2001)
[47] Yamada, S., Chomsuwan, K., Hagino, T., Tian, H., Minamide, K., Iwahara, M.: Conductive Microbead Array Detection by High-Frequency Eddy-Current Testing technique with SV-GMR Sensor. IEEE Transaction on Magnetics 41, 3622–3624 (2005)
[48] Somsak, T., Chomsuwan, K., Yamada, S., Iwahara, M.: Conductive Microbead Array Detection Based on Eddy-Current Testing Using SV-GMR Sensor and Helmholtz Coil Exciter. IEEE Transactions on Magnetics 42, 3572–3574 (2006)

High-Spatial Resolution Giant Magnetoresistive Sensors – Part II: Application in Biomedicine

C.P. Gooneratne[1], K. Chomsuwan[2], M. Kakikawa[3], and S. Yamada[3]

[1] King Abdullah University of Science and Technology,
Thuwal, Saudi Arabia
Chinthaka.Gooneratne@kaust.edu.sa
[2] King Mongkut's University of Technology Thonburi,
Bangkok, Thailand
[3] Kanazawa University, Kanazawa, Japan

Abstract. Magnetic fluid based hyperthermia therapy for treating cancerous tumors can be performed with a high success rate and minimal error given the possibility of detecting and estimating magnetic fluid weight density *in vivo*. In this chapter, a uniquely designed GMR needle probe is presented for the detection and estimation of magnetic fluid content density inside tumors. Experimental results showed that the proposed technique has a good potential to be implemented in hyperthermia therapy in the future.

1 Introduction

Magnetic fluid based hyperthermia has the potential to be an effective, non-invasive cancer therapy with negligible side effects [1-8]. Magnetic fluid is injected into the affected area and an external AC magnetic flux density is applied to exploit the self-heating properties of the magnetic beads (MBs) in the fluid. Temperatures in excess of 42 °C destroy tumors [9-12]. Generally, all parameters, except the magnetic fluid content density *in vivo*, are known in the specific heat equation which governs the heat given in hyperthermia therapy to destroy cancer cells. This is due to the fact that magnetic fluid injected into an affected area spreads to neighboring tissue thus, effectively reducing the magnetic fluid content density. Hence, accurate estimation of magnetic fluid content density *in vivo* is critical for successful cancer treatment by hyperthermia therapy. The purpose of this research is to develop a method and appropriate apparatus/tools to estimate magnetic fluid content density *in vivo* so that tumors can be destroyed without affecting healthy cells. The key feature of this research is the fabricated novel GMR needle probe. The GMR needle probe is designed in such a way so that it can be inserted *in vivo* in a minimally-invasive way to detect and estimate magnetic fluid content density.

2 Estimation of Magnetic Fluid Density Inside Tumors

2.1 Introduction

Recent cancer treatment has focused on the effectiveness of killing localized or deep seated cancer tumors. The use of magnetic materials to heat tumors was first proposed

C. Reig et al.: *Giant Magnetoresistance (GMR) Sensors*, SSMI 6, pp. 243–273.
DOI: 10.1007/978-3-642-37172-1_10

by Gilchrist et al in 1957 [13], who used iron oxides to heat lymph nodes. Since then much research has been done to heat tumors directly in a non-invasive way. In the past, heating methods were difficult, expensive and unsafe resulting in the whole procedure being unfeasible. However, hyperthermia therapy with the aid of magnetic fluid is now regarded as one of the most promising cancer therapies due to the progress made in the synthesis of superparamagnetic type beads in the last decade. Superparamagnetic type MBs have single domains, where the magnetization direction flips randomly due to temperature [14]. However, when an external magnetic flux density is applied the magnetization direction aligns in the direction of the applied flux; therefore, MBs can be controlled by an external magnetic flux density [15, 16]. In the case of magnetic fluid hyperthermia, MBs can be injected *in vivo* and moved to the target site, held there until treatment is completed and removed afterwards, by localized magnetic flux gradients. Furthermore, since MBs have large surface to volume ratios for binding of biological cells and are physiologically well tolerated there are limitless opportunities for their utilization in many biomedical applications such as biological cell tagging, targeted drug delivery and MRI [17-20]. Magnetite (Fe_3O_4) is the most widely used and promising MB available today for biomedical applications [21-23]. Fe_3O_4 beads have hydrophobic surfaces and when they interact with each other the beads agglomerate to form clusters which increase the bead size. Hence, to stabilize MBs and prevent agglomeration a stabilizer such as a surfactant or polymer is usually added during preparation [21-25].

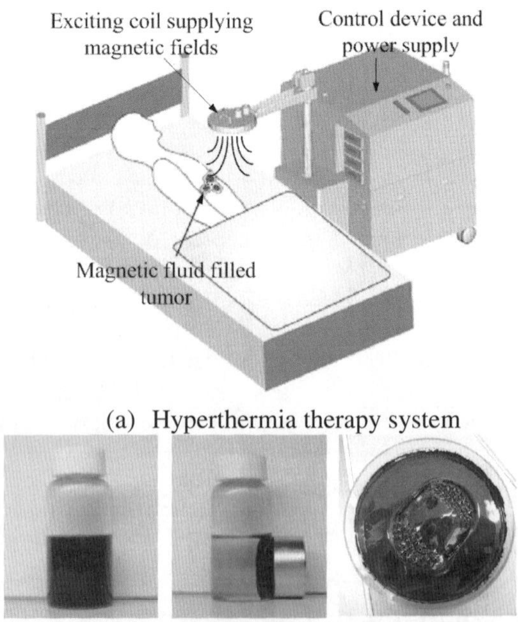

(a) Hyperthermia therapy system

(b) Magnetic fluid

Fig. 1. Magnetic fluid hyperthermia

Cancer cells are different to normal, healthy cells in many ways including how they react to heat [26]. Due to this difference it is possible for hyperthermia therapy to destroy without harming the healthy normal cells surrounding the tumor. Figure 1 (a) shows a magnetic fluid hyperthermia therapy system. As shown in Fig. 1 (b), magnetic fluids used in hyperthermia therapy are colloidal mixtures consisting of MBs suspended in a carrier fluid, usually an organic solvent or water; the fluid is injected directly into the tumor body or into an artery supplying the tumor. An external AC magnetic flux density of several kHz is then applied to the magnetic fluid filled area of the body. For biomedical purpose, the frequency has to be to 50 kHz to avoid neuromuscular electrostimulation and lower than 10 MHz for appropriate penetration depth of the radio frequency field [27]. Heat is produced in the magnetic fluid filled tumor since it is exposed to an external AC magnetic flux density. The heating losses of MBs are mainly due to Nèel relaxation and Brownian motion [28, 29]. Most types of cancer cells are more sensitive to temperatures in excess of 42° C than normal cells. Tumor apoptosis can be triggered if the temperature can be controlled at the therapeutic threshold of 42° C for a prolonged period of time, resulting in the tumor being destroyed or at least partly destroyed. Since the energy is coupled magnetically to the MBs, bone or boundaries of different conductive tissues do not interfere with power absorption as with electric field dominant systems used for regional hyperthermia for example [30]. Furthermore, there is also homogenous heating in the target region since even though there is a large number of MBs, each can be thought of as a separate hot source, giving temperature homogeneity during inactivation of cancer cells [31]. So, given that the fluid can be injected homogenously throughout the target region a homogenous cell inactivation could be expected. The AC magnetic flux density should also be homogenous because each bead has its specific power absorption, which is only constant if the applied magnetic flux density is also homogenous.

The specific heat capacity Q (W/ml), generated by magnetic fluid can be calculated as follows [32]:

$$Q = k_m f D_w B^2, \tag{1}$$

where k_m is a constant of 3.14×10^{-3} (W/Hz/(mgFe/ml)/T^2/ml), f is the exciting frequency of the applied magnetic flux density (kHz), D_w is the magnetic fluid weight density (mgFe/ml) and B is the amplitude of the applied magnetic flux density (T).

Currently, one of the main problems associated with magnetic fluid hyperthermia is that the magnetic fluid spreads inside tissue once injected, reducing D_w. From Eq. 1 it can be seen that Q is directly proportional to D_w. Inaccurate estimation of D_w has two major effects: i) if a low dosage is given the overall effect is thermal underdosage in the target region which often leads to recurrent tumor growth, ii) if heat given to a target region exceeds the therapeutic limit it may damage healthy cells. Hence, it can be stated that the quality of magnetic fluid hyperthermia treatment is proportional to the accuracy of estimating D_w *in vivo*. The purpose of this research is to develop a method and appropriate apparatus/tools to estimate D_w *in vivo* before hyperthermia therapy. The key feature of this research is the fabricated unique GMR needle probe which is designed in such a way so that it can be inserted *in vivo* in a minimally invasive way to detect and estimate D_w.

2.2 Analytical Estimation of Magnetic Fluid Parameters

2.2.1 Relationship between Relative Permeability, Magnetic Fluid Weight and Volume Density

The different variables used in this relationship are defined as follows:

D_v: (measured as a percentage); magnetic fluid volume density
D_w: (measured as weight per volume); magnetic fluid weight density
γ_f: (W-35 sample – Taiho Co. = 4.58); specific gravity of MBs

So then, the relationship between D_v and D_w can be expressed as follows:

D_w: weight of MBs in 1 ml volume : weight of combined MBs and water in 1 ml volume

$$D_w = \frac{(\gamma_f \times D_v)}{((1 - D_v) + \gamma_f \times D_v)}. \quad (2)$$

This can be simplified to,

$$D_v \approx D_w / \gamma_f . \quad (3)$$

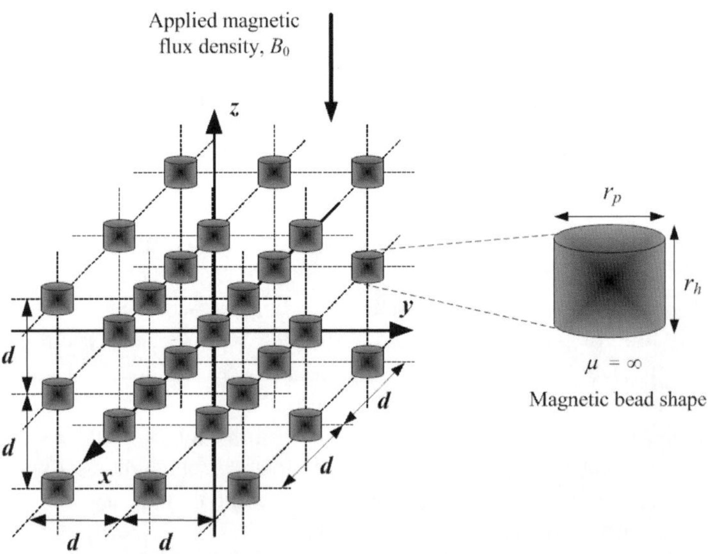

Fig. 2. Model of MBs uniformly distributed inside magnetic fluid

The relative permeability (μ^*) of magnetic fluid can be estimated by measuring the magnetic flux density in tissue injected with magnetic fluid. It is assumed that the MBs are cylindrical in shape with equal height (r_h) and diameter (r_p), and that they are uniformly distributed in the fluid as shown in Fig. 2. It is also assumed that μ^* of MBs is infinite and water is one. The permeance of the magnetic fluid based on an equivalent magnetic path is thus estimated from the assumptions [33]. Permeance generally refers to the degree to which a material admits a flow of matter or energy. D_w can be estimated based on the prediction of the magnetic flux path when magnetic fluid is placed under a uniform magnetic flux density. Due to the existence of MBs with infinite permeability the magnetic circuit for the external magnetic flux changes as shown in Fig. 3. When considering the permeance through MB magnetic flux lines will converge as shown in Fig. 3 (a). Hence, we take into account the surface area of the magnetic flux path. It is assumed that the diameter of the magnetic flux path surface area is twice the cylindrical bead diameter. Two equivalent magnetic paths, with and without magnetic nabeads are considered. The volume of the cylindrical MB, given that $r_p = r_h$, is

$$V_m = \frac{\pi r_p^3}{4}. \tag{4}$$

D_v = volume of MB : total volume

$$D_v = \frac{\pi r_p^3}{4} \div d^3 = \frac{\pi r_p^3}{4d^3}. \tag{5}$$

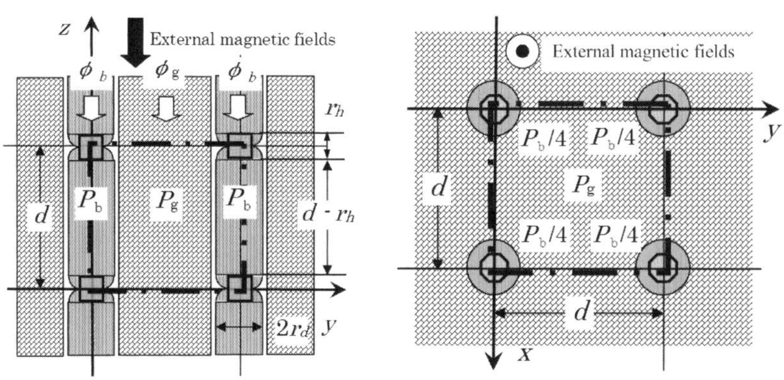

(a) Side view of magnetic circuit path.

(b) Top view.

Fig. 3. Equivalent magnetic circuit path of magnetic liquid under a *z* direction external magnetic flux density

Equation (5) can be rearranged to obtain

$$d = \left(\frac{\pi}{4D_v} \right)^{1/3} r_p. \tag{6}$$

The permeances of two magnetic paths are expressed as follows:
The permeance of the magnetic path through MB is given by

$$P_b = \frac{\mu_0 \pi r_p^2}{d - r_p} \approx \frac{\mu_0 \pi r_p^2}{d} \left(1 + \frac{r_p}{d} \right). \tag{7}$$

The permeance of the magnetic path through liquid is given by

$$P_g = \frac{\mu_0 \left(d^2 - \pi r_p^2 \right)}{d}. \tag{8}$$

Based on equations (7) and (8), the permeance per unit volume can be derived as

$$P = \frac{\mu S}{l} = \frac{\mu_0 \mu_s 1^2}{1} = \mu_0 \mu_s = \left(P_b + P_g \right) \left(\frac{1/d^2}{1/d} \right) = \frac{P_b + P_g}{d}. \tag{9}$$

Equation (9) can be expanded as follows:

$$\mu_0 \mu^* = \frac{1}{d} \left[\frac{\mu_0 \pi r_p^2}{d} \left(1 + \frac{r_p}{d} \right) + \frac{\mu_0 \left(d^2 - \pi r_p^2 \right)}{d} \right],$$

$$\mu_0 \mu^* = \mu_0 \left[\frac{\pi r_p^3}{d^3} + 1 \right]. \tag{10}$$

Relative permeability of magnetic fluid can then be expressed as

$$\mu^* = 1 + \frac{\pi r_p^3}{d^3}. \tag{11}$$

Substituting equation (6) into (11), for $D_v << 1$, we obtain

$$\mu^* = 1 + 4D_v. \tag{12}$$

Then, substituting equation (3) into (12) the following equation, for $D_w << 1$, is obtained.

$$\mu^* = 1 + 4D_w / \gamma_f. \tag{13}$$

Equation (13) shows that μ^* of magnetic fluid is directly proportional to D_v and D_w. Furthermore, the shape and/or size of the MBs have no effect on μ^* of magnetic fluid.

Assuming the same equivalent path the equivalent μ^* has the same expression even though the bead could be of spherical shape. This expression holds on the condition that the cavity includes a small amount of MBs.

The electron microscopy of the magnetic fluid shows that the MBs have a cluster structure as shown in Fig. 4 (a). It was then assumed that the cluster of magnetite is distributed uniformly as shown in Fig. 4 (b). It can be seen that there is some space between the magnetite beads in the cluster model, so we considered the space factor of spherical magnetite h_s. So then the effective specific gravity can be expressed as

$$\gamma_f' = h_s \gamma_f . \tag{14}$$

where h_s is 0.523.

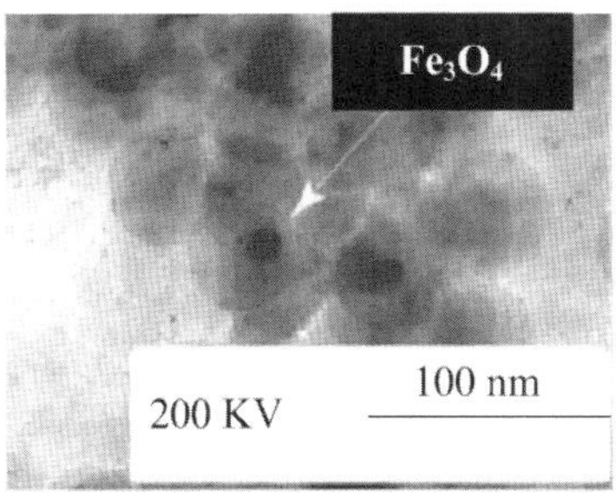

(a) Electron microscopy image of magnetic fluid.

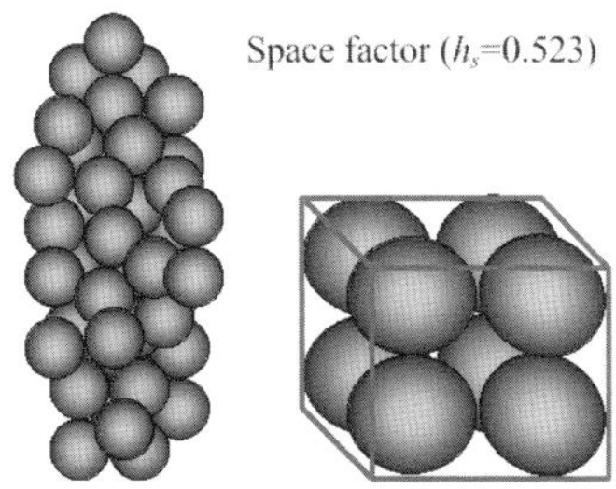

(b) Spherical type model of magnetite.

Fig. 4. Cluster structure of magnetic nanoparticles

Equation (13) can then be written as

$$\mu^* = 1 + 4D_w / h_s \gamma_f . \tag{15}$$

which, in general terms can be written as

$$\mu^* = 1 + C_d D_w / h_s \gamma_f . \qquad (D_w \ll 1) \tag{16}$$

where C_d is a coefficient which is theoretically 4 [34-36].

Equation (16) relates μ^* of magnetic fluid to D_w. To confirm equation (16) experimental analysis was carried out with the aid of a vibrating sample magnetometer (VSM). Hysteresis curves were obtained for magnetic fluid samples with different weight densities. Figure 5 shows the comparison of the experimental and theoretical results. It can be seen that μ^* is proportional to D_w. The theoretical and experimental results are also in good agreement.

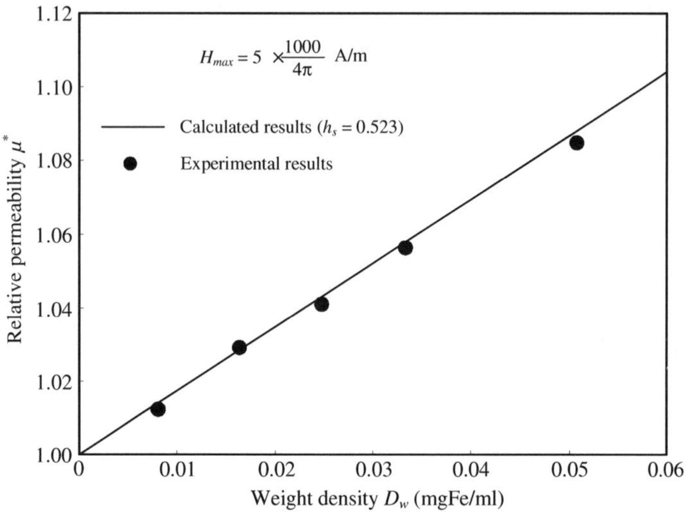

Fig. 5. Relationship between relative permeability and magnetic fluid

2.2.2 Estimation of Magnetic Fluid Weight Density by Measuring Magnetic Flux Density Inside and Outside a Magnetic Fluid-filled Cavity

Consider the situation shown in Fig. 6. An ellipsoidal cavity filled with magnetic fluid is placed under a uniform magnetic flux density. Given that the outside environment is air with $\mu^* = 1$, and magnetic fluid has μ^* slightly greater than 1, magnetic flux lines will converge and concentrate at the magnetic fluid filled ellipsoidal cavity. If a magnetic flux density B_0, is applied then the magnetic flux density in the cavity can be assumed as B_1. The magnetic flux density inside the magnetic fluid filled cavity B_1, will change according to D_w. The magnetic flux density at the center of the cavity B_1, can be expressed according to the following equation [37].

$$B_1 = \frac{B_0}{\frac{\mu_0}{\mu} + N\left(1 - \frac{\mu_0}{\mu}\right)}, \tag{17}$$

$$B_1 - B_0 = \frac{\mu^*}{1 + N\left(\mu^* - 1\right)} B_0 - B_0 = B_0\left(\frac{\mu^*}{1 + N\left(\mu^* - 1\right)} - 1\right), \tag{18}$$

$$\frac{B_1 - B_0}{B_0} = \left[(1 - N)\left(\mu^* - 1\right)\right] \qquad (\mu^* \approx 1) \tag{19}$$

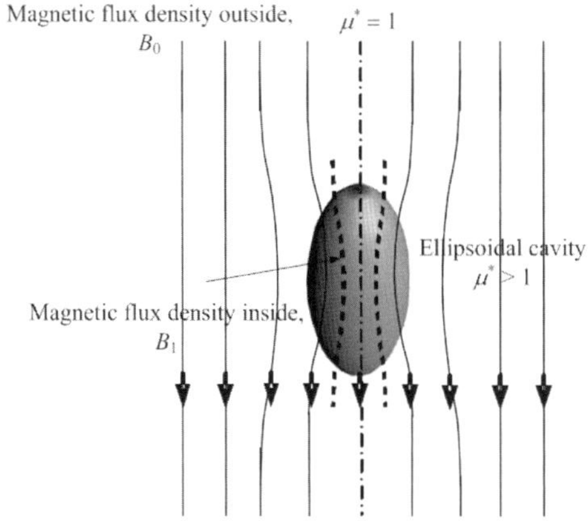

Fig. 6. Magnetic fluid filled ellipsoidal cavity under the influence of a uniform magnetic flux density

Substituting equation (16) into (19) we obtain the change in magnetic flux density (δ) as shown below.

$$\delta = \frac{B_1 - B_0}{B_0},$$

$$\delta = \frac{C_d (1 - N) D_w}{(h_s \gamma_f)}. \quad (20)$$

It can be seen from equation (20) that D_w can be effectively calculated from the difference between B_1 and B_0. The change in magnetic flux density is directly proportional to D_w. However, the demagnetizing factor N, which depends on the shape and size of cavity, influences the estimation of D_w.

2.3 GMR Needle Probe

2.3.1 Design of the GMR Needle Probe

The fabricated GMR needle probe as shown in Fig. 7 (a) is unique in the sense that it can be applied inside the body in a low-invasive way. The needle detection part is shown in Fig. 7 (b). The needle length is 20 mm, where 15 mm is available for insertion inside the body, and approximately 310 μm in diameter. Generally, such a fine needle can be expected to break easily due to its lack of rigidity. However, since the substrate itself is cut into a needle shape, a hard material such as aluminum titanium carbide (AlTiC), a sintered material of aluminum oxide (Al_2O_3) and titanium carbide (TiC), can be used as the base material to make the needle strong. The needle-shaped detecting part consists of a substrate to which a cutting process is applied to have a needle shape, four GMR elements formed of thin films on the surface of the substrate, four connection/bonding pads, lead conductors for electrically connecting the GMR elements to the connection/bonding pads and a protection film for covering the GMR elements and lead conductors, except parts of the connection/bonding pads. The GMR sensors, connection/bonding pads and lead conductors are formed on a wafer by a wafer process utilizing thin film photolithography techniques. After, a machining process is used to cut the wafer to a needle shape. The substrate is made of AlTiC. The GMR elements used in the needle have a spin-valve structure. The spin-valve structure consists of an antiferromagnetic layer that is used to fix or pin the magnetization of the pinned ferromagnetic layer, a non-magnetic space layer and a free layer of ferromagnetic material where the magnetization is free to move in response to an applied magnetic flux density. The magnetization direction of the pinned layer is perpendicular to the free layer and same in all of the four spin valve GMR elements. The connection/bonding pads and the lead conductors are made of copper (Cu). These connection/bonding pads are formed by a bump layer of Cu, and a bonding pad layer of gold (Au) that is laid on the bump layer.

The novel idea of the GMR needle probe is the GMR sensing area (75 ×45 μm) present at the tip of the needle. MBs are used *in vivo* as self-heating agents for hyperthermia treatment; accurate measurement of MB density is very important for successful treatment. When a uniform external magnetic flux density is applied to an area filled with MBs the flux lines will converge to this area (because the MBs make the relative permeability (μ^*) of this area slightly more than one) resulting in a change in magnetic flux between the applied flux and the flux at the target area. In such a case the GMR needle probe can be inserted into the target area in a minimally invasive

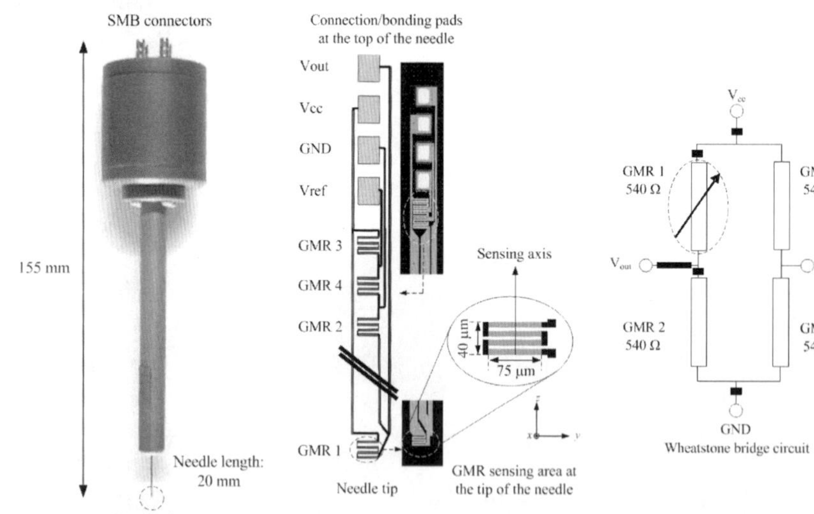

(a) Fabricated GMR needle probe.

(b) Needle design.

(c) Wheatstone bridge design of sensors.

Fig. 7. GMR needle probe

pattern to measure the magnetic flux density inside as well as outside the area. One of the main features of the GMR needle probe is the way the Wheatstone bridge circuit is designed as shown in Fig. 7 (c). There is a sensing area at the tip which is part of a Wheatstone bridge circuit with four other GMR sensing areas 20 mm further up, near the connecting/bonding pads. This means that rather than measuring the magnetic flux density inside and outside separately the GMR needle probe has the ability to measure both these quantities simultaneously.

2.3.2 Characterization of GMR Needle Probe

The DC characteristics of the GMR needle probe are shown in Fig. 8 for a magnetic flux density range of -12 to 12 mT. The DC characteristics of the GMR probe show that the maximum magnetic ratio is approximately 13.30 %. The linear region sensitivity of the GMR needle probe is around 2.5 %/mT. There is a very low hysteretic loop. It can be seen from the figure that the GMR needle probe has high sensitivity for applied magnetic flux density in the range of -1 to 1 mT.

Two types of experiments were performed to obtain the AC small signal characteristics of the GMR needle probe. Since the research performed involves AC magnetic flux the investigation of small signal AC characteristics of the GMR needle probe is very important.

Experiment 1: Measurement with Lee-Whiting coil

A Lee-Whiting coil was used to produce a magnetic flux density of 0.1 mT at 100 Hz as shown in Fig. 9 (a). The current to the coils are provided by the function generator through the high speed power amplifier. The output through V_{out} in the GMR needle probe Wheatstone bridge was sent to the oscilloscope which in turn was sent to a computer by GPIB for analysis. During experiments a current of 5 mA was fed to the

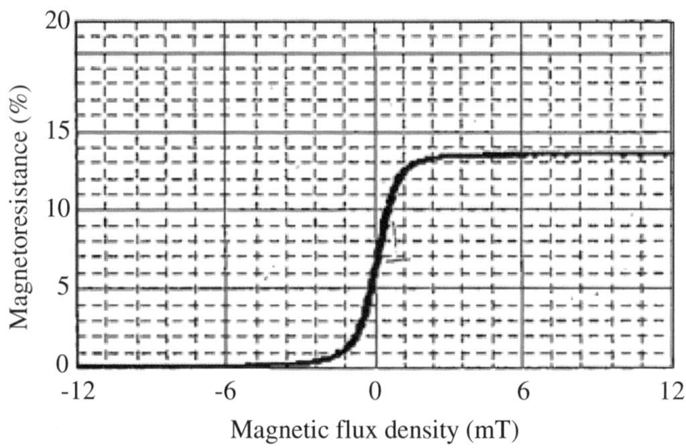

Fig. 8. DC characteristics of the GMR needle probe

GMR needle probe. The GMR needle probe was also tested for three different directions, *x*, *y* and *z*. Fig. 9 (b) shows the sensitivity results for all three directions. It can be seen that the response in the *x* and *y* direction is very low compared to the *z* direction. This can be explained by the fact that the sensing axis of the GMR sensing element at the needle tip is parallel to the magnetic flux density, which is in the *z* direction. The sensitivity in the *z* direction is approximately 13 mV/mT.

Experiment 2: Measurement with gradient distribution of magnetic flux density

The experimental setup is shown in Fig. 9 (c). Two coils are used with current flowing in parallel with respect to each other. This results in a magnetic flux density of zero at the center of the coils. The methodology for these experiments is based on the distance between the GMR sensing area at the tip of the needle and the other three GMR sensing areas near the connection/bonding pads. The distance between the GMR sensing area at the tip and the other three sensing areas is 20 mm. The magnetic flux density that is produced by parallel coils follows a gradient pattern. So the gradient at 20 mm can be obtained and used to calculate the AC sensitivity of the sensor. The gradient can be adjusted by the current to the coil which is proportional to the magnetic flux density. Since the magnetic flux density is zero at the center the GMR sensing areas near the connection/bonding pads can be placed there. The GMR sensing area at the tip can be placed where there would be a magnetic flux density and the Wheatstone bridge output (V_{out} - V_{ref}) can be obtained. The change of the signal will be solely due to the GMR sensing area at the tip. A current of 5 mA was fed to the GMR needle probe and a gradient of approximately 0.003.6 mT/20 mm was given by the two parallel coils. The frequency of the exciting current fed to the two parallel coils by the function generator through the high speed power amplifier was 100 Hz. The V_{out} and the V_{ref} signals were connected to a digital lock-in amplifier (NF electronics LI5640). The lock-in amplifier was used to analyze the results. The results showed that for AC small signal characterization at 100 Hz the sensitivity of the sensor is 15.3 mV/mT. Hence, for small signal characterization at 100 Hz the sensitivity of the sensor is between 13-15 mV/mT.

2.4 Estimation of Magnetic Fluid Weight Density Using the GMR Needle Probe

2.4.1 Experimental Methodology

In explaining the experimental method of estimating D_w (theoretically outlined in section 2.2.2), Fig. 10 is taken into account. Consider the event where the tip of the needle is inserted into the center of a tumor cavity under a uniform magnetic flux density (B_0). The four GMR sensors are exposed to B_0, assuming that the cavity is empty ($\mu^* = 1$ can be assumed inside and outside the cavity). So there is no change in magnetic flux density inside and outside the cavity since B_1 is equal to B_0. However, when the tumor cavity is filled with magnetic fluid, μ^* inside is greater than outside the cavity. Hence, the GMR sensing area at the tip of the needle is exposed to a

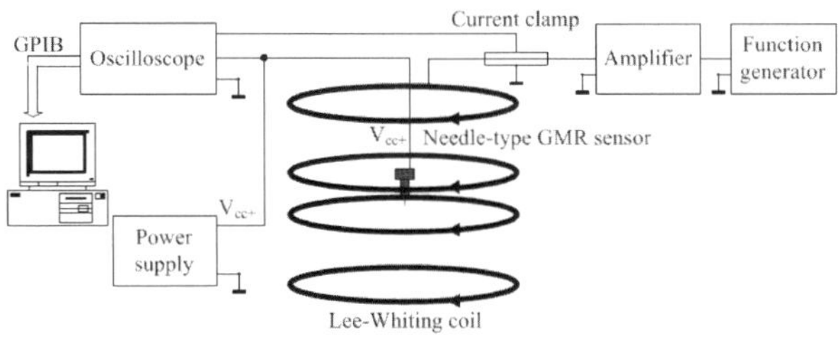

(a) Experimental set up for AC small signal characteristics.

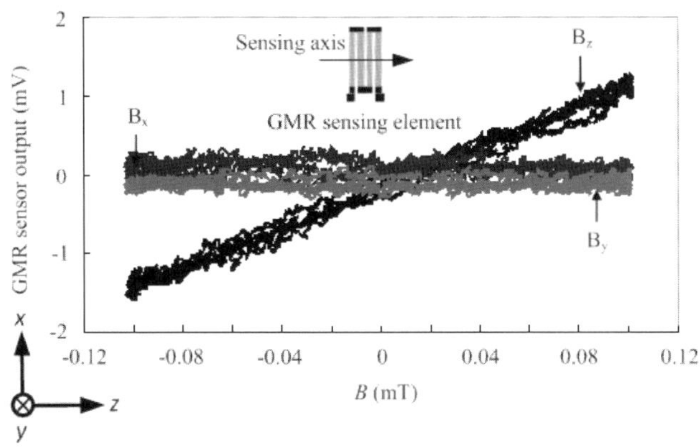

(b) AC small signal characteristics of the GMR needle probe (Section 2.3.2, Experiment 1).

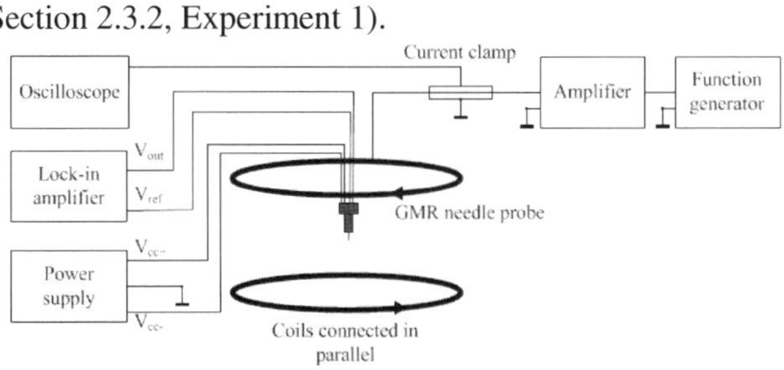

(c) Experimental setup for small signal AC characterization with gradient distribution of magnetic flux density (Section 2.3.2, Experiment 2).

Fig. 9. Small signal AC characterization at 100 Hz

magnetic flux density B_1, which is higher than the applied magnetic flux density B_0. However, since the other three sensors are located further up near the bonding pads, and hence outside the magnetic fluid filled cavity, they will still be exposed to the applied magnetic flux density. This way, B_1 and B_0 can be measured simultaneously. As explained in section 2.2.2, D_w is proportional to the differential magnetic flux density inside and outside a magnetic fluid filled cavity; thus, D_w can be estimated from the change in B_1 and B_0.

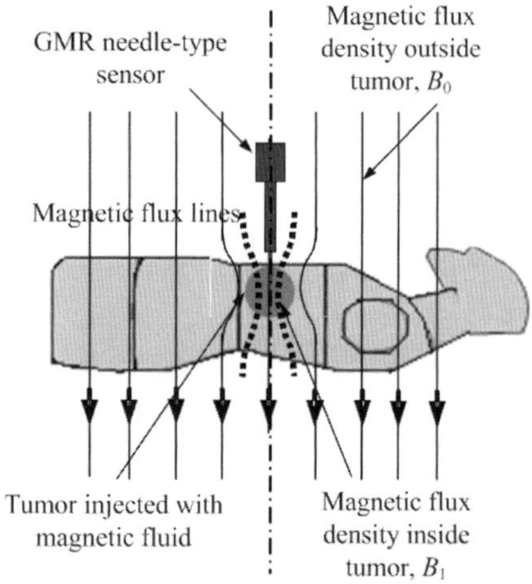

Fig. 10. Magnetic fluid filled tumor under a uniform magnetic flux density

One of the main requirements for the accurate estimation of D_w using the GMR needle probe is a uniform magnetic flux density. Helmholtz coils are used in a variety of applications, primarily due to its ability to produce a uniform magnetic flux configuration, ease of construction and flexibility. The accuracy of the relative magnetic flux density produced depends on how precise the Helmholtz coils are constructed and how accurate the current through them is maintained [38]. A lot of research has been done to find optimum parameters for designing super uniform Helmholtz coil systems [39-42]. A Helmholtz coil system consists of two coils, either circular or square, of equal radius and equal number of turns along an axis through the center of the coils, separated by a distance equal to the radius of the coils. The total magnetic flux produced is the sum of the two coils. A large volume of magnetic flux uniformity based around the mid-point between the two coils can be explained by the good deal of cancellation for the off-axis flux components generated by the coil. Magnetic flux density is generated by currents. A static flux will be produced if the currents are unchanging (DC). However for a changing current the flux will vary and will not only have magnetic, but also electric and electromagnetic components. Influence of fields other than magnetic increases when the operating frequency rises. However high frequency effects can be ignored if the operating frequency is kept low enough. In this region generally known as the "quasi-static" region, the differences between the flux generated by DC and AC can be assumed negligible. The Helmholtz coil has many applications in a variety of areas such as, calibration of magnetic instruments and probes, biomedical/bioelectromagnetic studies, diagnostic studies on electron beams, and in the study of the magnetic properties of materials [43-46].

In this research it is essential that the specifications required for the area of uniformity for experiments is explicitly met. A common error when designing a Helmholtz coil system is the assumption that the magnetic flux density will be uniform. This is true only for a certain volume around the center. Improvements and modifications are needed to improve the region of uniformity. A Helmholtz coil system was designed to obtain a uniform magnetic flux density with an error ≤ 0.01 % for 0.03m in the radial

(r) and axial (z) axes, for the experiments performed in sections 2.4.2 and 2.4.3. If a standard one coil pair is used the common radius of the system would be quite large for the required specifications. A large number of coil designs such as Maxwell tri-coil, Lee-Whiting, Alldred-Scollar and Rubens coil, have been reported [40]. However, for our specifications and experimental setup a planar coil system consisting of three coil pairs was chosen as the ideal system [47]. The coil design is shown in Fig. 11 (a). The three coil pair system is used to produce a magnetic flux density of B_0. The flux density will concentrate at the magnetic fluid filled cavity. The GMR needle probe is placed at the center of the cavity, hence the center of the common axis between the three coil pairs. The magnetic flux in the cavity can be assumed as B_1. B_1 changes with D_w. As explained before the change in B_1 and B_0 is proportional to D_w. It is critical that the magnetic flux distribution in this region should have an error of less than or equal to 0.01 % with respect to the center of the coils, because the percentage change in magnetic flux density for the magnetic fluid weight densities used for experiments is in the order of 1/10. So it is essential that the experimental area is more uniform to eliminate ambiguity in experimental results. The analytical results for the Helmholtz tri-coil are shown in Fig. 11 (b); it can be seen that the fluctuation of magnetic flux density is less than or equal to 0.01 %, 0.03 m in the axial and radial direction from the midpoint. The darker the region in the contour plot, the more uniform it is. The fabricated Helmholtz tri-coil is shown in Fig. 11 (c).

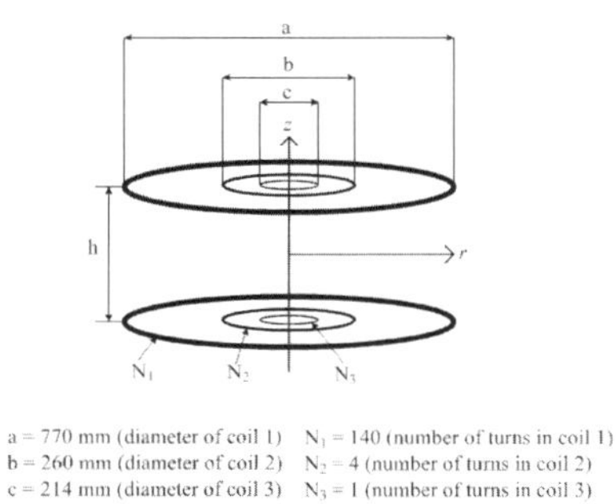

(a) Helmholtz tri-coil design.

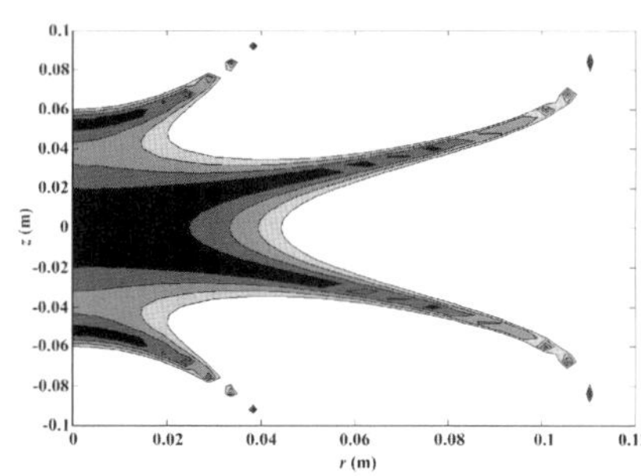

(b) Contour error plot ($\leq$ 0.01 % variation of magnetic flux density from the center of the coil).

(c) Fabricated Helmholtz tri-coil.

Fig. 11. Helmholtz tri-coil

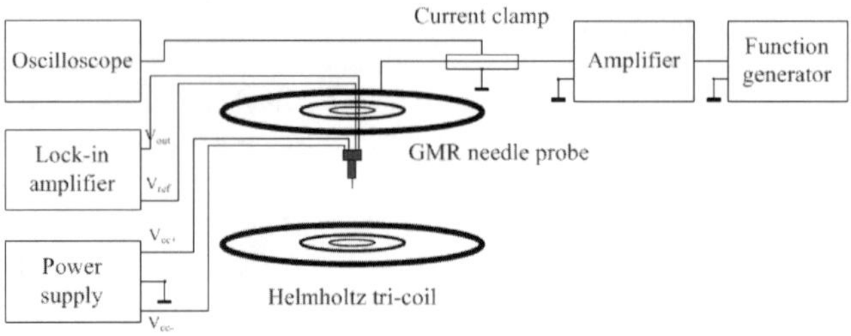

(a) Block diagram.

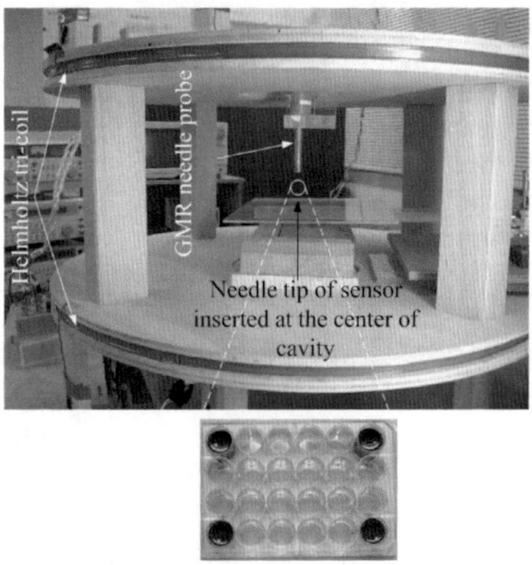

(b) Experimental apparatus.

Fig. 12. Experimental setup

Figure 12 shows the experimental setup for estimating low-concentration magnetic fluid inside plastic cylindrical containers. Low-concentration magnetic fluid can be defined by the content densities used in clinical applications. Magnetic fluid weight densities used for clinical applications are typically less than 2.8 % and potentially decreases even more when injected *in vivo*, due to spreading inside tissue. Hence, magnetic fluid of original D_w 40 % was thinned by mixing with distilled water. Plastic trays with embedded cavities (s = 0.625) were filled with thinned fluid of various densities. The fabricated Helmholtz tri-coil was used to produce a uniform magnetic flux density of 0.1 mT at 100 Hz (0.01 % fluctuation 0.03 m in the axial and radial direction from center of the coil). Current of 267 mA was provided to the coils of the Helmholtz tri-coil by a function generator (Sony Tektronix AFG310) through a high speed power amplifier (NF Electronics 4055). A current clamp (Hioki 3274) was clamped to the coils carrying current to the Helmholtz tri-coil and connected to an oscilloscope (Yokogawa DL4100) to analyze the current waveform. Additionally, a Gauss meter (MTI mm-340) was also used to confirm the magnetic flux density at the center of the Helmholtz tri-coil. A constant current of 5 mA was given to the GMR needle probe by a DC power supply (Matsusada Precision Instruments).

The GMR needle probe was inserted as shown in Fig. 12 (b). The GMR needle probe tip was placed in the middle of the Helmholtz coil and the cylindrical cavity was moved so that the needle tip was at the center of the cavity. The GMR needle probe was then used to estimate the D_w of the thinned magnetic fluid, by measuring the applied magnetic flux density ($B_0 = 0.1$ mT) and the magnetic flux density inside (B_1) thinned magnetic fluid filled cavities. The bridge output voltages across the GMR needle probe were measured by a lock-in amplifier (NF Electronics LI5640). By applying the results to equation (20), the change in magnetic flux density was obtained for all the cavities with different thinned magnetic fluid weight densities. The experimental results are shown in Fig. 13. The figure denotes the relationship between D_w and the change ratio of magnetic flux densities. When the cavity is thin and long ($N = 0$), the relationship shows the upper limit. The demagnetizing factor N for an elliptic body depends on the shape ratio of the cavity s, as shown in the figure. For spherical shaped cavities $s = 1$ and $N = 1/3$, and for flat shaped cavites $s = 0.5$ and $N = 0.527$. It can be seen from the experimental results that D_w is proportional to change in magnetic flux density and the results fall between theoretical lines for long and flat ellipsoidal cavities.

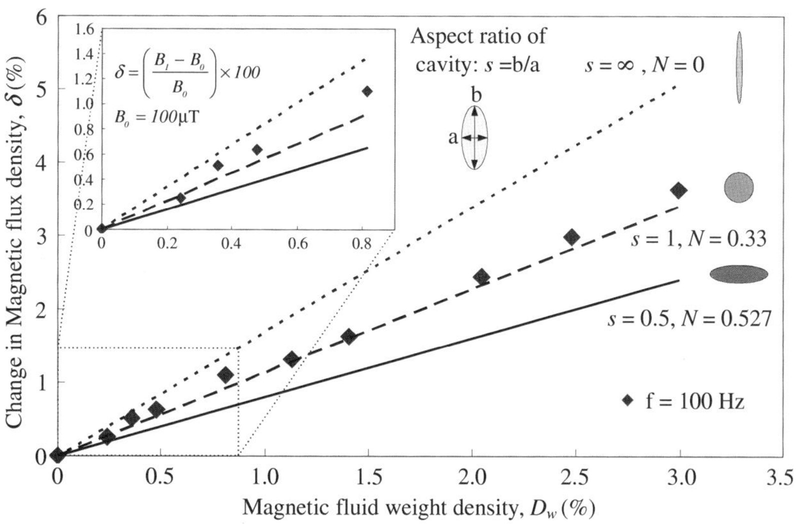

Fig. 13. Estimation of magnetic fluid weight density in a tray with embedded cavities

2.4.2 Detection and Estimation of Low Concentration Magnetic Fluid Inside Tumor-Simulating Cylindrical Agar Cavities

The evaluation of cancer staging is generally done with respect to the size of the tumor. The tumor, node, metastasis (TNM) staging system is commonly accepted for the evaluation of pathological stage [48]. Assuming that tumors are spherical in shape, the cutoff diameters for the different stages depend on the type of tumor. Generally tumors are classified into 5 stages (T0 – T5). In the T0 stage there is no evidence of a primary tumor but is defined as carcinoma in situ (CIS). This means that malignant cells that arise from epithelial cells have not invaded surrounding tissue. Recent studies have shown that the general cut off point between T1/T2 and T3 tumors is 50 mm [49, 50]. In T1 tumors diameters are not more than 20 mm and localized to one part of the body. In T2 tumors cancers are locally advanced and diameters vary between 20 and 50 mm. In later stage (T3 and T4) tumors the diameters are more than 50 mm, and also the cancer has started to spread into surrounding tissue, lymph nodes and body organs. Ideally hyperthermia therapy is performed on non-invasive, in situ tumors which are normally detected when they are small and confined [51]. In these cases (tumor diameter less than 20 mm) the cancer has not spread to other organs. Low-concentration magnetic fluid (less than 2.8 % D_w) is generally used in hyperthermia therapy, to keep the dose *in vivo* as low as possible. However, once injected the magnetic fluid tends to spreads inside tissue, further decreasing the low-concentration D_w. The specific heat capacity required to destroy a tumor is proportional to AC magnetic flux density amplitude, frequency and D_w. Specific heat capacity is also clearly correlated with therapeutic outcome since it can only be increased up to a certain critical value to avoid heating of healthy tissue. Hence, it is vital that D_w be known *in vivo* before as well as after treatment (to check for remnant density). This section provides details about a novel GMR needle probe that can be inserted *in vivo* in a low-invasive way to detect and estimate low-concentration D_w in T1 tumors for successful implementation of hyperthermia therapy.

To simulate the situation of detecting magnetic fluid inside the body, cylindrical agar pieces (simulating tumors) were injected with thinned magnetic fluid of various densities and immersed in potato starch, which acted as a reference medium. Agar is widely used in microbiology as a culture medium. Agar powder with jelly strength 400 – 600 g/cm^2 by Wako Company was used for experiments. The diameters of the agar pieces were chosen to be 4 – 14 mm (s = 1, N = 0.33) to simulate T1 tumors, which are less than 20 mm. A uniform magnetic flux density of 0.1 mT was applied by the Helmholtz tri-coil and the needle tip of the sensor was inserted at 10 mm intervals hence, to the middle of agar pieces along the length (225 mm) of the magnetic fluid filled cavity tray as shown in Fig. 14. The change in signal corresponds to the difference between the signal obtained inside the magnetic fluid filled agar and the reference medium (potato starch). It can be seen from Fig. 15 that the GMR needle probe can detect magnetic fluid injected into agar pieces with a diameter as low as 4 mm. Fig. 15 shows that for a given D_w of thinned magnetic fluid the change in signal does not vary so much between the four samples (since s and N is the same) and that the signal is proportional to the weight density of thinned magnetic fluid. This means that even though the size of the cavity may change (s and N constant), the signal will only change with D_w, thus verifying equation (20). Furthermore, detection of magnetic fluid in samples with diameters as low as 4 mm shows that the GMR needle probe has a potential to be used effectively as a tool for detecting drug coupled MBs, in targeted therapy for tumors.

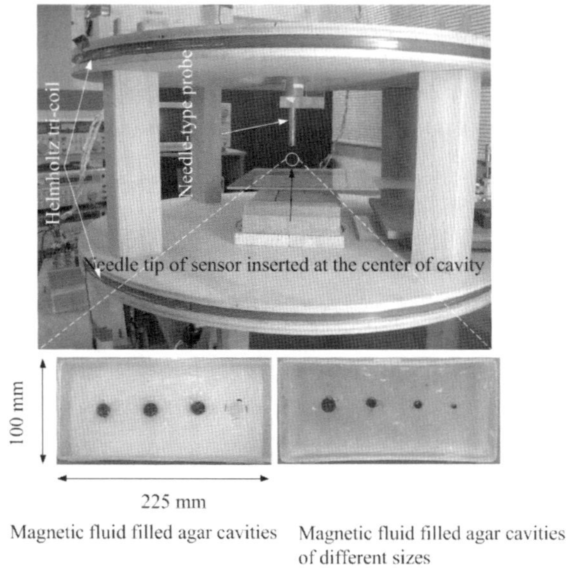

Fig. 14. Experimental setup for detecting and estimating magnetic fluid weight density inside agar cavities

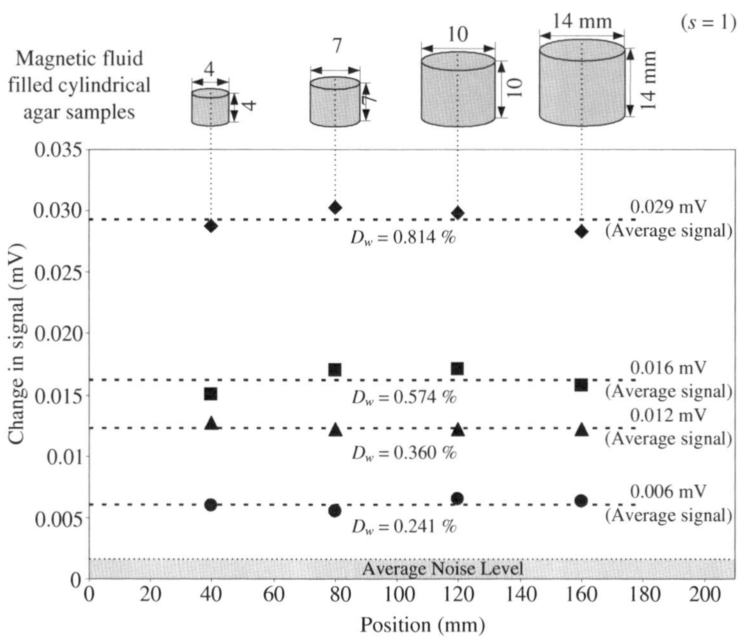

Fig. 15. Detection of magnetic fluid inside agar cavities

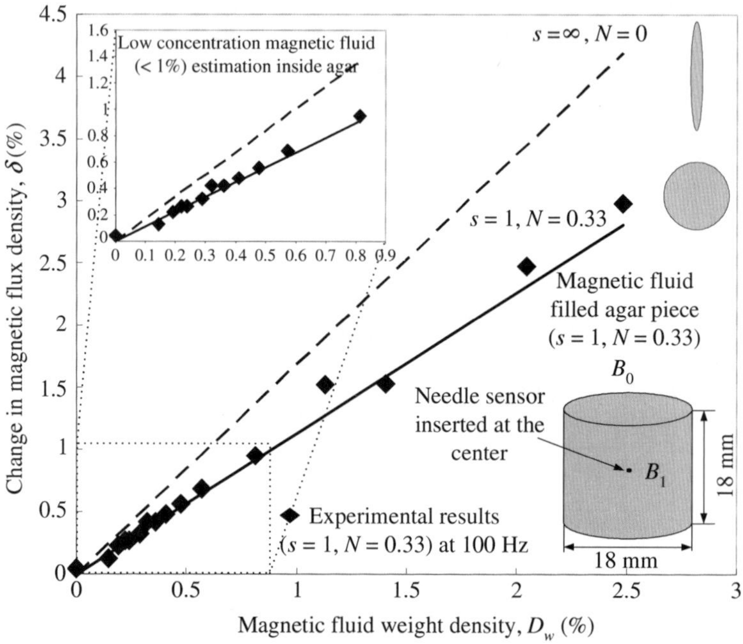

Fig. 16. Estimation of magnetic fluid weight density inside agar cavities

Since the GMR needle probe was used to successfully detect magnetic fluid inside cylindrical agar pieces of different sizes, experiments were performed to accurately estimate D_w inside cylindrical agar pieces simulating T1 tumors. One of the major obstacles for implementation of hyperthermia therapy as an effective cancer treatment is the retention of injected magnetic fluid without spreading to neighboring tissues and organs. Coupling magnetic fluid to tumor specific ligands such as antibodies, slow infiltration and repeated multi-site injections are some of the methods used to increase the retention of magnetic fluid in tumors. Experiments are performed with the GMR needle sensor to estimate D_w in 18 mm diameter agar cavities (s = 1, N = 0.33) since, to provide adequate heat to kill the tumor without affecting surrounding healthy cells, D_w needs to be confirmed before and after treatment (to check remaining density). The GMR needle was inserted at the center of the 18 mm agar cavities and B_1 and B_0 was measured simultaneously due to the bridge circuit design of the GMR needle probe. It can be seen from Fig. 16 that D_w is proportional to the change in magnetic flux density and agrees well with theoretical results obtained based on ellipsoidal cavities. Concentrations as low as 0.145 % weight density can be successfully estimated [52, 53].

2.4.3 Estimation of Magnetic Fluid Weight Density Inside Large Cylindrical Agar Cavities

In section 2.4.2 experiments were performed with agar cavities simulating T1 cancer tumors. Different size agar cavities were injected with magnetic fluid to simulate fluid filled tumors. The GMR needle probe was then used to measure the change in magnetic flux density inside and outside the agar cavities simultaneously. The size of the tumor should be considered since out of the 20 mm needle of the GMR probe 15 mm is available to be inserted inside the tumor. Since the needle tip should be inserted at the center of the tumor where the magnetic flux density is greatest (given that the tumor is exposed to a uniform magnetic flux density), the diameter of a given tumor should be less than or equal to 30 mm. Since diameters of the cylindrical agar cavities simulating tumors in the T1 stage of cancer were less than 20 mm the needle of the GMR sensor, which is 20 mm in length, was easily inserted into the center of the agar cavities to measure magnetic flux density inside. However, in T3 and 4 stages of cancer, commonly called the "later/advanced stages" of cancer, the diameters of tumors are more than 50 mm. So, the needle cannot be inserted at the center. Hence, a new experimental method was developed to estimate D_w inside large tumors. Taking advantage of the fact that the distance between the GMR sensing area at the tip and the three other sensors near the bonding pads is 20 mm, magnetic flux density is obtained at 20 mm steps as shown in Fig. 17. The total change in magnetic flux density is calculated by summing the change in magnetic flux densities at each step as shown in the equation below.

$$B_{TotalChange} = \sum_{i=0}^{n} \left(B_{i+1} - B_i\right). \tag{21}$$

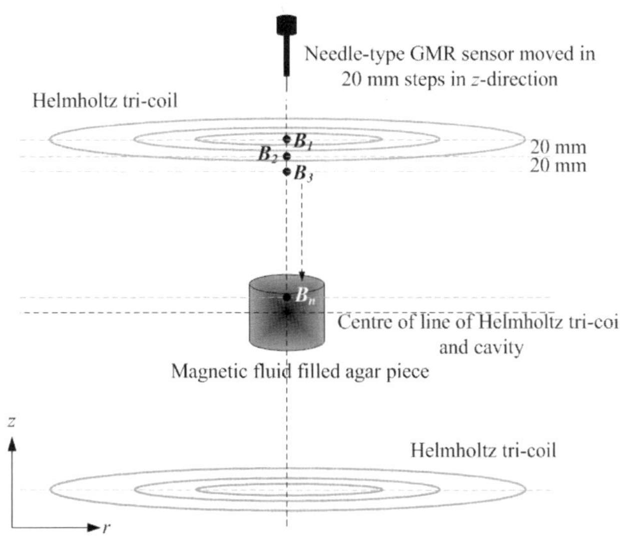

Fig. 17. Method of estimating magnetic fluid weight density inside large cavities

Figure 18 shows the comparison between the numerical and experimental results. It can be seen that if magnetic flux density can be measured at the center of the cavity, the change in magnetic flux density is approximately equal to the theoretical results. However, since the needle of the GMR probe is only 20 mm in length, the change in magnetic flux density is obtained by equation (21) for large cavities, where the needle cannot be inserted into the center of the cavity. The number of steps used in this case was 7. This meant that the needle was fully inserted in the magnetic fluid filled cavity after 7 steps, however the tip was not at the center of the cavity. It can be seen from Fig. 18 that the summing method proposed for large cavities gives a good approximation when the sensor needle cannot be inserted at the center of the cavity.

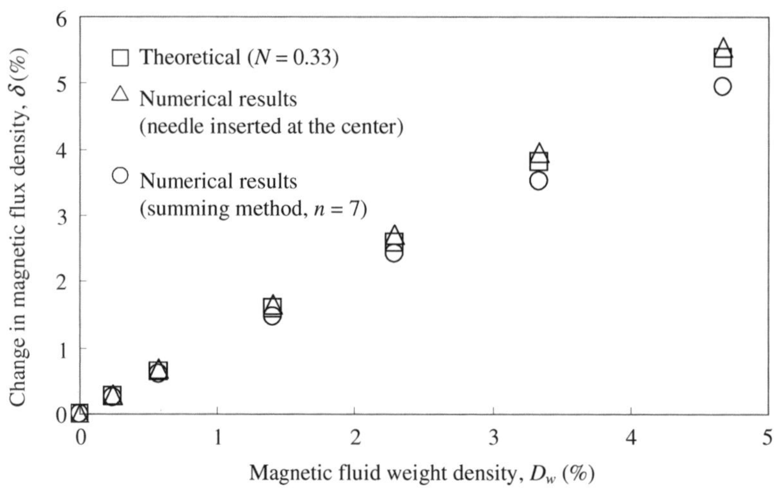

Fig. 18. Comparison of theoretical and numerical results for large cavities

The summing method was utilized to measure the change in magnetic flux density in magnetic fluid filled cylindrical agar cavities of 63 mm diameter (s = 1). 30 results were obtained in 30 seconds from the lock-in amplifier by MATLAB GUI and saved in the PC for further analysis. This method was more competent to handle the rapidly changing values of the lock-in amplifier, especially at points further away from the cavity. The average of 30 values was then taken. The experimental results are shown in Fig. 19. It can be seen that the total change in magnetic flux density increases with D_w. Moreover, the experimental results for large cavities compare favorably with theoretical results based on ellipsoidal cavities, numerical results obtained by numerical modeling and experimental results obtained for small cavities [54]. Figure 20 shows the results obtained for different size cylindrical agar cavities (s = 1) for a weight density of 2.29 %. Experiments done on smaller cavities (section 2.4.2) verified equation (21) for a range of sizes. It was shown that the change in magnetic flux density did not vary so much between different size cavities as long as s and hence N remained the same. The change in magnetic flux density only increased with D_w.

However, it must be noted that these experiments were performed with cavities smaller than 30 mm in diameter, hence allowing the insertion of the needle tip to the center of the cavity. In the experiments performed in this section the larger the cavity the further away it would be from the center of the cavity (where the magnetic flux density is the greatest) as shown in table 1. Thus, as diameter and/or height of a cavity increases (for a constant *N*) the total change in magnetic flux density decreases. Figure 20 compares the experimental results to numerical results. Even though the experimental results increase with the size of cavity they do not fluctuate so much compared to the numerical results.

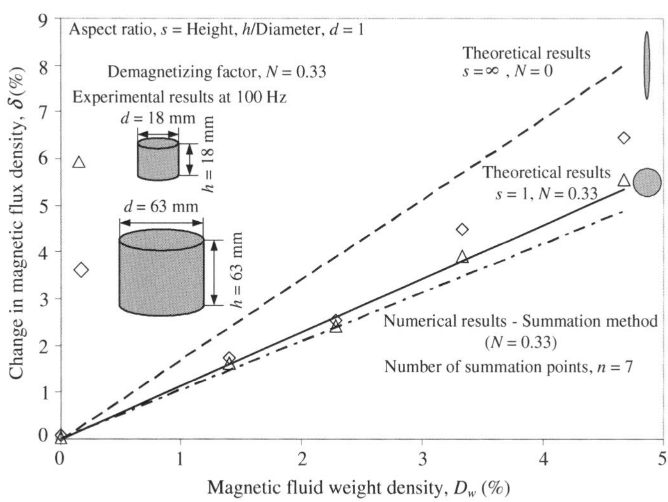

Fig. 19. Estimation of magnetic fluid weight density inside large cavities

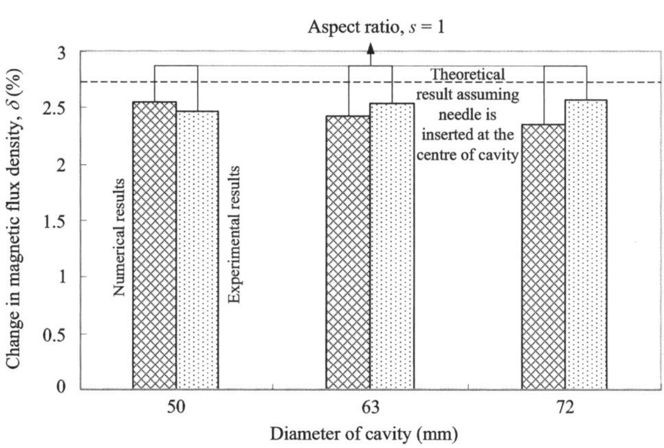

Fig. 20. Estimation of magnetic fluid weight density inside different sized cavities for D_w = 2.29 %

Table 1. Comparison of different sized cavities when needle is completely inserted inside cavity

Cavity	Diameter (mm)	Height (mm)	Distance from center of cavity to needle tip (mm)
1	50	50	10.0
2	63	63	16.5
3	72	72	21.0

2.4.4 Estimation of Very Low Concentration Magnetic Fluid Weight Density Inside Cylindrical Cavities

In section 2.4.2 experiments were performed to estimate low-concentration D_w inside cylindrical agar cavities; the limit of estimation was 0.145 %. However, estimation of very low-concentration magnetic fluid (less than 0.1 % D_w) is significant in hyperthermia therapy since the dosage is kept to a minimum *in vivo*, magnetic fluid spreads to neighboring tissue after injection which further reduces the dosage and some fluid may remain after treatment. Since successful treatment is directly proportional to D_w it is essential to estimate *in vivo*. Furthermore, current implementation of magnetic fluid based hyperthermia is in conjunction with other established forms of treatment such as radiotherapy, surgery and chemotherapy. It has been shown that cancer treatment with combination treatment is more effective compared to a given treatment on its own. However, if magnetic fluid hyperthermia therapy is to be more effective in combination treatment and also to be a feasible, stand alone, treatment there are several questions that need to be addressed. One of the main issues that need to be addressed involves the fate of magnetic fluid once used as self-heating agents in hyperthermia therapy. While it is safe to assume that only a small amount of magnetic fluid remains after treatment there is no conclusive evidence due to the novelty of magnetic fluid hyperthermia treatment. Even though, magnetic fluid is biocompatible it may be influenced by other bodily fluids or functions if it were to remain in the body for a prolonged period of time. The possible fact that magnetic fluid remains in the body for a long duration can be exploited for further treatment or for utilizing it for other applications. Remnant magnetic fluid can be moved to another part of the body by external magnetic flux gradients and reused for further/new treatment. The common factor that stands to benefit all these issues is accurate estimation of very low- concentration D_w. This section discusses the experimental apparatus fabricated and the experimental results estimating very low-concentration D_w inside agar cavities, using the GMR needle probe.

The aim of the experiments is to measure very low D_w of magnetic fluid in a cylindrical agar cavity, by measuring B_l. For this purpose a super uniform magnetic flux density generator is required for eliminating ambiguity of measurements. Since, for very low densities the percentage change in magnetic flux density is in the order of 1/10,000, it is important that the applied magnetic flux density is at least $1/10^{th}$ more uniform. For this purpose a Lee-Whiting type coil [40, 41] was designed and fabricated, producing a 0.001 % variation from the center of the coil in approximately 35 % of the outer coil spacing along the axial direction and 25 % of the diameter of

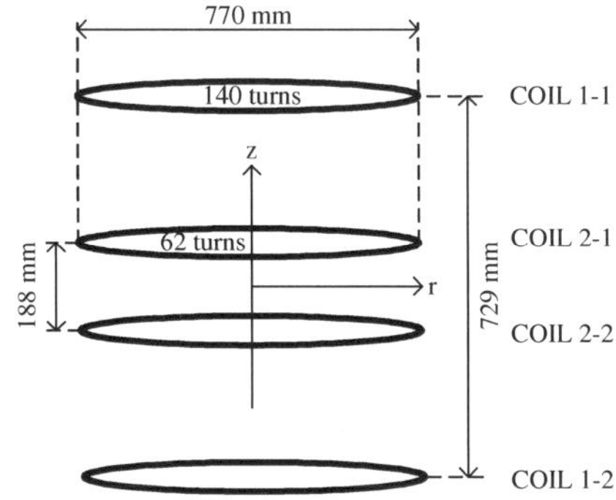

(a) Design of the Lee-Whiting coil.

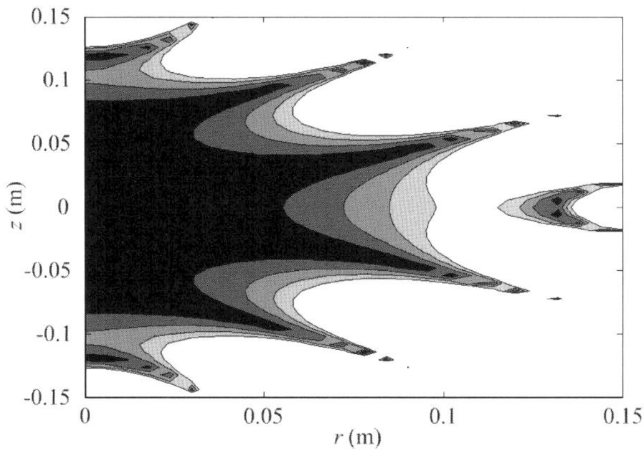

(b) Contour error plot (≤ 0.001 % fluctuation from center of the coil).

Fig. 21. Lee-Whiting coil

the coils in the radial direction. The Lee-Whiting coil design is shown in Fig. 21 (a) and the analytical results are shown in Fig. 21 (b).

Figure 22 shows the experimental setup where the fabricated Lee-Whiting coil was used to produce a uniform magnetic flux density of 0.09 mT at 50 and 100 Hz. Cylindrical agar pieces of d = 18 mm (s = 1) were injected with very low-concentration magnetic fluid (D_w = 0.03 – 0.2 %). The GMR needle probe was inserted to the center of magnetic fluid filled agar cavities. The differential magnetic flux density is in the order of nanotesla in the experimental situation. The bridge structure of the GMR needle probe measured the differential magnetic flux density simultaneously. The bridge output was amplified 100 times and sent to the lock-in amplifier. Then the data from the lock-in amplifier was transferred through GPIB to a computer for further analysis. For each low-concentration weight density, 5 values were taken and averaged. Experimental results shown in Fig. 23 indicate that the change in magnetic flux density is proportional to D_w [55]. However, the current limit of estimation has a good possibility to be influenced by the construction and coiling errors of the Lee-Whiting coil. Shown in table 2.2 are the error percentages at 0.02 m in the axial direction if the coils or diameters are altered by ±1/2 mm as shown in Fig. 24. Also, the error percentage increases if current distribution is considered as a square as shown in Fig. 24 instead of a point (as assumed in analytical analysis).

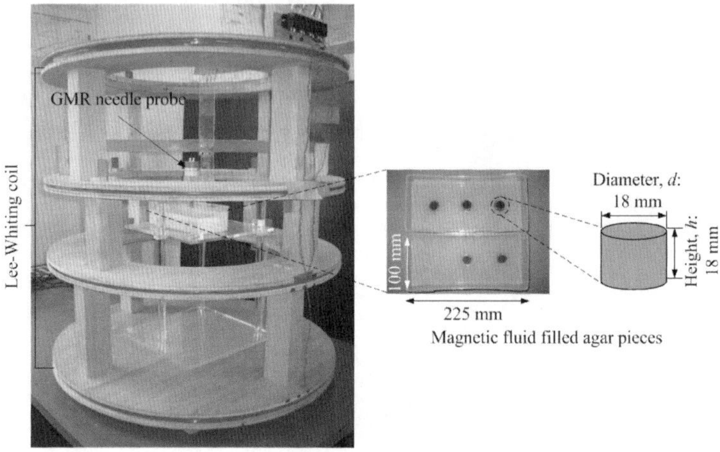

Fig. 22. Experimental setup

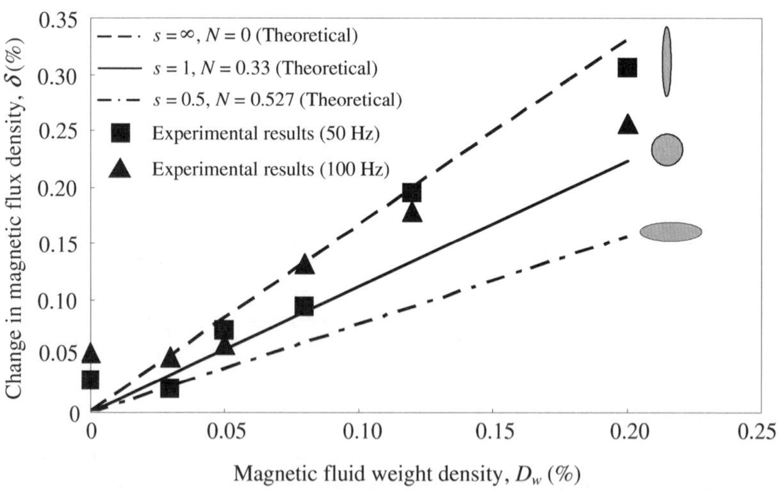

Fig. 23. Estimation of very low-concentration magnetic fluid weight density inside agar cavities

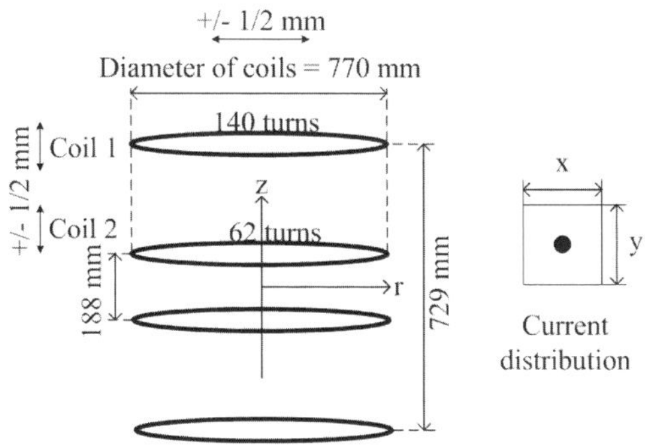

Fig. 24. Possible errors due to construction and coiling.

Table 2. Analysis of error percentages due to coiling and construction

Coil number	+1 mm (%)	-1 mm (%)	+2 mm (%)	-2 mm (%)
1_Distance	0.007	0.007	0.015	0.015
2_Distance	0.008	0.008	0.015	0.015
1_Radius	0.0037	0.0037	0.0075	0.0075
2_Radius	0.006	0.006	0.012	0.012

3 Conclusions

A novel GMR needle probe that is utilized to detect and estimate magnetic fluid weight density was reported in this section. The unique design of the fabricated GMR needle probe is especially made for application *in vivo* in a low-invasive way. A theoretical basis was obtained for detecting and estimating D_w *in vivo* based on relationships between relative permeability, weight density of magnetic fluid and magnetic flux density inside and outside a magnetic fluid filled cavity. An experimental setup (including a novel GMR needle probe, Helmholtz tri-coil and Lee-Whiting coil) and procedure with agar injected with magnetic fluid to simulate actual clinical process was developed.

Experiments were performed to detect and estimate D_w inside a variety of mediums simulating tumors, using the GMR needle probe, with the long term objective of estimating *in vivo*, especially in the area of hyperthermia therapy, a form of cancer treatment. Experiments were performed initially by inserting the GMR needle probe in a tray with magnetic fluid filled embedded cavities for confirmation of theoretical analysis. Cylindrical agar cavities simulating 1st and 2nd stage tumors were injected with magnetic fluid and the GMR needle probe was used to measure the magnetic flux density inside and outside the agar cavity. By measuring the differential magnetic flux density, D_w was estimated. The lowest D_w that could be estimated inside 18 mm diameter cylindrical agar cavities was 0.03 %. To estimate D_w in large tumors commonly found in later stages of cancer, a summing method was developed, taking into account the distance between the sensing element at the tip of the needle and the sensing elements near the bonding pads. The lowest D_w that could be estimated inside 63 mm cylindrical agar cavities was 1.414 %. The research performed in this section shows that in the future the GMR needle probe could be used not only in hyperthermia therapy but also in other advanced medical applications.

Acknowledgment. The authors acknowledge Prof. Masayoshi Iwahara (Retired) from the Institute of Nature and Environmental Technology, Kanazawa, Japan, for his help in this research. The authors would like to thank Professor Ichiro Sasada of Kyushu University for his kind help in designing and building the Helmholtz tri-coil system. Professor Subhas Chandra Mukhopadhyay of Massey University deserves the gratitude of the authors for his valuable advice. The authors also acknowledge Mrs. Agnieszka Łekawa-Raus and Dr. Adam Kurnicki for their contributions to this research. The authors gratefully acknowledge research support from the Ministry of Education, Culture, Sports, Science and Technology (MEXT) and the Japan Society for Promotion of Science (JSPS).

References

[1] Dennis, C.L., Jackson, A.J., Borchers, J.A., Ivkov, R., Foreman, A.R., Hoopes, P.J., Strawbridge, R., Pierce, Z., Goerntiz, E., Lau, J.W., Gruettner, C.: The influence of magnetic and physiological behaviour on the effectiveness of iron oxide nanoparticles for hyperthermia. Journal of Physics D: Applied Physics 41, 134020 (2008)

[2] Hiergeist, R., Andrä, W., Buske, N., Hergt, R., Hilger, I., Richter, U., Kaiser, W.: Application of magnetic ferrofluids for hyperthermia. Journal of Magnetism and Magnetic Materials 201, 420–422 (1999)

[3] Ivkov, R., DeNardo, S.J., Daum, W., Foreman, A.R., Goldstein, R.C., Nemkov, V.S., DeNardo, G.L.: Application of High Amplitude Alternating Magnetic Fields for Heat Induction of Nanoparticles Localized in Cancer. Clinical Cancer Research 11, 7093s–7103s (2005)

[4] Jordan, A., Scholz, R., Wust, P., Schirra, H., Schiestel, T., Schmidt, H., Felix, R.: Endocytosis of dextran and silan-coated magnetite nanoparticles and the effect of intracellular hyperthermia on human mammary carcinoma cells in vitro. Journal of Magnetism and Magnetic Materials 194, 185–196 (1999)

[5] Sincai, M., Ganga, D., Ganga, M., Argherie, D., Bica, D.: Antitumor effect of magnetite nanoparticles in cat mammary adenocarcinoma. Journal of Magnetism and Magnetic Materials 293, 438–441 (2005)

[6] Şincai, M., Gânga, D., Bica, D., Vékás, L.: The antitumor effect of locoregional magnetic cobalt ferrite in dog mammary adenocarcinoma. Journal of Magnetism and Magnetic Materials 225, 235–240 (2001)

[7] Wada, S., Tazawa, K., Suzuki, N., Furuta, I., Nagano, I.: Pulp ablation therapy by inductive heating: heat generation characteristics in the pulp cavity. Oral Diseases 13, 193–197 (2007)

[8] Wilhelm, C., Gazeau, F.: Magnetic nanoparticles: Internal probes and heaters within living cells. Journal of Magnetism and Magnetic Materials 321, 671–674 (2009)

[9] Storm, F.K., Harrison, W.H., Elliot, R.S., Morton, D.L.: Normal Tissue and Solid Tumor Effects of Hyperthermia in Animal Models and Clinical Trials. Cancer Research 39, 2245–2251 (1979)
[10] Atsumi, T., Jeyadevan, B., Sato, Y., Tohji, K.: Heating efficiency of magnetite particles exposed to AC magnetic field. Journal of Magnetism and Magnetic Materials 310, 2841–2843 (2007)
[11] Bae, S., Lee, S.W.: Applications of $NiFe_2O_4$ nanoparticles for a hyperthermia agent in biomedicine. Applied Physics Letters 89, 252503 (2006)
[12] Bae, S., Lee, S.W., Takemura, Y., Yamashita, E., Kunisaki, J., Zurn, S., Kim, C.S.: Dependence of Frequency and Magnetic Field on Self-Heating Characteristics of $NiFe_2O_4$ Nanoparticles for Hyperthermia. IEEE Transactions on Magnetics 42, 3566–3568 (2006)
[13] Gilchrist, R.K., Medal, R., Shorey, W.D., Hanselman, R.C., Parrott, J.C., Taylor, C.B.: Selective Inductive Heating of Lymph Nodes. Annals of Surgery 146, 596–606 (1957)
[14] Pankhurst, Q.A., Connolly, J., Jones, S.K., Dobson, J.: Applications of magnetic nanoparticles in biomedicine. Journal of Physics D: Applied Physics 36, R167–R181 (2003)
[15] Goodwin, S., Peterson, C., Hoh, C., Bittner, C.: Targeting and retention of magnetic targeted carriers (MTCs) enhancing intra-arterial chemotherapy. Journal of Magnetism and Magnetic Materials 194, 132–139 (1999)
[16] Jurgons, R., Seliger, C., Hilpert, A., Trahms, L., Odenbach, S., Alexiou, C.: Drug loaded magnetic nanoparticles for cancer therapy. Journal of Physics: Condensed Matter 18, S2893–S2902 (2006)
[17] Berry, C.C., Curtis, A.S.G.: Functionalisation of magnetic nanoparticles for applications in biomedicine. Journal of Physics D: Applied Physics 36, R198–R206 (2003)
[18] Ramanujan, R.V.: Clinical Applications of Magnetic Nanomaterials. In: Proceedings of the First International Bioengineering Conference, Singapore, pp. 174–177 (2004)
[19] Andrés Vergés, M., Costo, R., Roca, A.G., Marco, J.F., Goya, G.F., Serna, C.J., Morales, M.P.: Uniform and water stable magnetite nanoparticles with diameters around the monodomain-multidomain limit. Journal of Physics D: Applied Physics 41, 134003 (2008)
[20] Gupta, A.K., Gupta, M.: Synthesis and surface engineering of iron oxide nanoparticles for biomedical applications. Biomaterials 26, 3995–4021 (2005)
[21] Tartaj, P., del Puerto Morales, M., Veintemillas-Verdaguer, S., González-Carreño, T., Serna, C.J.: The preparation of magnetic nanoparticles for applications in biomedicine. Journal of Physics D: Applied Physics 36, R182–R197 (2003)
[22] Park, S.I., Kim, J.H., Kim, C.G., Kim, C.O.: Size-controlled magnetic nanoparticles with lecithin for biomedical applications. Journal of Magnetism and Magnetic Materials 312, 386–389 (2007)
[23] Dutz, S., Andrä, W., Hergt, R., Müller, R., Oestreich, C., Schmidt, C., Töpfer, J., Zeisberger, M., Bellemann, M.E.: Influence of dextran coating on the magnetic behaviour of iron oxide nanoparticles. Journal of Magnetism and Magnetic Materials 311, 51–54 (2007)
[24] Uozumi, Y., Tonohata, N., Imamura, M.: Magnetic Fluid Behaviour under External Fields in View of Medical Applications. In: Second International Conference on Innovative Computing, Information and Control, September 5-7, p. 370 (2007), doi:10.1109/ICICIC.2007.373
[25] Andrés Vergés, M., Costo, R., Roca, A.G., Marco, J.F., Goya, G.F., Serna, C.J., Morales, M.P.: Uniform and water stable magnetite nanoparticles with diameters around the monodomain-multidomain limit. Journal of Physics D: Applied Physics 41, 134003 (2008)
[26] Bahadur, D., Giri, J.: Biomaterials and magnetism. Sādhanā 28(3,4), 639–656 (2003)

[27] Duguet, E., Vasseur, S., Mornet, S., Goglio, G., Demourgues, A., Portier, J., Grasset, F., Veverka, P., Pollert, E.: Towards a versatile platform based on magnetic nanoparticles for in vivo applications. Bulletin of Materials Science 29, 581–586 (2006)
[28] Józefczak, A., Skumiel, A.: Study of heating effect and acoustic properties of dextran stabilized magnetic fluid. Journal of Magnetism and Magnetic Materials 311, 193–196 (2007)
[29] Hergt, R., Dutz, S.: Magnetic particle hyperthermia-biophysical limitations of a visionary tumour therapy. Journal of Magnetism and Magnetic Materials 311, 187–192 (2007)
[30] Jordan, A., Scholz, R., Maier-Hauff, K., Johannsen, M., Wust, P., Nadobny, J., Schirra, H., Schmidt, H., Deger, S., Loening, S., Lanksch, W., Felix, R.: Presentation of a new magnetic field therapy system for the treatment of human solid tumors with magnetic fluid hyperthermia. Journal of Magnetism and Magnetic Materials 225, 118–126 (2001)
[31] Jordan, A., Scholz, R., Wust, P., Fähling, H., Felix, R.: Magnetic fluid hyperthermia (MFH): Cancer treatment with AC magnetic field induced excitation of biocompatible superparamagnetic nanoparticles. Journal of Magnetism and Magnetic Materials 201, 413–419 (1999)
[32] Yamada, S., Chomsuwan, K., Mukhopadhyay, S.C., Iwahara, M., Kakikawa, M., Nagano, I.: Detection of magnetic fluid volume density with a GMR sensor. Journal of the Magnetics Society of Japan 31, 44–47 (2007)
[33] Mukhopadhyay, S.C., Chomsuwan, K., Gooneratne, C.P., Yamada, S.: A Novel Needle-Type SV-GMR Sensor for Biomedical Applications. IEEE Sensors Journal 7, 401–408 (2007)
[34] Yamada, S., Gooneratne, C.P., Chomsuwan, K., Iwahara, M., Kakikawa, M.: Estimation of Low-Concentration Magnetic Fluid Density with GMR Sensor. Review of Progress in Quantitative Nondestructive Evaluation 975, 873–880 (2008)
[35] Gooneratne, C., Lekawa, A., Iwahara, M., Kakikawa, M., Yamada, S.: Estimation of Low Concentration Magnetic Fluid Weight Density and Detection inside an Artificial Medium Using a Novel GMR Sensor. Sensors and Transducers Journal 90, 27–38 (2008)
[36] Gooneratne, C., Chomsuwan, K., Łękawa, A., Kakikawa, M., Iwahara, M., Yamada, S.: Estimation of Density of Low-Concentration Magnetic Fluid by a Needle-Type GMR Sensor for Medical Applications. Journal of the Magnetics Society of Japan 32, 191–194 (2008)
[37] Bozorth, R.M.: Ferromagnetism, 5th edn. New Jersey, Van Nostrand (1951)
[38] Bronaugh, E.L.: Helmholtz Coils for Calibration of Probes and Sensors: Limits of Magnetic Field Accuracy and Uniformity. In: Proceedings of the IEEE International Symposium on Electromagnetic Compatibility, Atlanta, U.S.A (1995)
[39] Wang, J., She, S., Zhang, S.: An improved Helmholtz coil and analysis of its magnetic field homogeneity. Review of Scientific Instruments 73, 2175–2179 (2002)
[40] Kirschvink, J.L.: Uniform Magnetic Fields and Double-Wrapped Coil Systems. Bioelectromagnetics 13, 401–411 (1992)
[41] Gottardi, G., Mesirca, P., Agostini, C., Remondini, D., Bersani, F.: A Four Coil Exposure System (Tetracoil) Producing a Highly Uniform Magnetic Field. Bioelectromagnetics 24, 125–133 (2003)
[42] Caprari, R.S.: Optimal current loop systems for producing uniform magnetic fields. Measurement Science and Technology 6, 593–597 (1995)
[43] Schill Jr., R.A., Hoff, K.: Characterizing and calibrating a large Helmholtz coil at low ac magnetic field levels with peak magnitudes below the earth's magnetic field. Review of Scientific Instruments 72, 2769–2776 (2001)

[44] Lee, S.G., Kang, C.S., Chang, J.W.: A square-loop Helmholtz coil system for the evaluation of a single-layer second-order high- it T_c SQUID gradiometer. Physica C 460-462, 1472–1474 (2007)

[45] Alamgir, A.K.M., Fang, J., Gu, C., Han, Z.: Square Helmholtz coil with homogenous field for magnetic measurement of longer HTS tapes. Physica C 424, 17–24 (2005)

[46] Lüdke, J., Ahlers, H., Albrecht, M.: Novel Compensated Moment Detection Coil. IEEE Transactions on Magnetics 43, 3567–3572 (2007)

[47] Sasada, I., Nakashima, Y.: A planar coil system consisting of three coil pairs for producing uniform magnetic field. Journal of Applied Physics 99, 08D904-08D904-3 (2006)

[48] TNM classification of malignant tumors, http://www.uicc.org/tnm

[49] Miyagawa, T., Shimazui, T., Hinotsu, S., Oikawa, T., Sekido, N., Miyanaga, N., Kawai, K., Akaza, H.: Does Tumor Size or Microvas-cular Invasion Affect Prognosis in Patients with Renal Cell Carcinoma? Japan Journal of Clinical Oncology 37, 197–200 (2007)

[50] Lugli, A., Zlobec, I., Singer, G., Lugli, A.K., Terracciano, L.M., Genta, R.M.: Napoleon Bonaparte's gastric cancer: a clinicopathologic approach to staging, pathogenesis, and etiology. Nature Clinical Practice Gastroenterology and Hepatology 4, 52–57 (2007)

[51] Hilger, I., Hert, R., Kaiser, W.A.: Towards breast cancer treatment by magnetic heating. Journal of Magnetism and Magnetic Materials 293, 314–319 (2005)

[52] Yamada, S., Gooneratne, C.P., Iwahara, M., Kakikawa, M.: Detection and Estimation of Low-Concentration Magnetic Fluid Inside Body by a Needle-Type GMR Sensor. IEEE Transactions on Magnetics 44, 4541–4544 (2008)

[53] Gooneratne, C.P., Kakikawa, M., Iwahara, M., Yamada, S.: GMR Sensor Application in Detecting and Estimating Magnetic Fluid Weight Density inside Various Size Tumors. Journal of the Magnetics Society of Japan 33, 175–178 (2009)

[54] Gooneratne, C.P., Kurnicki, A., Iwahara, M., Kakikawa, M., Mukhopadhyay, S.C., Yamada, S.: A GMR Needle Probe to Estimate Magnetic Fluid Weight Density Inside Large Tumors. In: Mukhopadhyay, S.C., Gupta, G.S., Huang, R.Y.-M. (eds.) Recent Advances in Sensing Technology. LNEE, vol. 49, pp. 1–14. Springer, Heidelberg (2009)

[55] Gooneratne, C.P., Kakikawa, M., Ueno, T., Yamada, S.: Measurement of Minute Changes in Magnetic Flux Density by Means of a Novel GMR Needle Probe for Application in Hyperthermia Therapy. Journal of the Magnetics Society of Japan 34, 119–122 (2010)

Magnetoresistive Sensors for Surface Scanning

D.C. Leitão[1], J. Borme[2], A. Orozco[3], S. Cardoso[1,4], and P.P. Freitas[1,4]

[1] INESC-MN and IN, Rua Alves Redol 9, 1000-029 Lisboa, Portugal
dleitao@inesc-mn.pt
[2] INL – International Iberian Nanotechnology Laboratory,
Av. Mestre José Veiga, 4715-31 Braga, Portugal
[3] Neocera, Beltsville, MD, USA
orozco@neocera.com
[4] Instituto Superior Tecnico (IST), Av. Rovisco Pais, 1000-029 Lisboa, Portugal

Abstract. This chapter provides an overview on several techniques used for surface imaging, including SQUIDs, Hall-effect sensors, Giant magneto-impedance sensors, and magnetoresistive (MR) sensors. Among all magnetic field sensors, only SQUIDs and MR devices have the potential to localize buried and non-visual field sources (such as defects in integrated circuits or magnetic field sources in biological environments. In particular, we describe how MR sensors have been used with advantage for integrated circuit (IC) mapping, with resolution below 500 nm and sensitivity to detect currents as low as 50 nA and have been used for many applications requiring low magnetic field detection. Challenges and experimental considerations on integration of MR sensors on a commercial analysis tool are provided here. Examples obtained with real devices demonstrate how Scanning Magnetic Microscopy has become an established failure analysis technique for visualizing current paths in microelectronic devices.

1 Introduction

In recent years has become essential to ensure the reliability, integrity and safety of systems in industry (nuclear, petrochemical, gas) and transport (aeronautics, railway, automotive) sectors. The detection of generation, propagation and failure of a defect in metallic parts is highly desirable, enabling a significant reduction of maintenance costs and consequently improving profit and productivity.

Non destructive testing (NDT) meets this challenge gathering methods to provide information on the health of a structure without impairing its future usefulness. Currently, conventional NDT techniques rely on eddy current (EC) inspection, ultrasonic, and acoustic emission. EC inspection is widespread, ensuring subjacent or surface-breaking flaws detection. EC-NDT works by inducing electrical currents in the structure under test with an electromagnetic field, and thus cracks within the structure distort the EC flow enabling its detection. Nevertheless, EC NDT using traditional inductive sensors has shown difficulty in locating hidden buried defects, low spatial resolution and slow speed of inspection.

Nowadays, the EC inspection method is undergoing a rapid change to answer industry specifications in terms of product quality and cost saving, focusing on the

C. Reig et al.: *Giant Magnetoresistance (GMR) Sensors*, SSMI 6, pp. 275–299.
DOI: 10.1007/978-3-642-37172-1_11

detection of small (< 100 μm) and deep flaws (buried under several mm), while decreasing the inspection time.

Among the existent magnetic field detection techniques, only superconducting quantum interference devices (SQUID) [1-3] and magnetoresistive (MR) [4, 5] devices have the potential to localize buried and non-visual field sources. SQUID sensitivity translates to the following in terms of maximum depth and minimum current: 18 μA currents can be detected 1 mm away with a SNR ratio better than 5 [1-3]. Lowering the scanning distance lowers the minimum detectable current: ~ 200 nA can be detected at a distance of 300 μm. Alternatively, increasing the current facilitates detection at a greater distance; for example, 1 mA of current can be detected 54 mm away. However, the extremely high field sensitivity of the SQUID compromises the spatial resolution, and requires complicated apparatus for operation at low temperatures.

These drawbacks motivated the search for alternative NDT methods, such as MR sensors. The excellent spatial resolution of MR sensors has been used with advantage [5]. Their potential for NDT has been addressed in experimental systems namely the detection of cracks on 20 mm thick aluminum [2], to locate and characterize small surface cracks [6], in the evaluation of metal medical implants for invisible cracks [7], inspection of printed circuit boards [8] and detection of hidden corrosion [9]. The detection of very small magnetic fields with high spatial resolution, makes MR sensors ideal for NDT [10, 12, 13] but also for biosensors [11, 13], precision position sensing, document validation [12] and magnetic imaging [14].

Nowadays, the miniaturization levels required in the electronics industry has resulted in complex packaging solutions [15]. Therefore, the difficulty in imaging buried defects or flaws in complicated packaging schemes with multiple stacked devices has increased significantly [Figure 1]. Also, notice that intricate die level analysis is ever more challenging, targeting the accurate detection of buried metallization layers such as 8 to 10 levels down from the chip surface [1, 2, 16]. Currently used non-destructive techniques to detect electrical fails in such complex microelectronic package technologies include Scanning Acoustic Microscopy [16, 17], Scanning SQUID Microscopy [2] and magnetoresistive microscopes [18, 19].

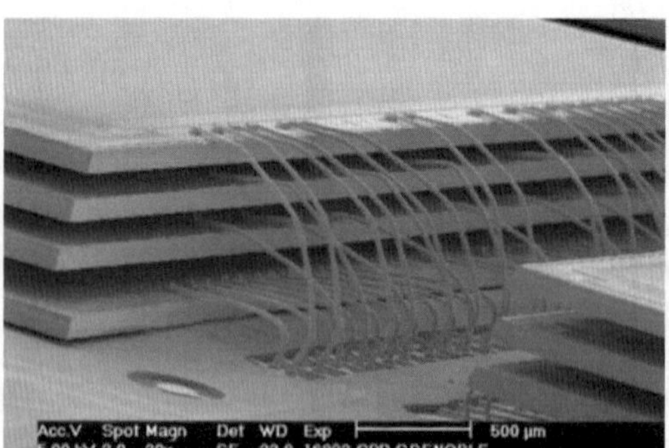

Fig. 1. Example of packaging inspection. Reprinted from Ref. [15] Copyright (2004), with permission from Elsevier.

Next sections will describe briefly some examples where these techniques have been used with advantage for mapping. Finally, section 3.4 will describe how MR sensors have been used for this purpose at an industrial level.

2 Imaging Sensors Devices: Overview

2.1 Flux Sensors

2.1.1 SQUIDs

A scanning superconducting quantum interference device (SQUID) incorporates a superconducting loop containing one or two Josephson junctions. The Josephson junction consists of a thin insulating layer (or constriction) between two superconductors [20, 21]. Overall, when the superconducting loop is submitted to a change in the applied flux applied, a phase difference across the junction will appear due to flowing currents.

In contrast to Hall probes or MR devices where the field sensitivity is nearly independent of the sensing area, in SQUIDs magnetic field sensitivity scales with $1/a^2$ (a is the diameter of the sensing area) [20, 22]. Typically, commercial DC SQUIDs exhibit a magnetic field noise of ~ 10 fT/√Hz [20, 21], still for NDT even small amounts of position noise during scanning can cause significant degradation of the obtained images [23]. Nevertheless, and although SQUIDs provide an extreme sensitivity, they bear the main disadvantage of operating at cryogenic temperatures, at least 77 K for high-T_C superconductors [20].

Several reports using particular designs for highly sensitive scanning SQUID microscope allowed direct observation flux quanta in high-T_C superconducting rings [24, 25]. SQUID microscopy has been used for visualization of magnetic structures at 77 K with a spatial resolution of about 30×10^{-6} m in the vertical component of the magnetic field [28]. Still, recent research on scanning SQUID microscopy focus on room temperature samples, where the main developments target improvements in hardware or software to achieve better spatial resolution [26, 27]. Fong *et al.* were able to image magnetic fields of room-temperature samples with sub-millimeter

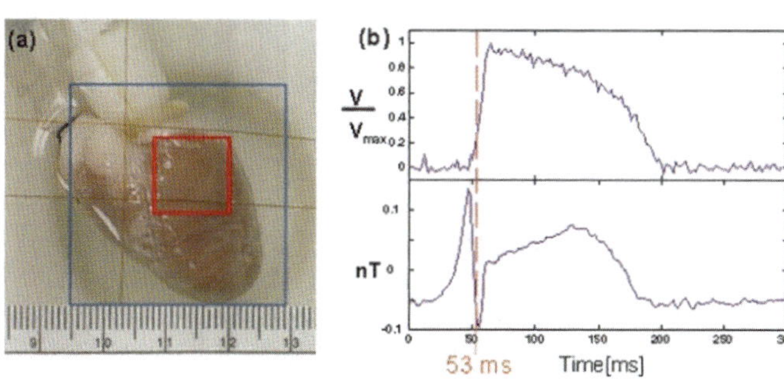

Fig. 2. Time trace of the transmembrane potential and the magnetic field recorded from a rabbit heart (Figure 10 of Ref. [26]). Reprinted with permission from Ref. [26] Copyright (2005), American Institute of Physics.

resolution. Their low-T_C multiloop SQUID sensor provided a field sensitivity of 1.5×10^{-12} T/√Hz, for frequencies above 100 Hz. Figure 2 displays action currents in cardiac tissue imaged by such device. The authors reported field sensitivities on the order of 180×10^{-15} T/√Hz for an optimized 1 mm multiloop SQUID sensor [26].

2.2 Field Sensors

2.2.1 Hall-Effect Sensors

Scanning Hall probe microscopy (SHPM) [29, 30] allows a noninvasive detection of surface magnetic fields. This technique is based on Hall Effect probes, where the changes of the external magnetic field translate into changes in the sensor output voltage [31]. With a theoretical field sensitivity of 2 nT/√Hz [32], SHPM offers a lower sensitivity than SQUID, but allows for nanometer-scale spatial resolution [30], works under variable temperature and variable magnetic field conditions. Figure 3 (left) shows an example of a nano-Hall sensor design [33]. Oral *et al.* described a low-noise SHPM with a magnetic field sensitivity of ~ 2.9×10^{-8} T/√Hz at 77 K and a spatial resolution of ~ 0.85 μm [30]. The same system was successfully operated at room temperature with a magnetic field resolution of ~ 3.8×10^{-6} T/√Hz [30, 34]. In addition, Howells *et al.* showed a high-resolution imaging of a magnetic media at 77 K with a SHPM system displaying a magnetic field sensitivity of 30×10^{-9} T/√Hz and spatial resolution of 0.8 μm [Figure 3(right)] [35].

Chang *et al.* introduced a hybrid Hall/STM microscope which provided a magnetic field sensitivity of ~ 10^{-5} T and a spatial resolution ~ 0.35 μm [36], whilst an optimized planar Hall effect magnetic sensor exhibited a field detection level down to 10^{-9} T [37].

Recently, nanometric Hall sensors with dimensions of ~ 50 nm [Figure 3 (left)] were incorporated into a room temperature SHPM exhibiting an optimum magnetic field sensitivity of 8.0×10^{-5} T/√Hz [33]; quantification of the spatial resolution was lacking. Although electron-beam lithography allows one to prepare sensors with lateral sizes down to the nanoscale range, effective lateral resolution of such a scanning sensor is limited by sensor-sample distance. When the sample is near the end of a measuring tip, its tilting can displace the sensor by hundreds of nanometers, being this the limiting factor of the resolution [38]. However, careful design of the sensor in recent systems claimed a minimum detection size of 80 nm [39].

SHPM systems can be built by attaching a suitable sensor to a commercial table-top AFM [40]. But many SHPM systems are developed to image superconducting materials and must work at cryogenic temperatures [42]. Several authors worked to overcome the small scan range of cooled piezoelectric crystals, either by using stepping motors effectively increasing the range to the centimeter scale [43-45] or by keeping piezoelectric scanner at room temperature while the sample is cooled [41]. Also, in order to reduce the infrastructure needed to operate the systems, others integrated a Stirling cycle refrigeration system allowing the operation at 35 K without the need for cryogenic fluids [46].

Most SHPM systems actually measure the component of the magnetic field perpendicular to the plane of the sample. Using three sensors patterned onto a pyramidal-shaped mesa, Fedor *et al.* [47] developed a SHPM system measuring the three components of the magnetic field.

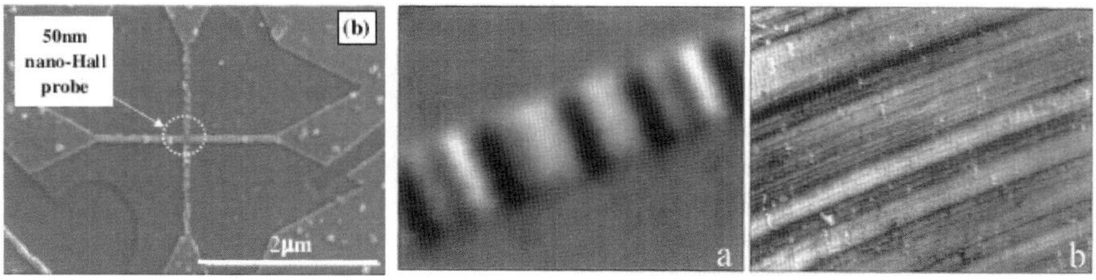

Fig. 3. (left) Nanometric Hall probe (from Ref. [33]). Copyright 2004 The Japan Society of Applied Physics. (middle; rigth) SHPM image at 77 K and simultaneous STM image. Reprinted from Ref. [35], Copyright (1999), with permission from Elsevier.

2.2.2 Giant Magneto Impedance Sensors

When a magnetic field is applied to a soft ferromagnetic conductor which is in turn subjected to a small alternating current (AC), a large change in the AC complex impedance of the conductor is visible [48]. This effect is known as Giant magneto-impedance (GMI) and is the base of GMI sensors, typically made of in metal-based amorphous alloys [48-50]. Although a field sensitivity of a typical GMI sensor can reach a value as high as 500 %/Oe [48, 52], the large size required for the sensing element restricts its spatial resolution [53, 54]. Nevertheless, since the GMI response strongly depends on the composition and shape of the sensor core a double-core GMI sensor based on Co-based amorphous magnetic wires, with reduced size has shown improved sensitivity [55]. The high magnetic field sensitivity of GMI sensors has already proved important in microstructural characterization at close proximity of the samples [56], being also used for detection surface of cracks [53], corrosion defects [56] and embedded flaws [57].

2.2.3 Comparison Table

In table 1 one compares figures of merit of selected sensors discussed above.

3 Magnetoresistive Sensors in Imaging and Scanning Microscopy

The spatial resolution of MR microscopy depends directly on the dimensions of the MR sensor, which in contrast to previously discussed sensors, is readily scalable through fabrication techniques. Furthermore, NDT imaging systems strongly profit from measuring distinctively the 3 components of the magnetic field along distinct directions. Such configuration is easily achieved with engineered MR sensors design. The latter, together with the relatively low cost and ease of implementation of such sensors and their ability of detecting very small magnetic fields give MR devices significant advantages over other magnetic imaging techniques such as SQUID, Hall sensors or Magnetic Force Microscopies.

Table 1. Details and figures of merit concerning NDT sensors, namely SQUID, GMI and Hall probes

Sensor	Particularities	Sensitivity/ Field Detection	Spatial Resolution	Ref.
Scanning Hall Probe Microscopy	AlGaAs/InGaAs/ GaAs	4×10^{-7} T/√Hz @ 300 K	0.08 μm	[39]
	5 μm patterned GaAs/AlGaAs	1×10^{-5} T/√Hz @ 4.2-300 K	5 μm	[44]
	Submicron Hall probe GaAs/AlGaAs	~ 2.9×10^{-8} T/√Hz @ 77 K ~ 3.8×10^{-6} T/√Hz @ 300 K	~0.85 μm	[30] [34]
	hybrid Hall-sensor/STM-positioning	~ 0.36×10^{-4} T/√Hz @ 4.2 K	~0.35 μm	[36]
	Planar Hall effect FeNi on Si(100)	300 V $T^{-1}A^{-1}$ @ 300 K	N/A	[37]
GMI sensors	Amorphous wire based	Up to 500 %/Oe @ 300 K	Low and size dependent	[52]
Scanning SQUID microscopy	high-T_C superconducting	N/A	30 μm (magnetic field vertical component)	[28]
	low-T_C niobium bare	1.5×10^{-12} T/√Hz	80 μm (sensor diameter)	[26]
	low-T_C niobium multiloop	$480 - 180 \times 10^{-15}$ T/√Hz	250 μm - 1 mm (sensor diameter)	[26]

The idea of using MR sensors as element to detect magnetic stray fields stems from the widely implemented hard-disk-heads, typically used to read information from hard disk media. In fact, the first reports concerning scanning MR microscopy used commercial MR record/playback heads as sensing elements which were raster scanned over particular magnetic samples [58-61]. Following these studies, the MR microscope system has been adapted to perform magnetic domain imaging on steel sheets [62] and nondestructive testing of materials [63]. Application of magnetic imaging to other research fields include the detection of defects in metallic plates [64, 65], in aircraft structures [66] and metallic containers used in nuclear power plants [67] showing ability to detect defects embedded 7 mm inside the metal, and land mine buried up to 8 cm inside the ground [68].

In particular, GMR based sensors have been successfully used in EC testing of nonmagnetic metals accurately detecting small surface-breaking cracks [7]. These probes had small dimensions and high sensitivities (220 mV/mT) with resolutions of 10 μT over a broad frequency range. Besides GMR, other types of magnetoresistive sensors are also used for the imaging applications such as anisotropic magnetoresistance (AMR) [69] and tunnel magnetoresistance (TMR) [70], although the first being the most used in current works. Still, AMR effect sensors are actively used in imaging either for scanning MR microscopy [71] or in multichannel scanning systems [72].

3.1 Arrays of Individual MR Sensors

When more than one individual sensor is located on a chip, the additional available information allows imaging and scanning of larger areas, and determining the location of magnetic objects. The use of arrays can provide substantial gains of scanning speed, or even remove the need for scanning and of moving parts altogether. Notice that, scanning over a surface or volume using a single sensor has the advantage of being an uniform measurement, while in arrays of magnetic sensors a dispersion in their properties of magnetic sensors will be reflected in the final result. Also, in the case of the scanning technique one is allowed to choose the spatial resolution by setting appropriate scanning speed and data acquisition rate, whereas an array provides a fixed value where the minimum distance is physically set by the individual sensor size, or by the size of each bridge arrangement of sensors. To compensate for the granularity and enhance the visual output, isolines of interpolated data can be used [73]. On the other hand, scanning systems often show line to line differences, which can be addressed during post-process data by a digital 3x3 low-pass convolution filter [74] or by other functions available in scanning-probe data analysis software. Another possibility to measure distances smaller than those provided by the sensor spacing resorts to the use of a fixed array together with a scanning system, the latter providing small displacements [75].

Due to fabrication reasons, most works use in-line or in-plane sensors. However, sensors can be disposed on a molded shape, allowing field imaging of industrial parts with known shapes [75]. The easiest way to obtain an array of sensors is to arrange discrete commercial sensors. Cano *et al.* [72] used 16 commercials in-line AMR sensors spaced by 5 mm, to scan in one direction over the surface of a magnetized sample with a magnetic field resolution of 0.1 μT. The same authors later used a scanning device equipped with 3 sensors with a smaller measurable field of 10 nT and with the capability to image the three components of the magnetic field by scanning over a surface. Other works extended the scanning method to the measurement of a volume using a three-axis stepper motor system [73]. As usually a MR sensor is sensitive to only one component of the field. However, the use of triple-axis magnetometers provides complementary information which can prove essential when a defect can be clearly evidenced in a specific field component [70].

When discrete sensors are used together, a higher data throughput can be obtained, but at the expense of complex wiring and control electronics. Moreover, and due to its compatibility with CMOS technology, a complete device can incorporate signal

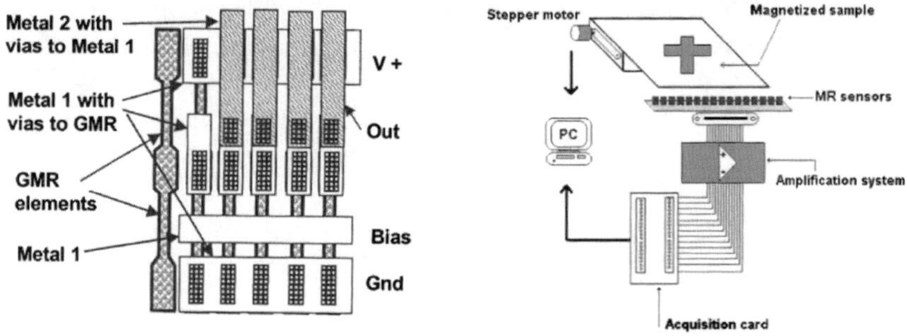

Fig. 4. (left) Example of a 16 element array. Reprinted with permission from Ref. [5]. Copyright (2003), American Institute of Physics. (right) Example of a sensor array measurement set-up. Reprinted with permission from Ref. [72]. Copyright (2005), American Institute of Physics.

conditioning and logic features besides the micrometer-scale MR sensor providing an optimized overall system performance [5]. In fact, large-scale integration of these sensors boosted a strong progress on the materials side, as smaller and denser elements required higher magnetic anisotropy [76]. Furthermore, intense research effort in large MR sensor arrays is devoted to assays [77], with an emphasis in compatibility with spotter technology [78, 79]. Additional works on biomedical applications focus also on simultaneous measurement of several sensors to provide imaging of biological samples or kinetics in solution [80].

3.2 Gradiometer Configuration

In a gradiometer, two magnetic elements are present and the signal measured depends on the difference of the magnetic fields at the two positions. Indeck *et al.* [81] introduced a magnetoresistive gradiometer composed of two MR sensors connected in a Wheatstone bridge, showing a reduction of thermal drift by one order of magnitude. In the case of a single MR sensor connected within a bridge in magnetometer configuration, magnetic shields are commonly used to render particular sensors insensitive. In the case of a gradiometer, these shields are often not required as two active sensors are used. Moreover, pairs of sensors in gradiometers configuration can be arranged in arrays for faster scanning or imaging.

Since two sensors are used for each measurement point, specific geometric considerations have to be taken into account, e.g. while scanning along a defect, the width of one sensor cannot exceed half the size of the smallest defect to be detected [82]. Such configurations can be used in the non-destructive control of metal parts to detect cracks [83] or inclusions of non-magnetic materials [84].

3.3 Cantilever MR Probes

MR sensors either AMR or GMR based, have already been used in scanning MR microscopy integrated in the back side of a cantilever [Figure 5]. In these systems, the sensor measures the magnetic fields while the cantilever detects the surface morphology [85, 86]. For hybrid AMR cantilever a lateral spatial resolution of a few mm and a field sensitivity of 0.17 mT were reported [85]. Recently, Sahoo *et al.* [87] used the MR sensor to detect the cantilever displacement, instead of the typical optical or piezoelectric detection modes. In this case, one aims to translate the cantilever displacement into a change in the magnetic field sensed by a MR sensor in close proximity. Here, MR sensors offer the advantages of high bandwidth (in excess of 1 MHz), small form factor, straightforward integration and scalability providing a route for high-throughput scanning probe microscope devices. The proposed solutions by Sahoo *et al.* rely on having a micromagnet attached to the end of the moving cantilever and a MR sensor placed at a fixed position relative to the cantilever [Figure 5]. Hence, a change in the cantilever position will induce a change in the magnetic field created by the micromagnet on the sensing layer of the MR probe, thus changing its resistance. Figure 5 (bottom) shows MR scanning probe microscopy contact-mode imaging of a surface using one of the architecture suggested in Ref. [87]. The authors envisaged a resolution of 84 pm achieved over a bandwidth of 1 MHz, overcoming the 200 pm (over 1 MHz) achievable using optical means in state-of-the-art AFMs.

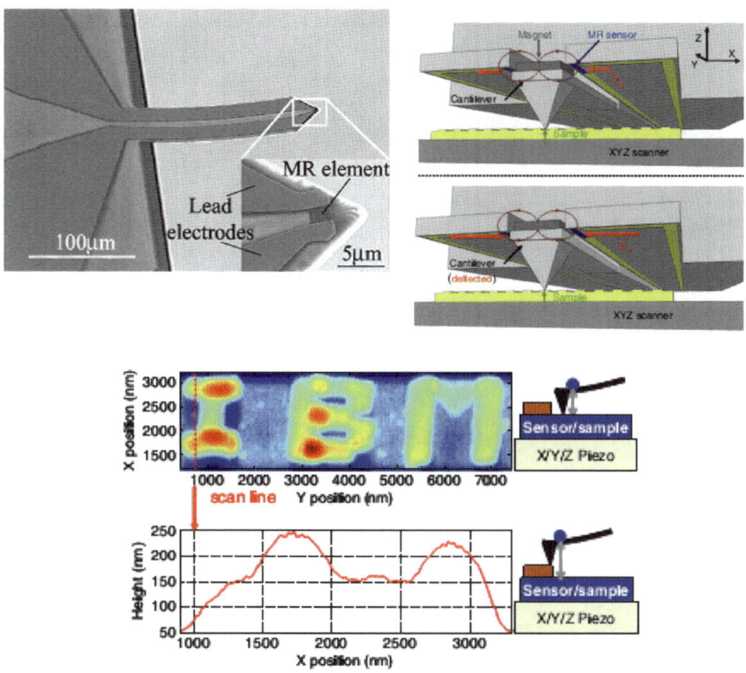

Fig. 5. (top) Examples of hybrid cantilever/MR sensor designs. (top-left) Reprinted with permission from Ref [85]. Copyright (2002), American Institute of Physics. (top-rigth) Reprinted with permission from Ref [87]. Copyright (2011), Institute of Physics (IOP). (bottom) Output of a MR-SPM system operating in contact mode. Reprinted with permission from Ref [87]. Copyright (2011), Institute of Physics (IOP).

3.4 Comparison of MR Sensors

The following table compares the distinct particularities and performance of selected MR based devices for NDT.

Table 2. Details and figures of merit multiple MR sensors

	Particularities and details		Sensitivity	Field Detection	Ref.
Magnetoresistive Sensors	Isolated Sensor Multilayer	conventional	0.04 %/G	N/A	[5]
		low hysteresis	0.07 %/G		
		high sensitivity	0.2 %/G		
	half-bridge Multilayer	conventional	0.2 mV/V/G	20 nV/nT @10 V	
		low hysteresis	0.35 mV/V/G	25 nV/nT @10 V	
		high sensitivity	1.0 mV/V/G	100 nV/nT @10 V	
	TMR	40 % TMR; 1.5 nm Al_2O_3 barrier	30 mV/V/G	3 mV/nT @10 V	[103]
	Phillips KMZ10A (16 channel)	AMR thin film NiFe	0.1 μT	N/A	[72]
	Bridge configuration of 4 barber pole 256 channel;	AMR; Ta/Ru/IrMn/Ru/NiFe/Ta	0.1 mT@ 8 kscans/s.	20 nT/√Hz	[102]
	-	GMR	10 μT	220 mV/mT	[7]
	Hybrid Sensor/ Cantilever	AMR; 40 nm-thick NiFe	0.17 mT	N/A	[85]

4 Magnetoresistive Microscopy for Die Level Fault

Test and Failure Analysis (FA) of integrated circuits (IC) and micro-devices is becoming an increasingly complex science. The ever-shrinking nature of the silicon technology and increased complexity of design and packaging has been making the search of root cause failure a difficult problem. This has forced the industry over the years to search for new techniques and tools to find defects that require increased sub-micron resolution with lower power consumption and more difficult access to circuits and transistors because of increased levels of metallization involved. To complicate things even more, in addition to the miniaturization trend, the need to expand functionality, form factor and real-state management and faster connections have pushed the industry to develop complex 3D package integration. This 3D circuit

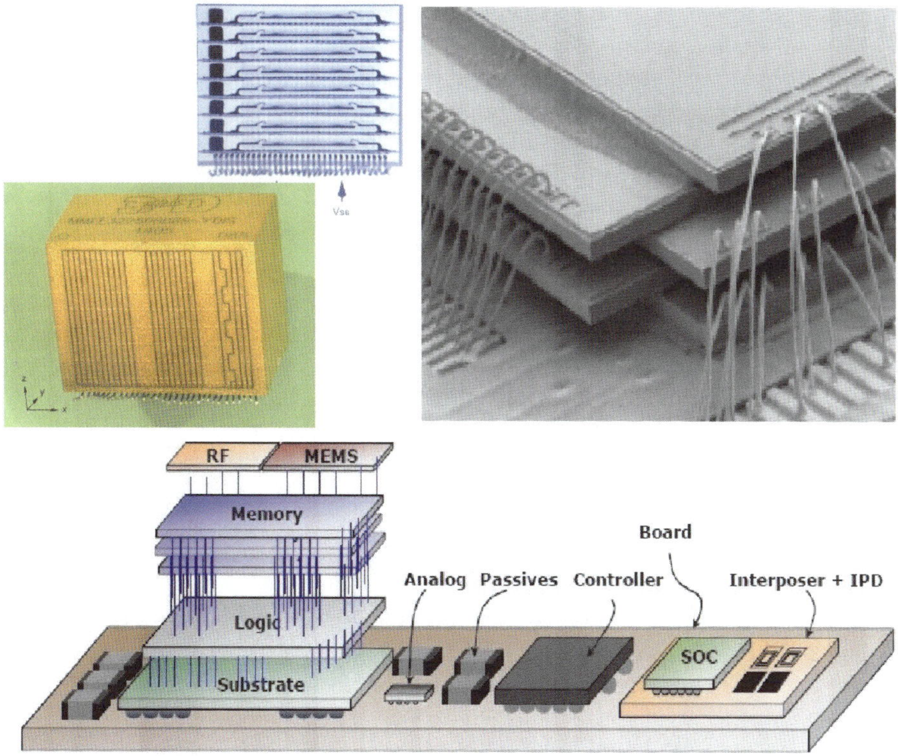

Fig. 6. Several examples of complex packaging

technology includes System-in-Package (SiP), wafer-level packaging, through-Silicon-vias (TSV), stacked-die and flex packages among others that calls for vertical stacking of multiple dies, fully integrated interconnects within Silicon and multiplication of opaque and metal layers, as exemplified in Figure 6.

The presence of buried, non-visual defects has forced scientist and engineers to develop new techniques to locate the failure location and to innovate in existing ones in order to keep pace with the trends and needs of the semiconductor industry and consequently, there is a myriad of acronyms right now for all the different techniques being currently used by the FA community. Complexities of present-day and future designs require use of nondestructive, non-contact fault isolation tools and techniques capable of guiding the FA engineer accurately, reliably, and quickly through multiple layers to the failure location.

Among them, magnetic field imaging, using SQUIDs as the sensing element, was introduced in 1998 [88] as a way to detect short circuit failures in ICs. The principle is very simple: the circuit of interest in the device under test (DUT) is powered up. The current generates a magnetic field around it and this magnetic field is detected by a sensor above the device. The sample is raster scanned and magnetic field is acquired at determined steps providing a magnetic image of the field distribution.

This magnetic field data is typically processed using an standard inversion technique [27, 89] to obtain a current density map of the device. The resulting current map can then be compared to a circuit diagram, an optical or infrared image, or a non-failing part to determine the fault location. In this sense, magnetic field imaging allows the Engineer to "see" what the current is doing in the DUT. Recent advances expanding the bandwidth of operation of SQUID sensors has allowed the localization of open circuit failures by detecting radio-frequency (RF) magnetic fields [1, 90].

The main advantage of magnetic imaging is that is a fully non-destructive, non-invasive, contact-less technique. In addition, magnetic signals generated by the current in the device under test pass unaffected by virtually all materials used in modern packaging technologies. It can thus see through multiple layers of metal, insulator and even multiple dies vertically stacked. All this makes magnetic imaging an excellent technique for Fault Isolation (FI) of defects and it is currently established as a standard technique for isolating shorts, leakages and high resistance defects in packages and die, as well as distinguishing between package-level and die-level defects [2, 16, 92-96].

As SQUIDs operate at cryogenic temperatures (typically at 77 K for high-temperature superconductors), they need to be isolated from the environment by an enclosure under vacuum while the DUT is raster scanned, at room temperature, under the sensor. The presence of the enclosure limits how close to the surface one can scan a sample. Ultimately, resolution is limited by the scanning distance or the sensor size and thus, SQUID imaging resolution is limited to about 25 μm lateral resolution for typical scanning distances and device thicknesses.

In FA, though, resolution is linked to failure localization accuracy, being this a better descriptor of a technique's capability. Failure localization accuracy is defined as the ability to correlate a defect through physical deprocessing results with the signature obtained from the fault isolation tool to meet or exceed some expected resolution. These two terms, lateral resolution and defect localization accuracy are often used to refer to the same concept: how good can a tool or technique physically locate the defect.

Magnetic imaging as applied to FI normally uses the processed current density image to pinpoint the failure location. Because of this, a peak localization technique allows, by use of software, to improve defect localization to about 3 μm.

Using magnetic imaging for FA imposes demanding restrictions on the type of sensor that can be used. There are three main requirements that are linked: need of high resolution, high magnetic field sensitivity (and thus, low noise figures) and the geometry of the sensor itself.

As stated before, SQUID sensors provide very high sensitivity and resolution in the few microns, making them ideal for packaged devices. They also provide coarse localization for die-level fault isolation. They lack, however, the submicron resolution required for die-level analysis. MR-based sensors on the other hand can be manufactured in small sizes, they operate at room temperature and noise figures have improved substantially in the recent years to provide enough sensitivity to be usable for FA [3, 12, 97-101]. Furthermore, they can be fabricated in a tapered-tip shape which allows for getting close to wirebonds or probe needles in front-side scanning situations as well as scanning in milled-out cavities [12].

4.1 Sensitivity

The key figure of merit for sensitivity in FA is minimum current detection. The reason is that the own nature of the defect can make it very sensitive to applied current or voltage and in some situations, too large voltage or current can result in the destruction of the defect or "healing" of the device. Besides that, different devices have different requirements in terms of limiting voltage and current. In general, FA Engineers tend to be on the safe side and restrict the current levels to < 1 mA with 20 mA to be typically considered too high. In addition, the need to locate defect leakage currents which can be as low as 20 nA imposes additional restrictions to the sensor sensitivity.

On the other hand, for magnetic imaging, the key figure is magnetic field sensitivity. There are two components to determine the magnetic field sensitivity when it comes to magnetic field scanning for FA: the intrinsic noise figure of the magnetic sensor used and the overall signal-to-noise ratio (S/N) for the image acquisition equipment. We will consider the second to be optimized for the application and identical for different type of sensors allowing us to just focus on intrinsic sensor noise figures.

The link between magnetic field sensitivity and minimum detectable current is obviously the distance from the current source to the sensor. For fixed scanning distance (that is, sensor-to-current distance), the sensor with better magnetic field sensitivity (lower noise figure) will allow for the lowest detectable current, making it the most desirable.

MR sensors used for magnetic field imaging in FA are typically based on a spin-valve (SV) or a magnetic tunnel junction (MTJ) device. Although MTJ sensors have larger response as compared to SV, the comparatively simple design, low voltage noise, high yield, and robust physical characteristics of SV sensors make them attractive for initial development and integration in commercial magnetic field microscopes [12, 101].

For practical imaging applications, the field noise of the sensor provides the best figure of merit for a sensitivity comparison and can be measured the same way for all sensors, including SQUIDs. Field noise measurements represent the minimum field that can be measured at a given frequency above the noise floor of the sensor and associated electronics. Figure 7 shows noise data for a SQUID, MTJ, commercial SV (hard-drive head), and an optimized SV sensor for failure analysis.

The SV and MTJ sensors only reach a white noise response beyond the upper end of this frequency range, so there is a steady improvement as a function of frequency. At 10 kHz, the MTJ sensor has a noise level of approximately 1 $nT/\sqrt{Hz}$. The optimized SV sensors have a noise level of approximately 6 $nT/\sqrt{Hz}$, 2 orders of magnitude better than SV hard-drive heads, which have a noise level of 900 $nT/\sqrt{Hz}$ at 10 kHz. These noise figure improvements directly translate into improvements in minimum current detectability in magnetic field imaging. For identical image S/N, the amount of current required scales directly with the noise level of the sensor. For example, an image of a circuit carrying 2 μArms imaged with an optimized SV sensor would require, under identical configuration and conditions, approximately 300 μArms to obtain the same quality image with a SV hard-drive sensor.

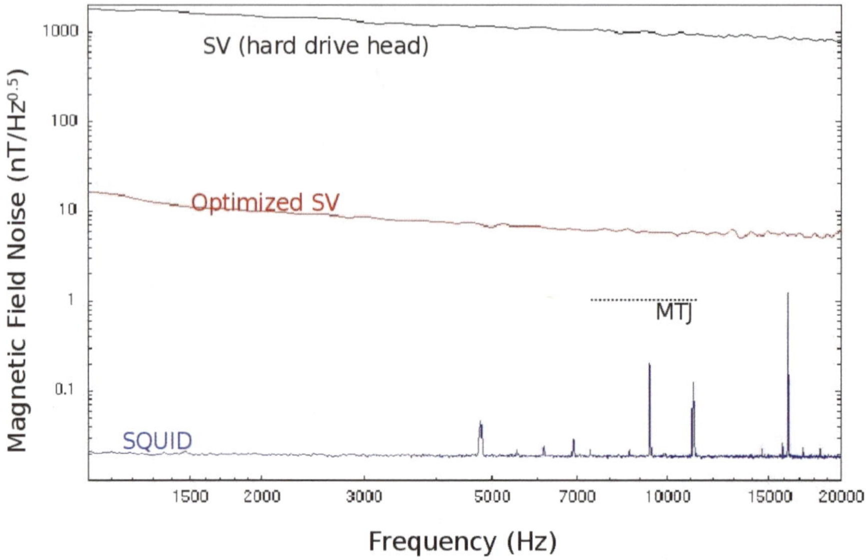

Fig. 7. Magnetic field noise as a function of frequency for SQUIDs, SV hard-drive heads, optimized SV sensors, and MTJs. Optimized SV sensors show 2 orders of improvement over conventional MR hard-drive heads.

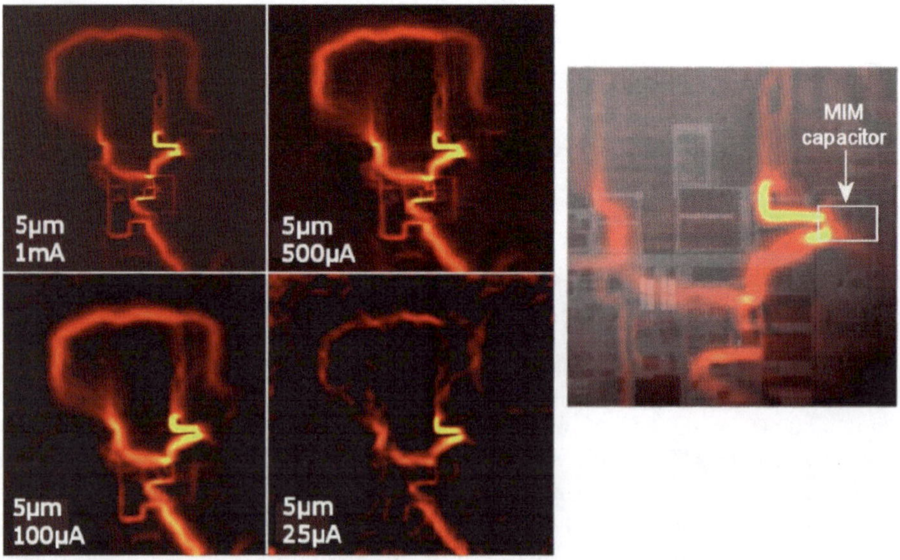

Fig. 8. (left) Current density image from backside scanning decreasing the current for a 5 μm thinned down die. (right) Zoom in detail of the current path at the defect MIM capacitor. The current image is overlaid with the infra-red (IR) image of the die. Images courtesy of M. Hechtl, Infineon.

The effect of current into the resulting images acquired with SV MR sensors has been widely studied. An excellent example was presented in the paper by M. Hechtl [100] and can be summarized in Figure 8(left) showing current density images from identical scans from the back side of a flip-chip mounted IC. A 5 μm-thick thinned down die was scanned using a commercial magnetic field imaging system. Identical scans were acquired except for varying the current for each scan from 1 mA to 25 μA. The impact of reducing the current on the sharpness, resolution and clarity of the images is evident.

Figure 8(right) shows the actual defect location, a defective MIM capacitor that was shorted. Two things are worth noticing: first, the SV sensor has enough sensitivity to locate the defect when applying just 25 μA of current to the device, as seen by the presence of the bright current paths produced by the short. The intensity of the current seen on the zoomed image on the MIM capacitor actually reveal that even lower current could have been used and the defect could have still been detected. Although details on the circuit parts that carry less current are vanishing progressively as the current decreases, just 1 mA of current is enough to reveal many details of the circuitry. Second, sensitivity and resolution are linked together. The lower the current, the less details of the circuit can be obtained.

The same author also did a study on how distance impacts current detectability and resolution. Figure 9(left) shows current density images for the same device scanned from the back side. In this case, the die was thinned down from 200 μm to 5 μm while all other conditions were kept identical. Obviously, the thinner the die, the closer the sensor-to-current distance.

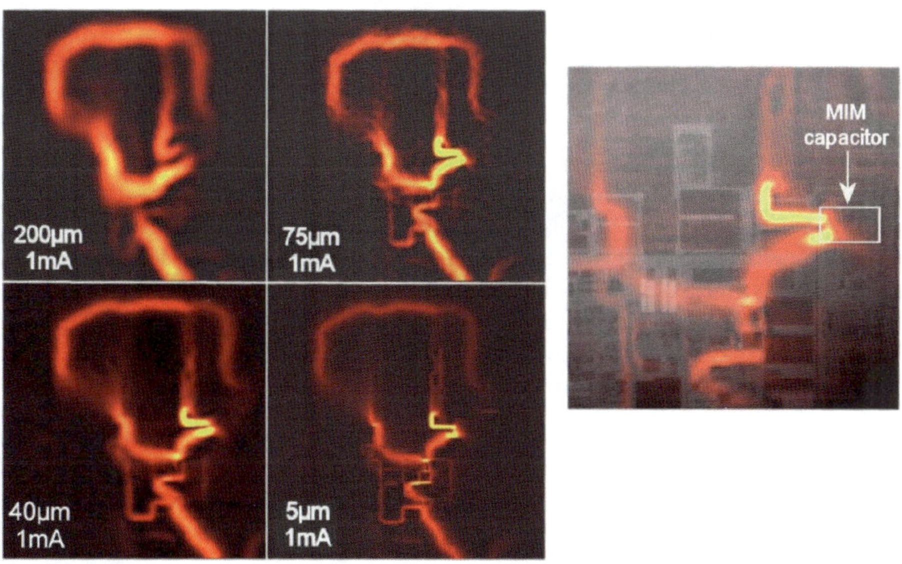

Fig. 9. (left) Current density image detected by the SV sensor at a current of 1 mA for different die thickness. (right) Zoom in detail of the current path at the defect MIM capacitor. The current image overlaid with the infra-red (IR) image of the die. Images courtesy of M. Hechtl, Infineon.

A large portion of the path structure can be identified by comparison with the top metallization layout of the interposer and only some short metal line of the bottom interposer layer is involved in the current path shown, appearing as blurred, as it is farthest away from the scanning sensor. At about 75 μm, the rectangular shape of the on-die current paths are distinct enough, a clear sign of resolution improvement.

It is remarkable that the image reveals not only on-die currents but also interposed current paths that are further away.

4.2 Resolution and Sensor Geometry

The resolution, d, for a sensor is proportional to the sensor-to-current distance, z, and the lateral dimensions, w, of the sensing area as $d \sim (4\ z^2 + w^2)^{0.5}$. MR sensors, with active sensing dimension along the scanning direction smaller than 50 nm, are thus capable of sub-micron resolution if allowed to scan in close proximity with the circuit under investigation. In principle, this makes them ideal candidates for front-side scanning providing nano-scale resolution but for actual device scanning, equally important to sensitivity and resolution is probe tip geometry. The reason for this is that real life device scanning requires electrical connection to the circuit that is done by either using probe needles, or, in the case of packaged devices, wirebonding to the package interconnect. Furthermore, for backside or packaged devices with no direct access to the circuitry, sometimes it is necessary to mill out a cavity on a previously identified coarse region that allows closer proximity to achieve higher resolution. These cavities are normally small in dimensions as the laser milling or focused-ion-beam (FIB) time required to machine them are very costly.

To illustrate this problem, let's consider, for example a typical MR hard drive sensor. These sensors are mounted on a rigid cantilever and have a footprint of roughly 1.5 mm × 1 mm. The size and the mounting geometry can make access to the die in a decapsulated chip difficult at best.

Figure 10 shows a hard-drive MR sensor probe positioned on a 1 cm × 1 cm die. In addition to the size of the sensor substrate itself, the cantilever supporting the sensor, as noticed on the left inset and right insets, extends well beyond the sensor block, thus risking contact with and damage to wirebonds or needles (Figure 10, right inset). They are also difficult or impossible to operate on very small dies.

SV MR sensors for FA has been designed with a body that tapers to a point, in a pencil-like shape with a footprint less than 50 μm × 50 μm. This allows easy access into etched packages, milled out cavities and close proximity (about 5 μm) to wirebonds and probe needles. Figure 11(left) shows a SV MR sensor scanning in a cavity (500 mm × 500 mm and 250 mm deep) of a small die while the image on the right illustrates the capability of the optimized MR sensors to scan in close proximity to wirebonds on a packaged die.

There is an advantage of using a dual sensor approach, like a SQUID/MR combination. The SQUID sensor can do coarse localization of the defect on thick packaged devices. This allows to determine if the problem is in the package, interconnect or die. If the failure location is found on the die, it is relatively easy to

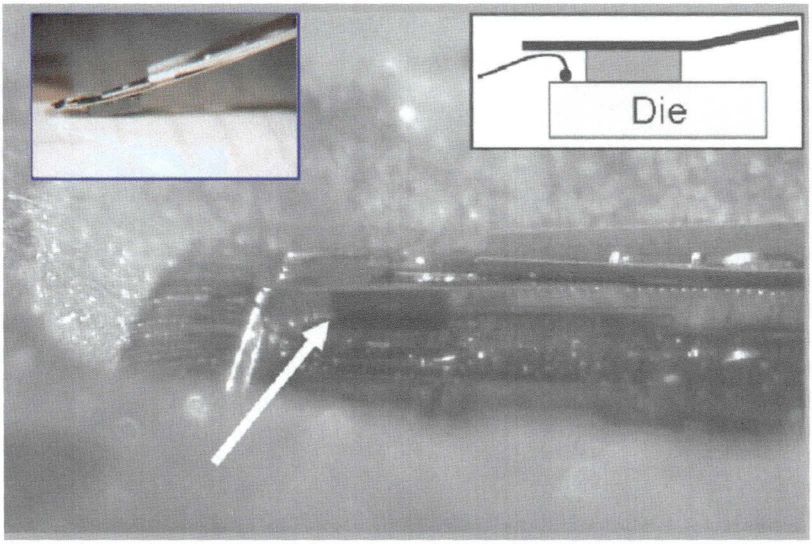

Fig. 10. Conventional MR hard-drive sensor on a die surface, with an illustrated inset. An arrow points to the sensor body mounted under the cantilever. The cantilever supporting the sensor protrudes on all sides, limiting access to the edges of the die.

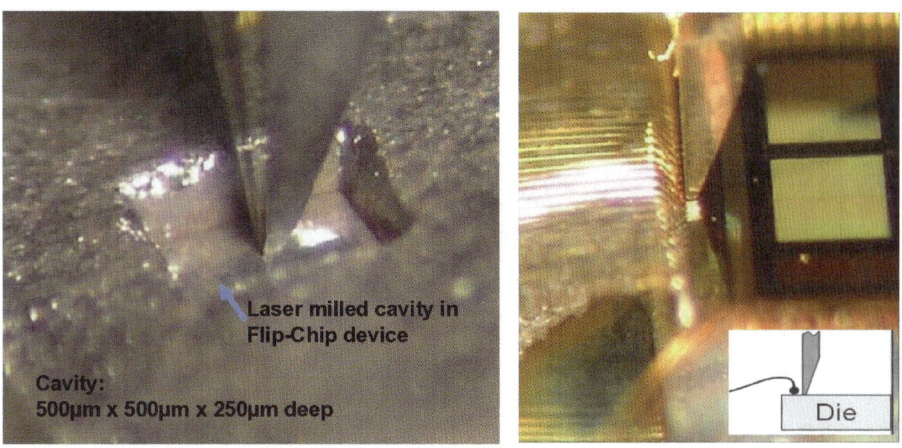

Fig. 11. (left) Local thinning through laser milling provides an opportunity to get closer to an area of interest for high resolution scanning. The SV MR sensor is shown in a cavity that is 500 µm × 500 µm and 250 µm deep. (right) SV MR sensor on a die surface, with an illustrated inset. A high-aspect-ratio tip provides easy access to the die and no restrictions for approaching wirebonds or probe needles.

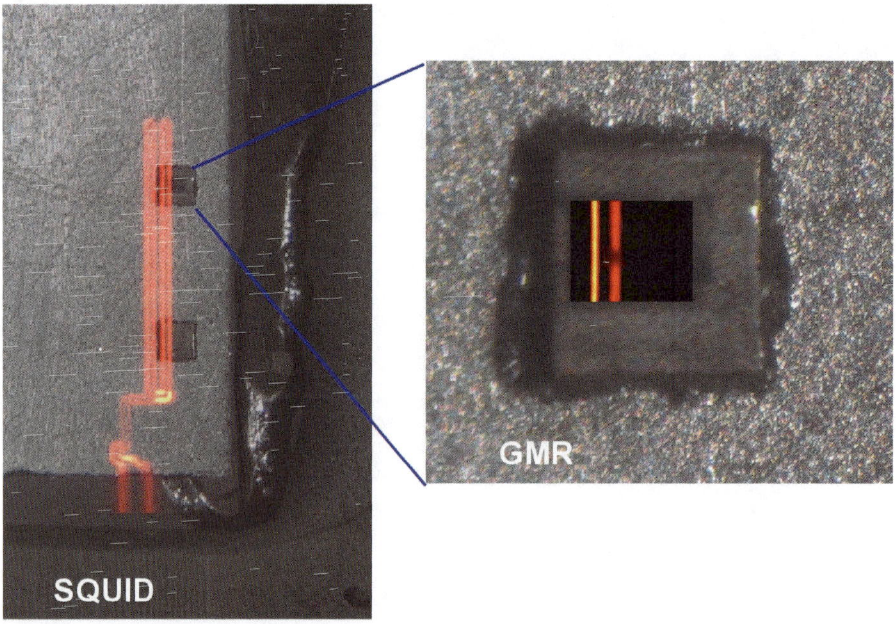

Fig. 12. (left) Current density image of a flip-chip device acquired with SQUID at > 250 mm scanning distance. Current density image is overlaid on optical image. Two FIB milled-out pockets are visible. (right) Current density image inside the cavity (same as shown on Figure 11) acquired with SV MR sensor. Additional details can be seen including a change in level of the current path on the right current path.

partially remove material (like thinning down the die or etching a cavity) to allow the MR sensor to provide the high resolution location of the defect. This is illustrated in Figure 12 where a flip-chip device was first scanned using a SQUID for global current localization [Figure 12(left)]. Two pockets were milled-out using FIB to allow the SV MR sensor tip to go in close proximity and provide better resolution and details [Figure 12(right)]. It can be noticed the difference in metal trace width of the two current lines and also a decrease in current level in the right current path.

The best resolution MR sensors can provide is when having direct access to the circuit, like is the case for front-side accessible devices (exposed die, wafer) as there is only a very thin (few hundreds of nm) passivation layer between the active metal circuit and the sensor. As an example of nano-scale resolution that can be achieved with MR sensor, Figure 13 shows current density images from SV scans of a die serpentine process monitor with 250 nm lines and 400 nm spacing. The two insets show zoom-in current density images of specific areas overlaid on scanning electron microscope (SEM) images to show how the current density matches perfectly the geometry of the metal traces as well as the spacing.

As discussed, there is a trade-off between sensitivity and resolution. The combined use of SQUID/MR sensors is the best way to optimize the data acquired by the system. A magnetic field microscope designed for FA should allow seamless scanning under one sensor or the other and under an optical camera for physical registration of locations. In this manner, all three images—optical, SQUID, and MR—are registered to each other. This permits easy selection of a region of interest (ROI) to re-scan with higher resolution or to move from the coarse, more sensitive SQUID sensor to the nanoscale-resolution-capable MR sensor. Experience has shown that a coarse scan with the SQUID reveals enough details about the current in the device for initial FI. In a quick and effortless manner, it facilitates identification of ROIs and isolation of defective components (for example, package, die, interconnect level, a particular die in a stacked-die configuration, etc.). With this information, the engineer can select an ROI and increase resolution of the image or even switch to the MR to improve defect localization. Thinning of the die, decapsulation, or other minor sample preparation may be desirable at some point to make it possible for the MR sensor to scan closer to the circuit in the ROI and achieve maximum resolution.

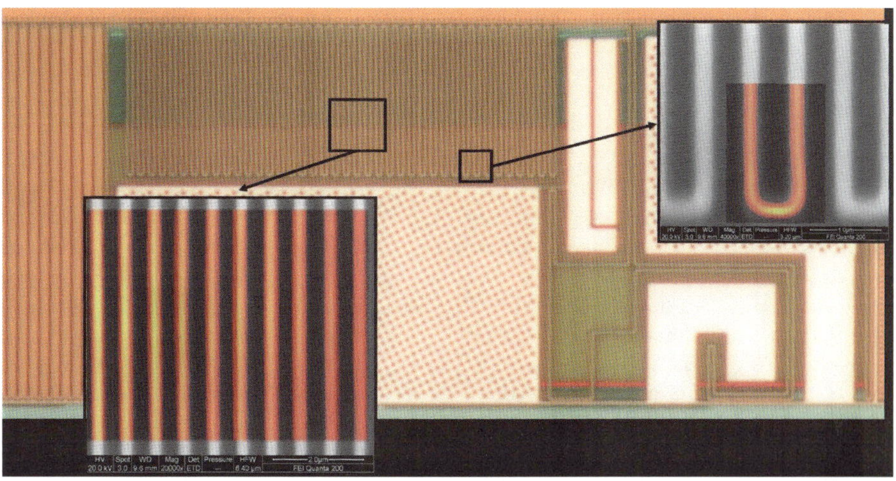

Fig. 13. Nanoscale-resolution current-density image overlaid on the optical image, using the MR sensor, of a test serpentine structure on a wafer. Metal trace width is 250 nm, while the spacing is 400 nm. Higher-resolution images in the insets were overlaid on scanning electron microscopy images.

5 Conclusion

Non destructive testing (NDT) resorting to magnetic imaging is a valuable a non-invasive and non-contact technique to detect either surface or deeply-buried defects in a wide range of structures and devices. In particular using magnetoresistive (MR) sensors for NDT has attracted a considerable attention in recent years as it fulfills the sensitivity and resolution requirements to localize buried and non-visual defects, in particular for complex package integration. Compared to the widely used SQUID devices for NDT, MR sensors have the advantage of being relatively low cost and of easier implementation. In this chapter, we have shown how MR sensors were successfully used in Scanning Magnetic Microscopy as a failure analysis tool to map current paths in microelectronic devices at an industrial level.

References

[1] Dias, R., Skoglund, L., Wang, Z., Smith, D.: Integration of SQUID Microscopy into FA Flow. In: Proc. 27th Int. Symp. Test. and Failure Analysis (ISTFA), pp. 77–81 (2001)

[2] Pacheco, M., Wang, Z.: Scanning SQUID Microscopy for New Package Technologies. In: Proc. 30th Int. Symp. Test. and Failure Analysis (ISTFA), pp. 67–72 (2004)

[3] Felt, F., Knauss, L., Gilbertson, A., Orozco, A.: Istfa Proc. 33rd Int. Symp. Test. and Failure Analysis, pp. 197–205 (2007)

[4] Ikeda, S., Miura, K., Yamamoto, H., Mizunuma, K., Gan, H.D., Endo, M., Kana, S., Hayakawa, J., Matsukura, F., Ohno, H.: Nat. Materials 9, 721 (2010)

[5] Smith, C.H., Schneider, R.W., Pohm, A.V.: High-resolution giant magnetoresistance on-chip arrays for magnetic imaging. J. Appl. Phys. 93, 6864 (2003)

[6] Sikora, R., Chady, T., Gratkowski, S., Komorowski, M., Stawicki, K.: Eddy Current Testing of Thick Aluminum Plates with Hidden Cracks. In: AIP Conference Proceedings, vol. 657, p. 427 (2003)

[7] Dogaru, T., Smith, S.T.: Giant magnetoresistance-based eddy-current sensor. IEEE Trans. Magn. 37, 3831 (2001)

[8] Dalichaouch, Y., Singsass, A.L., Putris, F., Perry, A.R., Czipott, P.V.: Low frequency electromagnetic technique for nondestructive evaluation. In: Proc. SPIE, vol. 3994, pp. 2–9 (2000)

[9] Crepel, O., Poirier, P., Descamps, P., Desplats, R., Perdu, P., Haller, G., Firiti, A.: Magnetic Microscopy for IC Failure Analysis: Comparative Case Studies using SQUID, GMR and MTJ systems. Microelectronics Reliability 44, 1559 (2004)

[10] Bajjuri, S., Hoffman, J., Siddoju, A., Meyendorf, N.: Development of GMR eddy current sensors for high temperature applications and imaging of corrosion in thick multi-layer structures. In: Proc. SPIE, vol. 5392, pp. 247–255 (2004)

[11] Kataoka, Y., Wakiwaka, H., Shinoura, O.: Proceedings of Sensors 2002, vol. 2, p. 65. IEEE, Piscataway (2002)

[12] Woods, S.I., Orozco, A., Knauss, L.A.: Advances in Magnetic Current Imaging for Die-Level Fault Isolation. Electr. Device Failure Analysis 8(4), 26–30 (2006)

[13] Martins, V.C., Cardoso, F.A., Germano, J., Cardoso, S., Sousa, L., Piedade, M., Freitas, P.P., Fonseca, L.P.: Femtomolar limit of detection with a magnetoresistive biochip. Biosen. Bioelect. 24, 2690–2695 (2009)

[14] Smith, C., Schneider, B.: Sens. Mag. 18, 11 (2001)

[15] Tee, T.Y., Zhong, Z.: Board level solder joint reliability analysis and optimization of pyramidal stacked die BGA packages. Microelectronics Reliability 44, 1957–1965 (2004)

[16] Hechtl, M., Steckert, G., Keller, C.: Localization of Electrical Shorts in Dies and Packages Using Magnetic Microscopy and Lock-In-IR Thermography. In: Proc. 13th Int. Symp. Phys. and Failure Analysis of Integr. Circuits (IPFA) (2006)

[17] Martin, P., El Matouat, A., Legendre, S., Colder, A., Descamps, P.: Fast and accurate method for flaws localization in stacked die packages from acoustic microscopy echoes transients. In: IECON 2011 - 37th Annual Conference on IEEE Industrial Electronics Society (2011)

[18] `http://www.neocera.com`

[19] `http://www.micromagnetics.com`

[20] Jenksy, W.G., Sadeghiz, S.S.H., Wikswo Jr., J.P.: SQUIDs for nondestructive evaluation. J. Phys. D: Appl. Phys. 30, 293 (1997)
[21] Krause, H.-J., Kreutzbruck, M.V.: Recent developments in SQUID NDE. Physica C 368, 70–79 (2002)
[22] Kirtley, J.R., Wikswo Jr, J.P.: Scanning Squid Microscopy. Annu. Rev. Mater. Sci. 29, 117–148 (1999)
[23] Lee, S.-Y., Matthews, J., Wellstood, F.C.: Position noise in scanning superconducting quantum interference device microscopy. Appl. Phys. Lett. 84, 5001 (2004)
[24] Kirtley, J.R., Ketchen, M.B., Stawiasz, K.G., Sun, J.Z., Gallagher, W.J., Blanton, S.H., Wind, S.J.: High-resolution scanning SQUID microscope. Appl. Phys. Lett. 66, 1138 (1995)
[25] Tsuei, C.C., Kirtley, J.R., Chi, C.C., Yu-Jahnes, L.S., Gupta, A., Shaw, T., Sun, J.Z., Ketchen, M.B.: Pairing Symmetry and Flux Quantization in a Tricrystal Superconducting Ring of $YBa_2Cu_3O_7$-δ. Phys. Rev. Lett. 73, 593 (1993)
[26] Fong, L.E., Holzer, J.R., McBride, K.K., Lima, E.A., Baudenbacher, F., Radparvar, M.: High-resolution room-temperature sample scanning superconducting quantum interference device microscope configurable for geological and biomagnetic applications. Rev. Sci. Instrum. 76, 053703 (2005)
[27] Chatraphorn, S., Fleet, E.F., Wellstood, F.C., Knauss, L.A., Eiles, T.M.: Scanning SQUID microscopy of integrated circuits. Appl. Phys. Lett. 76, 2304 (2000)
[28] Crankshaw, D.S., Trias, E., Orlando, T.P.: Magnetic Flux Controlled Josephson Array Oscillators. IEEE Transactions on Applied Superconductivity II(I), 1223 (2001)
[29] Davidović, D., Kumar, S., Reich, D.H., Siegel, J., Field, S.B., Tiberio, R.C., Hey, R., Ploog, K.: Phys. Rev. Lett. 76, 815 (1996)
[30] Oral, A., Bending, S.J., Henini, M.: Real-time scanning Hall probe microscopy. Appl. Phys. Lett. 69, 1324 (1996)
[31] Heremans, J.: Solid state magnetic field sensors and applications. J. Phys. D Appl. Phys. 26, 1149–1168 (1993)
[32] Boero, G., Demierre, M., Besse, P.-.A., Popovic, R.S.: Micro-Hall devices: performance, technologies and applications. Sensors and Actuators A: Physical 106(1-3), 314–320 (2003)
[33] Sandhu, A., Kurosawa, K., Dede, M., Oral, A.: 50 nm Hall Sensors for Room Temperature Scanning Hall Probe Microscopy. Japanese Journal of Applied Physics 43(2), 777–778 (2004)
[34] Oral, A., Bending, S.J., Henini, M.: Scanning Hall probe microscopy of superconductors and magnetic materials. J. Vac. Sci. Technol. B14, 1202 (1996)
[35] Howells, G.D., Oral, A., Bending, S.J., Andrews, S.R., Squire, P.T., Rice, P., de Lozanne, A., Bland, J.A.C., Kaya, I., Henini, M.: Scanning Hall probe microscopy of ferromagnetic structures. Journal of Magnetism and Magnetic Materials 196-197, 917–919 (1999)
[36] Chang, A.M., Hallen, H.D., Harriott, L., Hess, H.F., Kao, H.L., Kwo, J., Miller, R.E., Wolfe, R., van der Ziel, J., Chang, T.Y.: Scanning Hall probe microscopy. Appl. Phys. Lett. 61, 1974 (1992)
[37] Nguyen Van Dau, F., Schuhl, A., Childress, J.R., Sussiau, M.: Magnetic sensors for nanotesla detection using planar Hall Effect. Sens. Act. A 53, 256–260 (1996)
[38] Dinner, R.B., Beasley, M.R., Moler, K.A.: Cryogenic scanning Hall-probe microscope with centimeter scan range and submicron resolution. Rev. Sci. Instrum. 76, 103702 (2005)

[39] Pross, A., Crisan, A.I., Bending, S.J., Mosser, V., Konczykowski, M.: Second-generation quantum-well sensors for room-temperature scanning Hall probe microscopy. J. Appl. Phys. 97, 096105 (2005)
[40] Schweinböck, T., Weiss, D., Lipinski, M., Eberl, K.: Scanning Hall probe microscopy with shear force distance control. J. Appl. Phys. 87, 6496 (2000)
[41] Sandhu, A., Masuda, H., Senoguchi, H., Togawa, K.: Nanotechnology 15, S410 (2004)
[42] Kirtley, J.R.: A novel variable temperature scanning nano-Hall probe microscope system for large area magnetic imaging incorporating piezoelectric actuators maintained at room temperature. Rep. Prog. Phys. 73, 126501 (2010)
[43] Kustov, M., Laczkowski, P., Hykel, D., Hasselbach, K., D-Bouchiat, F., O'Brien, D., Kauffmann, P., Grechishkin, R., Givord, D., Reyne, G., Cugat, O., Dempsey, N.M.: Magnetic characterization of micropatterned Nd-Fe-B hard magnetic films using scanning Hall probe microscopy. J. Appl. Phys. 108, 63914 (2010)
[44] Cambel, V., Fedor, J., Gregušová, D., Kováč, P., Hušek, I.: Large-scale high-resolution scanning Hall probe microscope used for MgB2 filament characterization. Supercond. Sci. Technol. 18, 417 (2005)
[45] Perkins, G.K., Bugoslavsky, Y.V., Qi, X., MacManus-Driscoll, J.L., Caplin, A.D.: High field scanning Hall probe imaging of high temperature superconductors. IEEE Trans. Appl. Superc. 11(1), 3186–3189 (2001), doi:10.1109/77.919740
[46] Gregory, J.K., Bending, S.J., Sandhu, A.: A scanning Hall probe microscope for large area magnetic imaging down to cryogenic temperatures. Rev. Sci. Instrum. 73, 3515 (2002)
[47] Fedor, J., Cambel, V., Gregušová, D., Hanzelka, P., Dérer, J., Volko, J.: Scanning vector Hall probe microscope. Rev. Sci. Instrum. 74, 5105 (2003)
[48] Phan, M.-H., Peng, H.-X.: Giant magnetoimpedance materials: fundamentals and applications. Prog. Mater. Sci. 53, 323–420 (2008)
[49] Beach, R.S., Berkowitz, A.E.: Giant magnetic field dependent impedance of amorphous FeCoSiB wire. Appl. Phys. Lett. 64, 3652 (1994)
[50] Panina, L.V., Mohri, K., Bushida, K., Noda, M.: Giant magneto-impedance and magneto-inductive effects in amorphous alloys. J. Appl. Phys. 76, 6198 (1994)
[51] Panina, L.V., Mohri, K.: Magneto-impedance effect in amorphous wires. Appl. Phys. Lett. 65, 1189 (1994)
[52] Ripka, P.: Magnetic sensors and magnetometers. Artech House Publishers (2001)
[53] Goktepe, M., Ege, Y., Bayri, N., Atalay, S.: Non-destructive crack detection using GMI sensor. Phys. Status Solidi (c) 1, 3436–3439 (2004)
[54] Hamia, R., Cordier, C., Saez, S., Dolabdjian, C.: Giant magneto impedance sensor for nondestructive evaluation eddy current system. Sens. Lett. 7, 437–441 (2009)
[55] Tehranchi, M.M., Ranjbaran, M., Eftekhari, H.: A Double core giant magneto-impedance sensors for the inspection of magnetic flux leakage from metal surface cracks. Sensors and Actuators A 170, 55–61 (2011)
[56] Kim, D.J., Park, D.G., Hong, J.H.: Nondestructive evaluation of reactor pressure vessel steels using the giant magnetoimpedance sensor. J. Appl. Phys. 91, 7421–7423 (2002)
[57] Vacher, F., Alves, F., Gilles-Pascaud, C.: Eddy current nondestructive testing with giant magneto-impedance sensor. NDTE Int. 40, 439–442 (2007)
[58] Yamamoto, S.Y., Schultz, S.: Scanning magnetoresistance microscopy. Appl. Phys. Lett. 69, 3263 (1996)
[59] Yamamoto, S.Y., Schultz, S.: Scanning magnetoresistance microscopy (SMRM): Imaging with a MR head. J. Appl. Phys. 81, 4696 (1997)

[60] O'Barr, Lederman, M., Schultz, S.: A scanning microscope using a magnetoresistive head as the sensing element. J. Appl. Phys. 79, 6067 (1996)
[61] Yamamoto, S.Y., Vier, D.C., Schultz, S.: High resolution contact recording and diagnostics with a raster-scanned MR head. IEEE Trans. Magn. 32, 3410 (1996)
[62] So, M.H., Nicholson, P.I., Meydan, T., Moses, A.J.: Non-destructive surface inspection system for steel and other ferromagnetic materials using magneto-resistive sensors. IEEE Trans. Magn. 31, 3370 (1995)
[63] Nicholson, P.I., So, M.H., Meydan, T., Moses, A.J.: J. Magn. Magn. Mater. 160, 162 (1996)
[64] Smith, C.H., Schneider, R.W., Dogaru, T., Smith, S.T.: Eddy-Current Testing with GMR Magnetic Sensor Arrays. In: AIP Conf. Proc., vol. 700, pp. 406–413.
[65] Postolache, O., Ribeiro, A.L., Ramos, H.: Uniform eddy current probe implementation using planar excitation coil and GMR sensor array. In: Proceedings IMEKO TC4 (2011)
[66] Yashan, A., Bisle, W., Meier, T.: Inspection of hidden defects in metal-metal joints of aircraft structures using eddy current technique with GMR sensor array. In: Proc. 9th ECNDT, Berlin (2006)
[67] Vacher, F., Gilles-pascaud, C., Decitre, J.M., Fermon, C., Pannetier, M.: Non destructive testing with GMR magnetic sensor arrays. In: Proc. 9th ECNDT, Berlin (2006)
[68] Vyhnánek, J., Janošek, M., Ripka, P.: AMR gradiometer for mine detection. Sensors and Actuators A 186, 100–104 (2012)
[69] Janosek, M., Vyhnanek, J., Ripka, P.: CW metal detector based on AMR sensor array. Sensors 28-31, 1515–1517 (2011)
[70] Holzl, C.H.P.A., Wiesner, T., Zagar, B.G.: Quality assurance for wire connections used in integrated circuits via magnetic imaging. In: 2012 IEEE International Instrumentation and Measurement Technology Conference (I2MTC) (2012)
[71] Phillips, G.N., Eisenberg, M., Draaisma, E.A., Abelmann, L., Lodder, J.C.: IEEE Transactions on Magnetics 38 (2002)
[72] Cano, M.E., Martínez, J.C., Bernal-Alvarado, J., Sosa, M., Córdova, T.: 16-channel magnetoresistive scanner for magnetic surface imaging. Rev. Sci. Instrum. 76, 086106 (2005)
[73] Tumanski, S., Liszka, A.: The methods and devices for scanning of magnetic fields. Journal of Magnetism and Magnetic Materials 242–245, 1253–1256 (2002)
[74] Cano, M.E., Pacheco, A.H., Cordova, T., Mazon, E.E., Barrera, A.: Superficial magnetic imaging by an xy-scanner of three magnetoresistive channels. Rev. Sci. Instrum. 83, 033705 (2012)
[75] Mook, G., Michel, F., Simonin, J.: Electromagnetic Imaging Using Probe Arrays. Journal of Mechanical Engineering 57(3), 227–236 (2011)
[76] Pohm, A.V., Beech, R.S., Bade, P.A., Chen, E.Y., Daughton, J.M.: Analysis of 0.1 to 0.3 micron wide, ultra dense GMR memory elements. IEEE Trans. Mag. 6(30), 4650–4652 (1994)
[77] Baselt, D.R., Lee, G.U., Natesan, M., Metzger, S.W., Sheehan, P.E., Colton, R.J.: Biosensors & Bioelectronics 13, 731–739 (1998)
[78] Han, S.-J., Xu, L., Yu, H., Wilson, R.J., White, R.L., Pourmand, N., Wang, S.X.: CMOS Integrated DNA Microarray Based on GMR Sensors. In: International Electron Devices Meeting (IEDM 2006) (2006)
[79] Freitas, P.P., Cardoso, F.A., Martins, V.C., Martins, S.A.M., Loureiro, J., Amaral, J., Chaves, R.C., Cardoso, S., Germano, J., Piedade, M.S., Sebastião, A.M., Fonseca, L.F., Pannetier-Lecoeur, M., Fermon, C.: Spintronic platforms for biomedical applications. Lab-on-Chip 12(3), 546–557 (2012)

[80] Gaster, R.S., Xu, L., Han, S.-J., Wilson, R.J., Hall, D.A., Osterfeld, S.J., Yu, H., Wang, S.X.: Quantification of protein interactions and solution transport using high-density GMR sensor arrays. Nature Nanotechnology 6, 314–320 (2011)
[81] Indeck, R.S., Judy, J.H., Iwasaki, S.: A magnetoresistive gradiometer. IEEE Trans. Mag. 6(24), 2617–2619 (1988)
[82] Pelkner, M., Neubauer, A., Reimund, V., Kreutzbruck, M., Schütze, A.: Routes for GMR-Sensor Design in Non-Destructive Testing. Sensors 12(9), 12169–12183 (2012)
[83] Kreutzbruck, M., Neubauer, A., Pelkner, M., Reimund, V.: Adapted GMR Array used in Magnetic Flux Leakage Inspection: 18th World Conference on Nondestructive Testing, Durban, South Africa, April 16-20 (2012)
[84] Kloster, A., Kröning, M., Smorodinsky, J., Ustinov, V.: Linear magnetic stray flux array based on GMR gradiometers. In: Indian Society for Non-Destructive Testing (NDE 2002), ISNT Chapters in Chennai, Kalpakkam & Sriharikota, December 5-7 (2002), http://publica.fraunhofer.de/documents/N-31528.html
[85] Nakamura, M., Kimura, M., Sueoka, K., Mukasa, K.: Scanning magnetoresistance microscopy with a magnetoresistive sensor cantilever. Appl. Phys. Lett. 80, 2713–2715 (2002)
[86] Takezaki, T., Yagisawa, D., Sueoka, K.: Magnetic Field Measurement using Scanning Magnetoresistance Microscope with Spin-Valve Sensor. Jap. J. Appl. Phys. 45, 2251–2254 (2006)
[87] Sahoo, D.R., Sebastian, A., Häberle, W., Pozidis, H., Eleftheriou, E.: Scanning probe microscopy based on magnetoresistive sensing. Nanotechnology 22, 145501 (2011), doi:10.1088/0957-4484/22/14/145501
[88] Knauss, L.A., Frazier, B.M., Christen, H.M., Silliman, S.D., Harshavardhan, K.S., Fleet, E.F., Wellstood, F.C., Mahanpour, M., Ghaemmaghami, A.: Power Shorts from Front and Backside of IC Packages Using Scanning SQUID Microscopy. In: ISTFA 1999: Proceedings of the 25th International Symposium for Testing and Failure Analysis (ASM International), pp. 11–16 (October 1999)
[89] Wikswo, J.P.: The Magnetic Inverse Problem for NDE. In: SQUID Sensors: Fundamentals, Fabrication and Applications, pp. 629–695. Kluwer Academic Publishers, The Netherlands (1996)
[90] Xie, M., Qian, Z., Pacheco, M., Wang, Z., Dias, R., Talanov, V.: Fault Isolation of Open Defects Using Space Domain Reflectometry. In: Proc. 38th Int. Symp. Test. and Failure Analysis, Phoenix, AZ (2012)
[91] Vallett, D.P., Bader, D.A., Talanov, V.V., Gaudestad, J., Gagliolo, N., Orozco, A.: Localization of Dead Open in a Solder Bump by Space Domain Reflectometry. In: Proc. 38th Int. Symp. Test. and Failure Analysis, Phoenix, AZ (2012)
[92] Eng, T.T., Lwin, H.E., Muthu, P., Chin, J.M.: Backside Deprocessing Technique & Its Novel Fault Isolation Application. In: Proc. 12th IPFA, pp. 110–113 (2005)
[93] Woods, S.I., Knauss, L.A., Orozco, A.: Current Imaging Using Magnetic Field Sensors. In: Microelectronics Failure Analysis Desk Reference, 5th edn., pp. 303–311. ASM International (2005)
[94] Orozco, A., et al.: Proc. 29th Int. Symp. Test. and Failure Analysis, Santa Clara, CA, pp. 9–13 (2003)
[95] Orozco, A.: Fault Isolation of Circuit Defects Using Comparative Magnetic Field Imaging, U.S. Patent 7,019,521
[96] Crepel, O., Descamps, P., Poirier, P., Desplants, R., Perdu, P., Firiti, A.: Proc. 30th Int. Symp. Test. and Failure Analysis, Worcester, MA, pp. 29–31 (2004)

[97] Baibich, M.N., Broto, J.M., Fert, A., Nguyen Van Dau, F., Petroff, F., Eitenne, P., Creuzet, G., Friederich, A., Chazelas, J.: Giant Magnetoresistance of (001)Fe/(001)Cr Magnetic Superlattices. Phys. Rev. Lett. 61, 2472 (1988)

[98] Binasch, G., Grünberg, P., Saurenbach, F., Zinn, W.: Enhanced magnetoresistance in layered magnetic structures with antiferromagnetic interlayer exchange. Phys. Rev. B 39, 4828 (1989)

[99] Schrag, B.D., Carter, M.J., Liu, X., Hoftun, J.S., Xiao, G.: Magnetic current imaging with tunnel junction sensors. In: Proc. ISTFA, p. 13 (2006)

[100] Hechtl, M.: Backside GMR Magnetic Microscopy for Flip Chip and Related Microelectronic Devices. IEEE Proceedings of 15th IPFA – 2008, 174–177 (2008)

[101] Orozco, A.: Magnetic Current Imaging in Failure Analysis. Electronic Device Failure Analysis 11, 14–21 (2009)

[102] da Silva, F.C., Halloran, S.T., Kos, A.B., Pappas, D.P.: 256-channel magnetic imaging system. Rev. Sci. Instrum. 79, 013709 (2008)

[103] Tondra, M., Nordman, C.A., Lang, E.H., Reed, D., Jander, A., Akou, S., Daughton, J.M.: Proc. SPIE, vol. 4393, p. 135 (2001)

Author Index